The deformation and processing of structural materials

Related titles from Woodhead's materials engineering list:

Corrosion in reinforced concrete structures (ISBN 1 85573 768 X)

An understanding of corrosion in concrete is an essential basis for enabling engineers to decide which structural techniques and building processes are most effective in improving the durability of new or existing concrete structures. In this authoritative new book the fundamental aspects of corrosion in concrete are analysed in detail. An overview of current monitoring techniques together with a discussion of practical applications and current numerical methods that simulate the corrosion process provides the civil and structural engineer with an invaluable guide to best practice when it comes to design aimed at minimising the effects of corrosion in concrete. The corrosion protective properties of concrete and modified cements are also discussed. The most frequently used stainless steels are examined together with an analysis of their reinforcement properties. Special attention is given to their handling and their welding requirements and the economics of their use. A comprehensive overview of surface treatments and corrosion inhibitors is presented alongside their practical applications as well as detailed coverage of electrochemical protection and maintenance techniques.

Durability of engineering structures (ISBN 1 85573 695 0)

Structures often deteriorate because not enough attention is given to the design stage. Most standards for structural design do not cover design for service life. Designing for durability is often left to the structural designer or architect who does not have the necessary skills, and the result is all too often failure. Knowledge of the long-term behaviour of materials, building components and structures is the basis for avoiding these costs. This book is aimed at degree students in civil engineering, engineers, architects, contractors, plant managers, maintenance managers and inspection engineers.

Advanced polymer composites for structural applications in construction
(ISBN 1 85573 736 1)

The 2nd International Conference on the application and further exploitation of advanced composites in construction was held at the University of Surrey from 20 to 22 April, 2004. With over 100 delegates the conference brought together practising engineers, asset managers, researchers and representatives of regulatory bodies to promote the active exchange of scientific and technical information on the rapidly changing scene of advanced composites in construction.

Details of these books and a complete list of Woodhead's materials engineering titles can be obtained by:

- visiting our web site at www.woodheadpublishing.com
- contacting Customer Services (e-mail: sales@woodhead-publishing.com; fax: +44 (0) 1223 893694; tel.: +44 (0) 1223 891358 ext. 30; address: Woodhead Publishing Limited, Abington Hall, Abington, Cambridge CB1 6AH, England).

The deformation and processing of structural materials

Edited by

Z. Xiao Guo

CRC Press
Boca Raton Boston New York Washington, DC

WOODHEAD PUBLISHING LIMITED
Cambridge England

Published by Woodhead Publishing Limited
Abington Hall, Abington
Cambridge CB1 6AH
England
www.woodheadpublishing.com

Published in North America by CRC Press LLC
2000 Corporate Blvd, NW
Boca Raton FL 33431
USA

First published 2005, Woodhead Publishing Limited and CRC Press LLC

British Library Cataloguing in Publication Data
A catalogue record for this book is available from the British Library.

Library of Congress Cataloging in Publication Data
A catalog record for this book is available from the Library of Congress.

The publishers' policy is to use permanent paper from mills that operate a sustainable forestry policy, and which has been manufactured from pulp which is processed using acid-free and elementary chlorine-free practices. Furthermore, the publishers ensure that the text paper and cover board used have met acceptable environmental accreditation standards.

Woodhead Publishing Limited ISBN 1 85573 738 8
CRC Press ISBN 0-8493-2578-1
CRC Press order number: WP2578

Project managed by Macfarlane Production Services, Markyate, Hertfordshire
(macfarl@aol.com)
Typeset by Replika Press Pvt Ltd, India
Printed by T J International Limited, Padstow, Cornwall, England

Contents

Contributor contact details

Preface

Professor Z. Xiao Guo
Department of Materials
Queen Mary, University of London
Mile End Road
London
E1 4NS
UK

Tel: +44 (0) 20 7882 5569
Fax: +44 (0) 20 8981 9804
E-mail: x.guo@qmul.ac.uk

Chapter 1

Professor Uday Chakkingal
Department of Metallurgical and Materials Engineering
Indian Institute of Technology Madras
Chennai 600 036
India

Tel: +91 44 2257 4775
Fax: +91 44 2257 4752
E-mail: udaychak@iitm.ac.in

Chapter 2

Professor A. Arslan Kaya
TUBITAK-MRC
Materials Institute
P.O.B. 21 Gebze
41470 Kocaeli
Turkey
E-mail: AliArslan.Kaya@mam.gov.tr

Professor M. Pekgüleryüz
McGill University
Department of Metals and Materials Engineering
3601 University Street
Montreal, QC, H3A 2B2
Canada

Professor D. Eliezer
Ben-Gurion University of The Negev
Dept. of Materials Science & Engineering
PO Box 653, Beer-Sheva, 84105
Israel

Tel: +972 8 6461 467
Fax: +972 8 6472 931
E-mail: deliezer@bgumail.bgu.ac.il

Chapter 3

Professor P. Manohar
Department of Engineering
Robert Morris University
6001 University Boulevard
Moon Township PA-15108
USA

E-mail: Manohar@rmu.edu

Professor M. Ferry
School of Materials Science and Engineering
University of New South Wales
UNSW Sydney 2052
Australia

Tel: +61 2 9385 4453
Fax: +61 2 9385 5956
E-mail: m.ferry@unsw.edu.au

Chapter 4

Professor S. Tin
Department of Materials Science and Metallurgy
Cambridge University
Pembroke Street
Cambridge CB2 3QZ
UK

Tel: +44 (0) 1223 334320
Fax: +44 (0) 1223 331956
E-mail: st298@hermes.cam.ac.uk

Chapter 5

Professor K. Xia
Department of Mechanical and Manufacturing Engineering
University of Melbourne
Victoria 3010
Australia

Tel: +61 3 8344 6664
Fax: +61 3 9347 8784
E-mail: k.xia@unimelb.edu.au

Chapter 6

Professor M. Ferry
School of Materials Science and Engineering
University of New South Wales
UNSW Sydney 2052
Australia

E-mail: m.ferry@unsw.edu.au

Chapter 7

Professor Z.Y. Fan and Dr A. Das
Department of Mechanical Engineering
Brunel University
Kingston Lane
Uxbridge
Middlesex UB8 3PH
UK

Tel: +44 (0) 1895 816391
Fax: +44 (0) 1895 816393
E-mail: mtstzzf@brunel.ac.uk
E-mail: Amitabha.Das@brunel.ac.uk

Chapter 8

Professor N. Ridley
Manchester Materials Science Centre
The University of Manchester
Grosvenor Street
Manchester M1 7HS
UK

Tel: + 44 (0) 161 200 3614
Fax: + 44 (0) 161 200 3586
E-mail:norman.ridley@manchester.ac.uk

Preface

Structural materials remain the backbone of many industrial technologies, ranging from aerospace, automobile, building, sports to medical devices, where mechanical properties of the materials are of particular importance. Mechanical properties are very sensitive to microstructural features or 'defects', such as vacancies, interstitials, dislocations, dislocation networks, precipitates, grain boundaries, phase boundaries and other interfaces. 'Materials are like people – it's their defects that make them interesting' (a statement often attributed to Professor Frank). Such defects in materials can be engineered by means of deformation to harness desirable mechanical properties. Advances and innovations in materials processing continue to improve such properties and/ or reduce the cost of engineering materials. It is both timely and fitting that a reference book on the deformation and processing of structural materials is compiled to reflect such advances for the benefit of engineers, technologists and research students.

General theories of deformation have been covered in several textbooks in the past. However, key aspects of microstructural development are very specific to the type of materials in question. Indeed, different engineering applications also exert varied emphases on the mechanical properties. Hence, this book aims to treat major structural materials or key processes as 'individuals'. Their characteristics are carefully and clearly examined by an international team of renowned researchers in their respective fields.

The book contains eight chapters, each of which was compiled with reference to up-to-date scientific literature and the state-of-the-art technologies. Important groups of structural materials and their specific processing techniques are covered in the first six chapters, including aluminium alloys, magnesium alloys, ferrous alloys, superalloys, intermetallics, and metal matrix composites. The last two chapters are devoted to two very important processing techniques, semi-solid processing and superplastic forming, and their applications to structural components.

This book is a valuable reference for research scientists and industrialists who would like to gain an insight into the deformation and processing characteristics of modern structural materials. It is also an indispensable source of information for final-year project students and postgraduates.

I would like to thank all the contributing authors who have devoted much time and effort in reviewing the literature and putting the chapters together. My appreciation also goes to staff at Woodhead Publishing Limited for assistance with most of the coordination.

Z. Xiao Guo
Queen Mary, University of London

1
Aluminium and its alloys

U CHAKKINGAL, Indian Institute of Technology Madras, India

1.1 Introduction

Aluminium and its alloys form an important class of structural materials. The most important characteristic of aluminum is its relatively low density compared to steels. By utilizing suitable alloying and proper heat treatment practice, specific strengths equivalent to those of high strength steels can be developed in aluminum alloys. In addition, aluminium alloys can be easily fabricated into useful engineering components through a variety of manufacturing processes. This has led to widespread use of aluminium alloys in automobile, aerospace, packaging and construction industries.

This chapter will discuss the important aspects of deformation and processing of aluminium alloys. The introductory section (section 1.1) contains a review of the classification of aluminium alloys, heat treatable and non-heat treatable alloys, wrought aluminium alloys, cast aluminium alloys, etc. Deformation behaviour of aluminium is covered in section 1.2. Section 1.3 covers the microstructural aspects of deformation of aluminium alloys during hot and cold deformation with particular emphasis on the development of dislocation structures, formation of cells and subgrains, etc. Section 1.4 covers the various hot and cold forming techniques applicable to the processing of aluminium alloys with particular emphasis on extrusion. Section 1.5 reviews various severe plastic deformation techniques, particularly the equal channel angular extrusion process that has the potential to produce a class of ultra fine grained aluminium alloys. Common defects that arise in the processing of aluminium alloys are discussed in section 1.6. The contents of the chapter are summarized in section 1.7. Sources for further information are listed in section 1.8. An extensive list of references is given in section 1.9.

1.1.1 Classification of aluminium alloys

Aluminium alloys are usually subdivided into wrought alloys and cast alloys. These alloy types can be either heat treatable or non heat treatable. The heat

Table 1.1 Designation system for wrought aluminium alloys

Main alloying element	Alloy designation
No major alloying element (Al > 99%)	1xxx
Copper	2xxx
Manganese	3xxx
Silicon	4xxx
Magnesium	5xxx
Magnesium and silicon	6xxx
Zinc	7xxx
Other elements	8xxx
Unused	9xxx

Source: R.B.C. Cayless, Alloy and temper designation systems for aluminium and aluminium alloys, Metals Handbook Vol. 2, 10th edn, *Properties and Selection of Non-ferrous Alloys and Special Purpose Materials*, OH, USA, ASM International, 1990, 15–28. Reprinted with permission.

treatable alloys develop high strengths after suitable ageing treatments. For wrought alloys, a four-digit designation system standardized by the Aluminium Association, USA is widely followed (Aluminium Association, 1997). This classification is based on the main alloying elements present (please see Table 1.1). In addition a temper designation is usually denoted next to the alloy type. This system consists of a letter indicating the basic temper. The four basic tempers are denoted by the letters: F as fabricated, O annealed, H strain hardened, and T heat treated. The H and T tempers are further defined by using one or more digits to specify the amount of strain hardening (for the H temper) or the specific thermal treatment followed (for the T temper). Cast aluminium alloys are usually designated using a three-digit notation.

1.1.2 Important alloy grades

The heat treatable wrought aluminium alloys are based on Al-Cu (the 2xxx series), Al-Mg-Si (the 6xxx series) and the Al-Zn-Mg (the 7xxx series). The heat treatable alloys develop high strengths after proper solutionizing and ageing through precipitation hardening. One of the earliest aluminium alloys developed was 2017 (duralumin) which contains 4% Cu. Widely used aluminium alloys in the 2xxx series include 2014, 2017, 2024 and 2219. The 2xxx series alloys are widely used in the aerospace and transportation industries. The 6xxx series alloys are based on Mg and Si as the main alloying elements. During ageing, high strengths are developed due to the precipitation of Mg_2Si. Some of the important alloys in this series are 6061, 6063, 6111 and 6016. These alloys have moderately high strengths, good corrosion resistance and good weldability. An important characteristic of 6061 and 6063 is their excellent extrudability enabling the production of complex structural shapes by hot extrusion. These alloys are widely used in automotive and construction applications. The 7xxx series of alloys are based on Zn as the main alloying element. Common alloys in the 7xxx series include 7005, 7055, 7075, 7150 and 7475. These alloys are characterized by their very high strengths compared to other aluminium alloys. These alloys are widely used in the aerospace industry.

The non heat treatable wrought aluminium alloys are based on Al-Mn (the 3xxx series) and Al-Mg (the 5xxx series). The most common alloys in the 3xxx series are 3003 and 3004. Even though they cannot be strengthened by precipitation hardening, they are strain hardenable and also possess excellent formability. Alloy 3004 is widely used in the manufacture of beverage cans. Commonly used alloys in the 5xxx series include 5052 and 5083. These alloys can be strengthened by strain hardening and have excellent weldability. These alloys are widely used in the building and construction industry. The important alloy types in each class are described in detail elsewhere (Avner, 1987; Higgins, 1993; Kaufman, 1999; Polmear, 1989).

1.2 Deformation behaviour of aluminium alloys

The deformation behaviour on a macroscopic scale can be characterized by the flow curve of the material. The flow curve is a true stress–true strain curve and is useful in analysis of forming loads. Typically it is determined using compression testing on cylindrical specimens carried out as per ASTM standards (American Society for Testing and Materials, ASTM, 2000). Even with use of proper lubricants, some amount of barreling and non-uniform deformation can occur which can introduce errors into the measured flow curve. The flow curve can be corrected for the effect of friction using various approaches (see e.g., Cooke and Larke, 1945).

The flow curve obtained at room temperature is usually represented by a constitutive equation of the form,

$$\sigma = K\varepsilon^n \tag{1.1}$$

where σ is the flow stress at a strain of ε, K is the strength co-efficient and n is the strain hardening exponent. A typical flow curve for an aluminium alloy determined at room temperature is shown in Fig. 1.1. The values of K and n depend upon the alloy composition and the heat treatment given to the material. For commercially pure aluminium in the annealed condition the value of K is approximately 110 MPa and n is 0.26 whereas for a higher strength alloy like AA 5754 the K values are obviously higher (K = 390 MPa and n = 0.19) (Lange, 1985). For a 6xxx series alloy like AA 6063 in the solutionized condition, values of K and n were found to be 186 MPa and 0.23 respectively (Khan *et al.*, 2003). For forming load estimation, the flow stresses

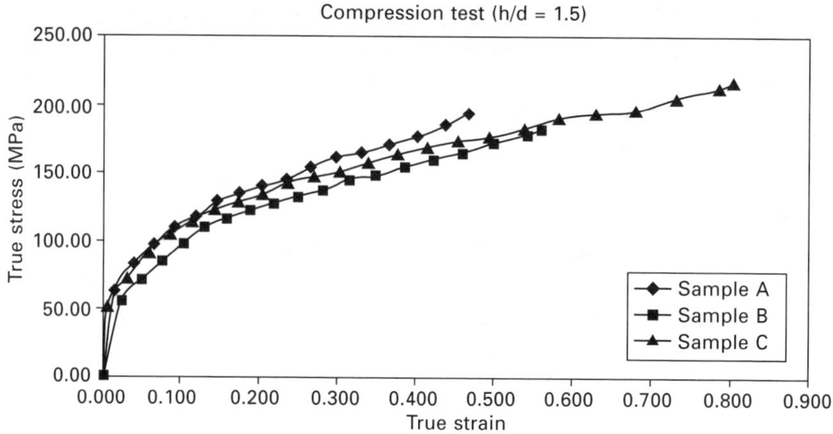

1.1 Typical experimentally measured flow curve for an aluminium alloy.

at various strains or the averaged value of flow stress over the strain range encountered during deformation can be used. The influence of strain rate can usually be neglected at room temperature.

Other constitutive equations can also be used especially for analysis of forming processes by numerical techniques. These include the Ramberg-Osgood equation and the Voce equation. The Ramberg-Osgood equation is especially useful for small strain deformation where the elastic strains cannot be ignored. The Ramberg-Osgood equation is given by the equation (Dowling, 1993)

$$\varepsilon = \frac{\sigma}{E} + \left(\frac{\sigma}{K}\right)^{\frac{1}{n}} \qquad 1.2$$

where σ is the flow stress at a strain of ε, K is the strength co-efficient, E is Young's modulus and n is the strain hardening exponent. At very high values of strain, a better prediction of the flow stress is obtained from the Voce equation (Airod *et al.*, 2003). The Voce equation is given by

$$\sigma = \sigma_0 + (\sigma_s - \sigma_0)(1 - \exp(-C\varepsilon)) \qquad 1.3$$

where σ is the flow stress at a strain of ε, σ_0 is the yield strength, σ_s is the steady state or saturation stress, and C is a constant co-efficient that is determined from experimental data.

At hot working temperatures, because of dynamic recovery processes, extensive strain hardening does not take place. The flow curves rise and reach a constant value which is characteristic of metals that undergo dynamic recovery. Typically strain rate effects on the flow stress have to be taken into account at elevated temperatures. The effect of strain rate on the flow stress (at some temperature, T and strain, ε) is usually expressed using the equation

$$\sigma = C(\dot{\varepsilon})^m \tag{1.4}$$

where σ is the flow stress, $\dot{\varepsilon}$ is the strain rate, C is a constant and m is the strain rate sensitivity of the material. Typically at elevated temperature, m is in the range 0.1 to 0.2. A more complete constitutive equation that can be used under hot working conditions has been proposed by Sellars *et al.* (1972). This equation is given by

$$\dot{\varepsilon} = A(\sinh \alpha\sigma)^{n'} \exp\left(\frac{-Q}{RT}\right) \tag{1.5}$$

where $\dot{\varepsilon}$ is the strain rate, σ is the flow stress, R is the gas constant A, α and n' are experimentally determined constants and Q is the activation energy for plastic deformation at the hot working temperature T. Values of these constants for some of the aluminium alloys have been listed by McQueen and Ryum (1985), McQueen and Ryan (2002). Additional mechanical property data at both room temperature and elevated temperatures has been compiled by Kaufman (1999).

1.3 Processing and microstructure

1.3.1 Microstructure development during deformation at room temperature

Development of the deformed microstructure in aluminium alloys is influenced by the fact that aluminium has high stacking fault energy. Because of the high stacking fault energy the spacing between the dissociated partial dislocations in aluminium is very small and this facilitates easy cross slip. Alloying elements, especially magnesium in aluminium alloys, can lower the stacking fault energy and hence the nature of the deformed microstructure.

Formation of dislocation structures in commercial-purity aluminium and in aluminium alloys containing only small amounts of magnesium

There is a rapid rise in dislocation density as a result of plastic deformation. The dislocations in a deformed metal are not randomly distributed. There is a driving force for the dislocations to re-arrange themselves into configurations that minimize the stored energy per unit length of the dislocation line. These dislocation configurations are referred to as low energy dislocation structures (LEDS) (Kuhlmann-Wilsdorf, 1982, 1989). With increasing amounts of plastic deformation a series of structures develop within an individual deforming grain. At small strains, dislocation tangles form that are simply regions of low dislocation density surrounded by regions of high dislocation density. With increasing strain, more and more dislocations are generated and these are trapped at the tangled walls. This results in what is called a cell structure, where the interior of the cell is relatively free of dislocations and the cell walls are composed of dislocations (Barker *et al.*, 1989). The dislocations within the cell wall cause a misorientation between two adjacent cells. This misorientation is usually of the order of 2 degrees.

From studies on high purity aluminium, it has been shown dislocation tangles begin to develop by a strain of 0.05 and a well developed cell structure develops by a strain of 0.15 (Young *et al.*, 1986). The amount of misorientation between two adjacent cells can be correlated with the amount of strain. With increasing deformation, different substructures can develop. Groups of cells can be bounded by dense dislocation walls which are dislocation boundaries; the difference being the misorientation across the dense dislocation walls is much higher than the misorientation between two adjacent cells. In effect, several groups of cells are bounded by the dense dislocation walls. Substructures referred to as microbands, which are composed of several smaller cells, can also form later on the dense dislocation walls (Bay *et al.*, 1989). A schematic diagram showing the development of these kinds of dislocation substructures is shown in Fig. 1.2.

According to Bay *et al.* (1989), all the cells and microbands enclosed by the dense dislocation walls share the same combination of active glide systems. These glide systems are, however, fewer in number than are required to satisfy the Taylor condition fully because the rate of work hardening increases with increasing number of active slip systems. (The Taylor condition states that the operation of at least five independent slip systems is required for a crystal to undergo general shape change by plastic deformation.) The development of the dislocation substructure is not necessarily uniform from grain to grain or even within one particular grain. These differences arise because of differences in the original orientations of the grains with respect

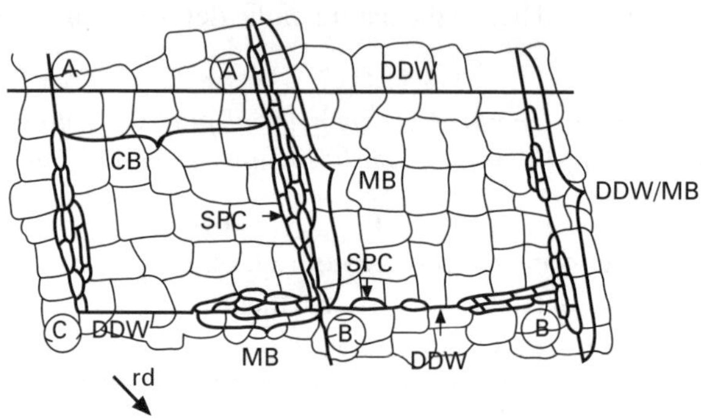

1.2 Schematic diagram showing the development of dislocation substructures in aluminium. (Reprinted from Materials Science and Engineering A, vol. 113, Bay B, Hansen N and Kuhlmann-Wilsdorf D, *Deformation Structures in Lightly Rolled Pure Aluminium*, pp. 385–399, 1989 with permission from Elsevier).

to the direction of externally applied loading. Some grains contain only dislocation cells, some contain dense dislocation walls in combination with dislocation cells, some contain both dense dislocation walls and microbands in addition to dislocation cells (Bay *et al.*, 1989). A typical deformed microstructure showing a dislocation cell structure with microbands is shown in Fig. 1.3. With further plastic deformation, the misorientations between the neighboring cells increases and they are referred to as subgrains. Typically the misorientation angle between two adjacent subgrains is less than 15 degrees. When the misorientation across the boundaries exceeds 15 degrees, the boundaries are regarded as high angle grain boundaries separating individual grains.

After large amounts of plastic deformation, microbands intersecting grain boundaries can also be observed. These are sometimes also referred to as

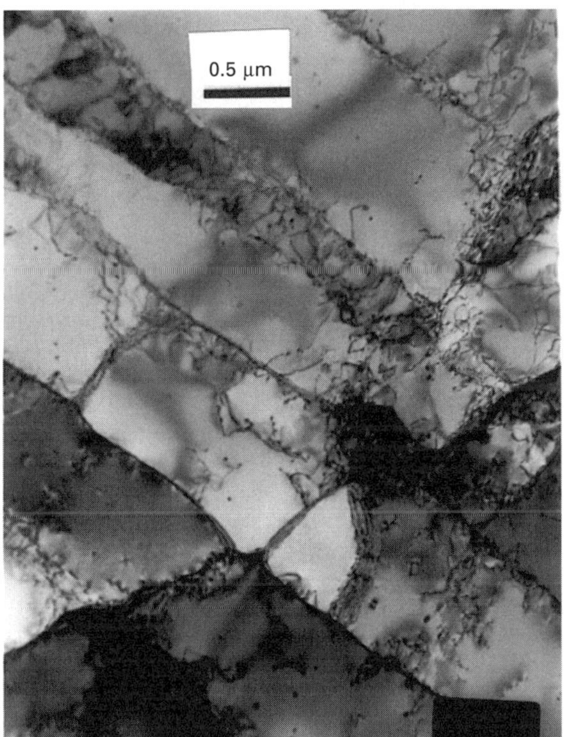

1.3 TEM micrograph showing a typical dislocation substructure in aluminium formed as a result of plastic deformation at room temperature. (Reprinted from *Scripta Materialia*, vol. 39, Chakkingal U, Suriadi A and Thomson P F, 'Microstructure development during equal channel angular drawing of pure aluminium at room temperature', pp. 677-684, copyright (1998), with permission from Elsevier).

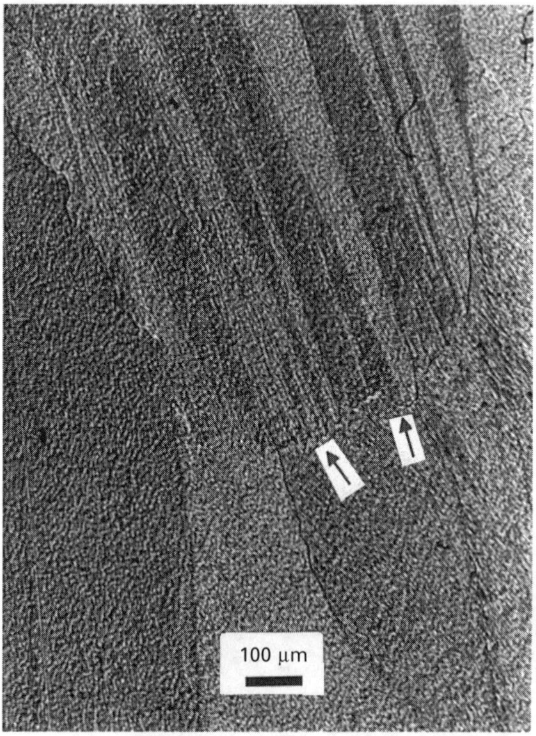

1.4 Optical micrograph showing the intersection of microbands with grain boundaries in aluminium after room temperature plastic deformation. (Reprinted from *Materials Science and Engineering A*, vol. 266, Chakkingal U, Suriadi A and Thomson P F, *The development of microstructure and influence of processing route during equal channel angular drawing of pure aluminium*, pp. 241–249, copyright (1999), with permission from Elsevier).

micro shear bands. A distinct shear offset is usually observed at locations where the microband intersects grain boundaries. Figure 1.4 shows a deformed microstructure in aluminium showing micro bands intersecting grain boundaries. For some plastic deformation conditions, as in equal channel angular drawing where the deformation is largely by shear, the microbands are found to lie parallel to the trace of the {111} plane in the grain (Chakkingal *et al.*, 1999). However, this is not always the case for deformation by cold rolling (Korbel *et al.* 1986a). Conversion of some of the microbands to macroscopic shear bands can also occur under suitable conditions. Shear bands usually lie along planes of maximum macroscopic shear stress and are formed after large amounts of plastic deformation. They cross several grain boundaries and their formation usually involves some amount of strain softening (Kuhlmann-Wilsdorf, 1999).

Influence of other alloying elements, especially magnesium, in influencing formation of dislocation structures in aluminium alloys

The presence of other alloying elements can also influence the evolution of the deformed microstructure in aluminium alloys. A typical example is in alloys containing large amounts of magnesium (approximately 5% Mg). Magnesium lowers the stacking fault energy in aluminium and hence these alloys do not easily form dislocation cells and subgrains. At small strains dislocations are arranged in nearly uniform arrays such that they form a Taylor lattice containing multiple Burgers vectors and having alternating misorientations along the {111} slip planes (Hughes, 1993). With increasing amounts of plastic strains, these alloys have a strong tendency to form microbands and shear bands. Because of formation and intersection of a large number of microbands, substructure that resembles the subgrains in other aluminium alloys begins to develop. Such microstructures have been shown to develop as a result of cold rolling and also by cyclic extrusion and compression (Korbel *et al.*, 1986b, Richert *et al.*, 2001).

The presence of particles and precipitates in the alloy can also influence the development of the deformed microstructure. Presence of small particles in commercial purity aluminium has been shown to promote formation of equiaxed structures while decreasing the formation of dense dislocation walls and microbands (Barlow *et al.*, 1989, Hughes *et al.*, 1998).

Static recovery and recrystallization

When cold worked high purity aluminium is annealed, some softening can occur as a result of static recovery without any change in the elongated structure of the original grains. The boundaries of the dislocation cell structures that were created as a result of plastic deformation become well defined and the cells eventually transform into subgrains. A considerable fraction of the stored energy of plastic deformation in the cold worked metal is removed as a result of recovery processes in aluminium. This release of stored energy occurs by the removal of point defects and dislocation rearrangement into low energy structures. The change in dislocation configuration during recovery occurs due to the annealing of dislocation loops and the formation of regular dislocation networks. Growth of subgrains under appropriate conditions can also occur by a process of subgrain coalescence and sub boundary coalescence (Young *et al.*, 1986). Static recovery of the deformed microstructure reduces the driving force for static recrystallization. Static recrystallization involves the nucleation of new stress-free grains (which can be aided by static recovery processes) and their growth thereby consuming the deformed microstructure. Static recrystallized microstructure usually shows equiaxed grains with high angle grain boundaries without extensive dislocation substructures within the grains. Recrystallization in aluminium also follows the Avrami rate equation.

1.3.2 Microstructure development during deformation at hot working temperatures

During hot deformation of metals, restoration processes like dynamic recovery and dynamic recrystallization can occur. In aluminium and its alloys, the usual restoration mechanism is dynamic recovery. As in cold working, the evolution of the microstructure during hot working of aluminium alloys is also influenced by the fact that aluminium has high stacking fault energy. This facilitates easy cross slip of screw dislocations that are required for the generation of recovered microstructures. Presence of alloying elements (especially magnesium) that lower the stacking fault energy and precipitates also influence the development of the microstructure. In commercial purity aluminium and in alloys where magnesium is present only in small amounts, dynamic recovery is the only restoration mechanism during hot working. Because of the high levels of dynamic recovery, dynamic recrystallization generally does not occur (McQueen *et al.*, 1991).

When dynamic recovery occurs, the flow curve ($\sigma - \varepsilon$ curve) rises monotonically to some saturation stress level that depends upon the deformation temperature and the strain rate, with no further increase with subsequent strain. Dislocations generated as a result of plastic deformation are continuously annihilated and rearranged to form low energy dislocation walls that define equiaxed polygonized subgrains. The hot worked microstructure shows deformed and elongated grains which are neatly subdivided into subgrains. The subgrain size usually reaches a limiting value of a few microns in aluminium alloys; with continued deformation usually there is no further decrease in the subgrain size. A typical dynamically recovered microstructure resulting from a hot working operation is shown in Fig. 1.5.

The critical value of the strain for the onset of dynamic recovery and the saturation value of the stress are lower as the deformation temperature is increased and the strain rate is decreased, or in other words as the value of the Zener-Hollomon parameter Z ($Z = \dot{\varepsilon} \exp(Q/RT)$, where $\dot{\varepsilon}$ is the strain rate, Q is the activation energy for dynamic recovery, R is the universal gas constant and T is the deformation temperature (McQueen *et al.*, 1996). The sizes of the stable subgrain that result from dynamic recovery also depend upon the hot working conditions. McQueen *et al.* (1996) have expressed this by the equation

$$d_s^{-1} = a + b \log Z \qquad 1.6$$

where d_s is the subgrain size, Z is the Zener-Hollomon parameter and a and b are material constants. The flow stress is also dependent on the subgrain size and can be expressed by the equation (Raj *et al.*, 1986)

$$\frac{d_s}{b} = K\left(\frac{G}{\sigma}\right)^m \qquad 1.7$$

1.5 TEM micrograph showing a typical dynamically recovered microstructure resulting from a hot working operation in aluminium. (Reprinted from *J. Materials Processing Technology*, vol. 117, Chakkingal U and Thomson P F, 'Development of microstructure and texture during high temperature equal channel angular extrusion of aluminium', pp. 169–177, copyright (2001), with permission from Elsevier).

where d_s is the subgrain size, b is the magnitude of the Burgers vector, G is the shear modulus, σ is the flow stress and K and m are constants.

Formation of grain boundary serrations

After considerable hot working, as a result of dynamic recovery, serrations can appear along the grain boundaries in aluminium. The serrations begin to appear as the subgrain structure develops and usually the apex of serrations of one grain occur where the grain boundary migration has taken place along a subgrain boundary into a neighbouring grain (McQueen *et al.*, 1995). This local migration of the grain boundary migration occurs to absorb dislocation walls as the subgrain boundaries are constantly being eliminated and reformed during dynamic recovery. The occurrence of serrations is dependent on the subgrain size (which is influenced by the temperature and strain rate) and the alloying elements present. Figure 1.6 shows a typical example of grain

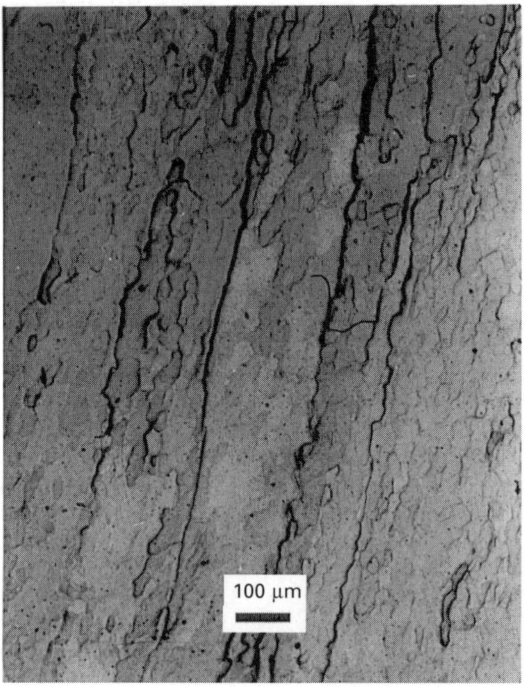

1.6 Optical micrograph showing serrations developed at grain boundaries in aluminium as a result of hot working. (Reprinted from *J. Materials Processing Technology*, vol. 117, Chakkingal U and Thomson P F, 'Development of microstructure and texture during high temperature equal channel angular extrusion of aluminium', pp. 169–177, copyright (2001), with permission from Elsevier).

boundary serrations formed during hot working of commercially pure aluminium.

The formation of grain boundary serrations can lead to what is referred to as geometric dynamic recrystallization after large strains (Blum *et al*., 1996). When serrated boundaries come close to each other as a result of thinning of the original grains, serrations of opposite boundaries can come into contact. This causes splitting or perforation of the original grains by high angle grain boundaries and the apparent appearance of equiaxed grains. Because this process does not occur by the nucleation of new stress free grains and their growth consuming the deformed microstructure, it does not correspond to dynamic recrystallization.

Even though a large body of work seems to suggest that dynamic recovery is the main restoration mechanism during hot working and that dynamic recrystallization does not occur in commercial-purity aluminium and in aluminium alloys with low alloy contents, there are some studies that have found evidence of dynamic recrystallization under specific deformation conditions (Ren *et al.*, 1995, Ravichandran *et al.*, 1991).

Influence of alloying elements

Alloying elements that lower the stacking fault energy can change the dynamic restoration mechanisms during hot working. Solutes, like magnesium in sufficient amounts which lower the stacking fault energy, can reduce the tendency for dynamic recovery leading to dynamic recrystallization after a critical strain is reached. In commercially used aluminium alloys, the presence of precipitates can also promote dynamic recrystallization during hot working due to particle stimulated nucleation. In general, dynamic recrystallization does not seem to occur in Al alloys that contain less than 4% Mg (McQueen *et al.*, 1991). In aluminium alloys containing solutes like Cu and Zr dynamic recrystallization does not occur; in these alloys the presence of precipitate particles lowers the grain boundary mobility and even lowers the rate at which static recrystallization proceeds (McQueen *et al.*, 1994).

In some alloys precipitates (with sizes that do not pin grain boundaries preventing grain boundary motion) can cause particle stimulated nucleation due to the creation of locally high dislocation densities in the vicinity of the particles. This has been reported for alloy 5083 (Al-4.5 wt% Mg-0.8wt%Mn) under hot working conditions due to presence of fine precipitates of $MnAl_6$ (McQueen, 1988).

1.4 Processing and properties

Aluminium and its alloys exhibit very high levels of formability and can therefore be processed by using all conventional forming techniques like rolling, extrusion, forging, drawing, etc., to produce engineering components. In addition, some aluminium alloys also exhibit superplasticity and can be superplastically formed to different near net shapes. Key aspects of forming of aluminium alloys are described below.

1.4.1 Rolling of aluminium and its alloys

Aluminium and its alloys can be rolled to produce flat rolled products like plates, sheets and foils. The cast ingots are hot rolled initially followed by cold rolling to produce flat products of the required thickness. The final mechanical properties can be controlled by manipulating the degree of cold working, partial or full annealing, ageing (for precipitation hardening grades), use of different stabilization treatments, etc. Commercially the 6xxx series of alloys are an important class of alloys which are produced in the form of sheets by rolling primarily for auto body applications. Important alloys in this category include AA 6009, AA 6010, AA 6016 and AA 6111; the higher strength alloy AA 6111 is primarily used in the US, while the higher formability alloy AA 6016 is preferred in Europe for auto body applications (Engler *et al.*, 2002).

For producing sheets, the direct chill cast alloys are scalped to remove surface defects and soaked at an elevated temperature (usually around 500–550 °C depending on the alloy grade) for extended periods of time to homogenize the billets and eliminate interdendritic segregation. Models that predict microstructural changes in various commercial grades of aluminium alloys during the homogenization process have been proposed (Dons, 2001). After initial breakdown hot rolling to break down the coarse as-cast microstructure, hot rolling is carried out in a multi-stand mill to produce sheet a few mm thick. This is then cold rolled to produce the final sheet with desired thickness. For auto body applications, the sheet is then continuously annealed at temperatures around 500–550 °C, followed by quenching, to promote recrystallization and dissolve precipitates that contribute to age hardening. The processing parameters have to be controlled to get optimum mechanical properties and crystallographic texture in the sheet as this influences subsequent formability (Engler *et al.*, 2002).

Some aluminium alloys can also be cast directly into thinner strips of thickness varying from 15 mm to 40 mm which can later be cold rolled to sheets or foils after an intermediate hot rolling step. Twin roll casting is another variation where a strip of thickness 6–12 mm is directly cast and cold rolled to produce sheet or foil of the desired gauge thickness. These processes have the advantage of greater energy efficiency. Alloys with a narrow freezing range are especially suited for twin roll casting. In this process, molten metal is fed into water cooled rolls where it solidifies and it is then rolled. Successful twin roll casting depends upon proper selection of processing parameters like roll diameters, selection and application of lubricant, etc. (Yun *et al.*, 2000).

1.4.2 Extrusion of aluminium and its alloys

Extrusion is a very important processing route for aluminium alloys. Various components and structural members like channels, rods, tubes, etc., can be easily and economically fabricated by extrusion. For example, alloys like 6061 and 6063 are widely used for extruding architectural members; alloys 2024, 7075, etc., are widely used for extruding aircraft structural shapes; while tubes and pipes can typically be extruded from a wide range of aluminium alloys (Saha, 2000).

Hot extrusion of aluminium alloys

The age-hardenable aluminium alloys are more commonly hot extruded compared to the non-age-hardenable alloy grades. The 6xxx series of aluminium alloys is especially suited for hot extrusion and is widely used for making extruded components. The ease with which aluminium alloys can be extruded is usually designated by a dimensionless number called the relative extrudability number. The relative extrudability is a measure of the maximum extrusion speed that can be employed during extrusion and is influenced by billet temperature and state of stress in the deformation zone (Saha, 2000). The relative extrudability is primarily limited by the temperature rise in the deformation zone due to deformation heating; at high extrusion speeds, the exit temperature can approach the solidus temperature of the alloy. Alloy AA 6063, which is a commonly extruded alloy, has been assigned a relative extrudability of 100 and extrudability of other alloys is indicated relative to this alloy. In general, alloys with relative extrudabilities in the range 50 to 150 (e.g. alloys AA 1060, 3003, 5052, 6061 and 6063) are considered easy to extrude; those with relative extrudabilities in the range 30–50 (e.g. AA 6066) are considered moderately difficult to extrude; those with values less than 30 (AA 2024, 7075, 7079, etc.) are considered difficult to extrude (Laue *et al.*, 1981).

Billets for hot extrusion are normally homogenized to remove grain boundary segregation and dissolve brittle intermetallic compounds. Hot extrusion is carried out using flat dies (in contrast to converging dies that are normally used in cold extrusion) and therefore a dead metal zone develops in the corners of the dies. Lubricants are generally not used and therefore sticking friction can be expected to prevail along the billet-container and billet-die interface. The extrusion pressure that is required depends upon the billet temperature, extrusion ratio, the length of the billet and the complexity of the shape of the extruded product. The extrusion pressure increases with decreasing billet temperature, increasing extrusion ratios and increasing billet length. The billet temperature depends upon the alloy grade to be extruded and is usually in the range of 420–520 °C. For example, billet temperatures range from 420 °C for a AA 1060 to 430–500 °C for AA 6061 (Laue *et al.*, 1981).

The actual billet temperature that is utilized is also influenced by other process parameters like the extrusion speed because these factors influence the exit temperature of the extruded product. The exit temperatures cannot be so high that incipient melting around precipitate particles occurs. Additionally, age hardenable alloys are usually press quenched on the run out table, therefore the exit temperature has to be such that the elements that contribute to precipitation hardening during subsequent ageing are taken into solution. This removes the need for an additional solutionizing step for the extruded products.

The temperature rise due to deformation during extrusion usually leads to an increase in the exit temperature. This rise in exit temperature leads to variations in the mechanical properties of the extruded product. Therefore various techniques have been devised to maintain a constant exit temperature during extrusion and this is referred to as isothermal extrusion. Isothermal extrusion is usually accomplished by taper heating the billet such that the front of the billet is at a higher temperature compared to the rear or by taper quenching the back end of the billet with a water spray before transfer to the container for extrusion. Alternatively, if the billet temperature is uniform, isothermal extrusion can be accomplished by reducing the ram speed in a controlled manner as the extrusion proceeds. The principle of isothermal extrusion is illustrated in Fig. 1.7.

Tubes and hollow sections can also be easily hot extruded from solid billets either by seamless extrusion, which uses a mandrel to define the inner section, or by using dies with welding chambers. This is illustrated in Figs 1.8 and 1.9 respectively. When welding chamber dies are used, the metal stream is split up as it enters the deformation zone where it enters into separate ports and flows around the central bridge which supports a short mandrel. The separate streams of metal that flow through the ports are welded together in the welding chamber and then the extruded product exits the die as a hollow section (Dieter, 1986). The temperature in the welding chamber

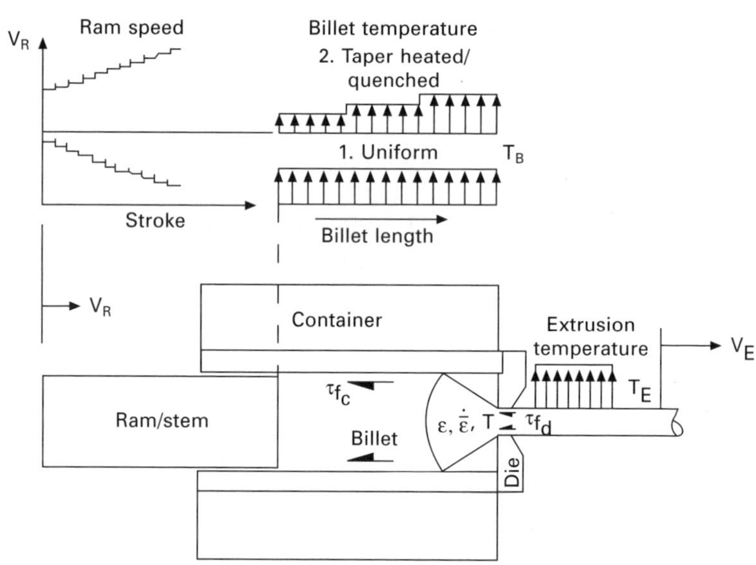

1.7 Schematic diagram illustrating the underlying principle behind the isothermal extrusion of aluminium alloys. (Source: Saha P K, *Aluminum Extrusion Technology*, ASM International, 2000. Reprinted with permission from ASM International).

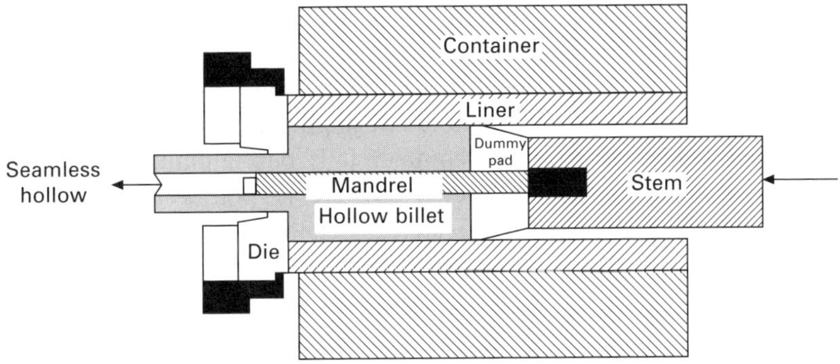

1.8 Diagram illustrating the principle of seamless tube extrusion. (Source: Saha P K, *Aluminum Extrusion Technology*, ASM International, 2000. Reprinted with permission from ASM International).

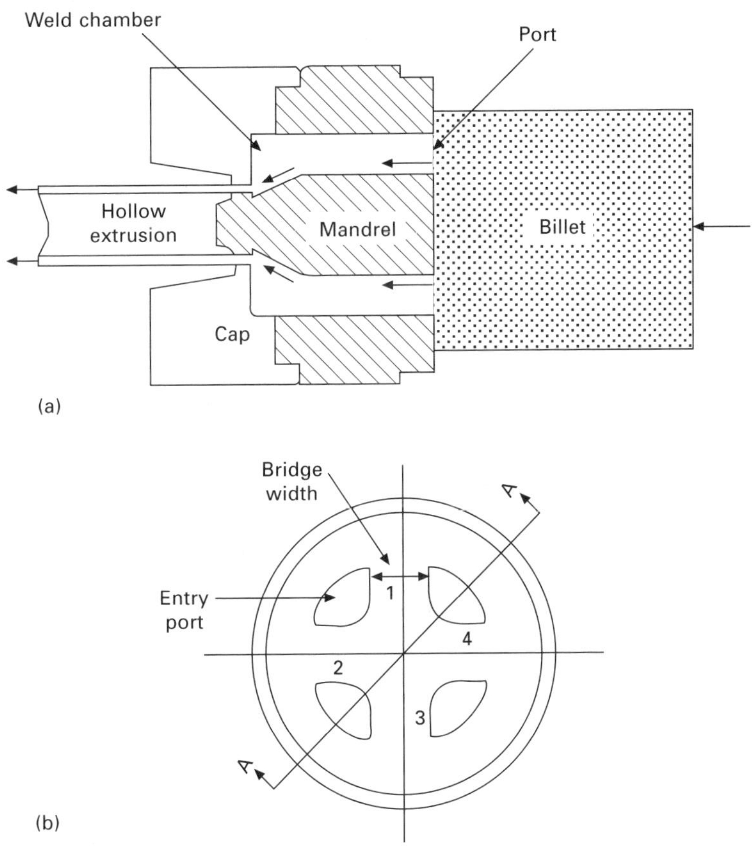

1.9 Schematic of welding chamber (porthole) die hollow extrusion. (a) Cross-section showing metal flow into port streams and around the mandrel. (b) Face of the die set through which the billet enters. (Source: Saha P K, *Aluminum Extrusion Technology*, ASM International, 2000. Reprinted with permission from ASM International).

has to be sufficiently high to ensure clean welding. The extrusion pressures are usually high (compared to solid extrusions with the same extrusion ratios) because of the greater frictional resistance to flow through the ports and around the mandrel. This technique can be applied to alloys that easily form good, clean welds. For example, commercially pure aluminium, AA 6063, AA 6061, AA 3103, AA 3003, etc., have very good welding properties whereas alloys like 2024, AA 7075, etc., have poor welding properties (Laue *et al.*, 1981).

Cold extrusion of aluminium alloys

Cold extrusion refers to extrusion carried out with billets at room temperature. Cold extrusion has some inherent advantages in that higher mechanical properties are obtained in the extruded product if heat generated due to deformation does not cause recovery or recrystallization. Cold extruded components also have better surface finish and are more dimensionally accurate. Because the forming loads are much higher, the proper choice of lubricant and die design are very important.

1.4.3 Other important forming techniques for processing aluminium alloys

Aluminium alloys can also be processed by other conventional forming techniques. Aluminium alloys can be closed die forged to produce parts with good surface finish and dimensional tolerances. Aluminium alloys are generally considered to be more difficult to forge compared to steels. Alloys 6061, 2014, 7075, 7010, etc., are among the alloys that are used for producing forged components for use in the automobile and aerospace industries. Forging temperatures for aluminium alloys are in the range of 400–550 °C and depend upon the alloy composition. Normally, heated dies and such lubricants as colloidal graphite in water are used (Bray, 1990). Metal temperature is normally controlled within a narrow range during forging.

Deep drawing forms another important technique for the processing of aluminium alloy components. Aluminium alloy beverage cans are produced in large numbers using this manufacturing route from cold rolled alloy sheets. Cups of required diameter are initially drawn from blanks. This is then redrawn and ironed to form a can of desired wall thickness. The processing of the sheets has to be controlled to get optimum mechanical properties and crystallographic texture for good drawability.

Stretch forming is another fabrication technique employed for the production of aluminium components because aluminium alloys have high ductilities. In this technique, sheet is gripped at the ends and stretched over a form block beyond its elastic limit to produce desired shapes. This technique is widely used in the aerospace industry to produce parts with a large radius of curvature (Dieter, 1986).

Some aluminium alloys exhibit superplasticity. This is a property by which alloys exhibit very large elongations before necking and fracture occur during a tensile test. Such alloys can also be formed into components by superplastic forming. For aluminium alloys to exhibit superplasticity, they should exhibit high values of the strain rate sensitivity, m and have a fine grained microstructure (grain size typically less than 10 microns) which is stable at the forming temperature (which is usually higher than $0.5T_m$). Low strain rates (lower than 10^{-2} per second) are also required. In contrast to conventional mechanisms of deformation at hot working temperatures, the deformation mechanism during superplastic forming is primarily grain boundary sliding. Commercially Al-Zn-Mg-Cu alloys belonging to the 7xxx series (AA 7475 and AA 7075) and alloys like AA 2091 are among the important aluminium alloys that exhibit superplasticity. In all cases controlled thermo-mechanical processing is required to obtain the desired microstructure. For example, the effect of different thermo-mechanical treatments and forming parameters on the superplastic behaviour of AA 7075 has been described by Smolej *et al.* (2001). Aluminium alloys can also be produced in the form of wires, rods, bars, etc., from cast ingots by a combination of rolling, extrusion, drawing, etc.

1.4.4 Precipitation hardening of aluminium alloys

Precipitation hardening is an important strengthening mechanism in aluminium alloys especially in the 2xxx, 6xxx and the 7xxx series. Strengthening is achieved by a two-step process: a solutionizing step at elevated temperatures followed by an ageing treatment at lower temperatures in order to increase the strength and hardness. The term natural ageing is used to refer to the increase in hardness observed in solutionized specimens after extended periods of time at room temperature. Normally ageing treatment, referred to as artificial ageing, is carried out at higher temperatures (which could range from 130–180 °C) for a few hours to develop peak strength and hardness. The time and temperature selected for the ageing treatment depends on the alloy composition and the final mechanical properties required.

Traditionally the Al – 4% Cu system has been used to illustrate precipitate hardening. The initial solutionizing step involves heating the alloy to a temperature of about 550 °C where all the Cu is taken into solution followed by a quenching step to obtain a supersaturated solution of Cu in Al. During ageing, a series of precipitation steps occurs, leading to an increase in strength and hardness. The equilibrium precipitate in this system is θ ($CuAl_2$); however the precipitates that form initially do not correspond to this phase. Instead a series of transition phases are formed during ageing before the equilibrium phase $CuAl_2$ forms. The first precipitate to nucleate is referred to as a GP (Guinier-Preston) zone. GP zones are little more than disc shaped clusters, a few atomic layers thick and around 100 A° in diameter. These zones form first because they are coherent with matrix leading to low interfacial energies. This lowers the activation energy barrier for formation of these precipitates. The activation energy barriers for the nucleation of various intermediate phases are much lower compared to those for the formation of the equilibrium precipitate.

Even though the interfacial energies are small, large coherency strains have to be accommodated in the vicinity of the GP zones. In the Al-Cu system, coherency strains are minimized because the GP zones are such that the discs are perpendicular to the elastically soft <100> direction in the aluminium matrix (Porter *et al.*, 1981). On continued ageing the GP zones disappear and are replaced by other transition phases (referred to as θ) before the equilibrium precipitate is formed. With the formation of these precipitates the degree of coherency is slowly lost and the final precipitate that forms is incoherent with the matrix. With the changing nature of the precipitates, the nature of dislocation interaction with the precipitate also changes leading to differences in mechanical properties. Usually by the time the equilibrium precipitates appear, the peak value of strength has been crossed and the alloy is in the overaged condition. The appearance of these precipitates also depends upon the ageing temperature. If ageing is carried out at a temperature higher than the GP zone solvus temperature but below the θ'' solvus temperature of a precipitate, the GP zones will not appear. Instead the first precipitate that forms will be θ'' (Porter *et al.*, 1981).

Precipitation hardening sequences in other aluminium alloys are shown in Table 1.2. The GP zones formed in other systems are not necessarily disc shaped. The equilibrium precipitate in the Al-Cu-Mg system is $CuMgAl_2$, in the Al-Mg-Si system it is Mg_2Si, in the Al-Zn-Mg system, it is $MgZn_2$. The precipitation hardening process is also influenced by the presence of vacancy

Table 1.2 Precipitation hardening sequence in some aluminium alloys

Alloy type	Precipitation sequence	Reference
Al – Ag	GP zones (spheres) $\rightarrow \gamma'$ (plates) $\rightarrow \gamma$ (Ag_2Al)	Porter *et al.* (1981)
Al – Cu	GP zones (discs) $\rightarrow \theta''$ (discs) $\rightarrow \theta'$ (plates) $\rightarrow \theta$ ($CuAl_2$)	Martin (1968)
Al – Cu - Mg	GP zones (rods) $\rightarrow S'' \rightarrow S'$ (laths) $\rightarrow$ S (Al_2CuMg) (laths)	Ber (2000a)
Al – Zn - Mg	GP zones (spheres) $\rightarrow \eta'$ (plates) $\rightarrow \eta$ ($MgZn_2$) (plates or rods)	Deschamps *et al.* (1999)
Al – Mg - Si	GP – I zones β'' (needles) $\rightarrow$ β' (rods) $\rightarrow \beta$ (Mg_2Si) (plates)	Bryant *et al.* (1998)

concentrations that are in excess of equilibrium concentrations as a result of quenching. Excess vacancies promote nucleation by forming clusters that act as heterogeneous nucleation sites and by enhancing diffusion rates in the matrix. The increase in dislocation density as a result of deformation before ageing also promotes heterogeneous nucleation and influences the nature of the precipitation process. For example, the Al-Zn-Mg family of alloys, which is widely used in the aerospace industry, is usually given a small plastic deformation to promote precipitation and to relieve internal stresses that arise due to quenching as otherwise the internal stresses would lead to dimensional instability during later machining operations (Deschamps *et al.*, 1999). The normal precipitation sequence in these alloys is of the form: supersaturated solid solution → GP zones → metastable η' → stable η (see Table 1.2). Plastic deformation prior to ageing changes the precipitation sequence and also the size and spatial distribution of the precipitates (Deschamps *et al.*, 1999).

The precipitation hardening process in a system can also be enhanced by small additions of other alloying elements and by pre-ageing treatments (which are essentially ageing treatments carried out at temperatures much lower than the recommended artificial ageing temperature). For example, in the 6xxx series alloys used in the automobile body panel applications, alloys that develop high strengths due to ageing during the paint bake cycle are particularly preferred. The paint bake cycle usually corresponds to a thermal treatment of about thirty minutes at temperatures in the range of 170–180 °C (Bryant *et al.*, 1998). The commonly used aluminium alloy for this application is AA 6111 which is essentially an Al–Mg–Si alloy with about 0.5–0.9% addition of Cu to enhance the rate at which ageing takes place. In AA 6111, in addition to the metastable β'', β' and the equilibrium β phase, a quaternary precipitate referred to as the *Q* phase can also form influencing the strength (Perovic *et al.*, 1999). Pre-ageing can also influence the ageing process. For example, for an AA 6016 alloy subjected to a pre-ageing treatment of 24 hours at 100 °C, the strength after the paint bake cycle was considerably increased compared to the case where only natural ageing had occurred prior to the paint bake cycle (Bryant *et al.*, 1998).

The ageing response of aluminium alloys can be represented using time–temperature–transformation (TTT) diagrams showing the formation of various precipitates or by using diagrams that show the mechanical properties resulting from different ageing treatments. Diagrams such as these have been plotted for various commercially important alloy systems like Al-Cu-Mg (Ber, 2000a), Al-Cu, Al-Zn-Mg–(Cu), Al-Mg-Si-(Cu)(Ber, 2000b), Al-Li-Mg-Sc-Zr (Davydov *et al.*, 2000) and Al-Li-Zn-Mg-Cu (Wei *et al.*, 2000). A typical TTT diagram for ageing in a 6061 type alloy is shown in Fig. 1.10. Typical mechanical properties of some aluminium alloys are shown in Table 1.3.

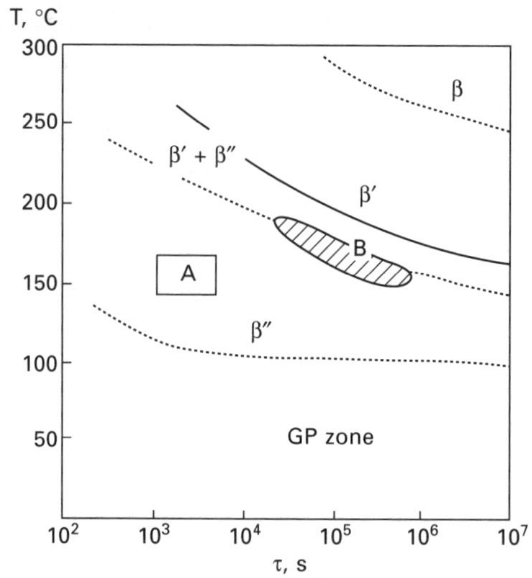

1.10 A TTT diagram (during ageing) for a 6061 alloy. (Reprinted from *Materials Science and Engineering* A, vol. 280, Ber L B, 'Accelerated artificial ageing regimes of commercial aluminum alloys', II – Al-Cu, Al-Zn-Mg-(Cu), Al-Mg-Si-(Cu), pp. 91–96, Copyright (2000), with permission from Elsevier).

Table 1.3 Typical mechanical properties of some aluminium alloys

Alloy	Condition	Yield strength (MPa)	Ultimate tensile strength (MPa)	% Elongation	Reference
2024	T6	409	451	10	Ber (2000a)
2219	T6	346	434	8.2	Ber (2000b)
3003	H18	185	200	10	Kaufman (1999)
5052	H34	215	260	16	Kaufman (1999)
6061	T6	280	315	12	Porter (1981)
6063	T6	215	240	18	Kaufman (1999)
7075	T6	570	620	10	Kaufman (1999)

1.5 Severe plastic deformation techniques

In recent years, much work has been carried out on using severe plastic deformation techniques to produce ultra fine grained structures in aluminium alloys. These processes can be used to develop high strengths, special textures, refine grain size, enhance superplastic behaviour in some alloys, etc. An important process is the equal channel angular extrusion process which was originally described by Segal *et al.* (1981) and Segal (1999). In this process, the die consists of two channels of equal cross-sectional area intersecting at some specified angle 2ϕ (which is usually between 90 and 120 degrees). A schematic diagram illustrating this process is shown in Fig. 1.11. A well lubricated specimen is placed in one of the channels and extruded into the other with a punch. Deformation experienced by the sample is basically one of internal shearing. The equivalent plastic strain imparted to the specimen as a result of this extrusion is given by Segal *et al.* (1981) as

$$\varepsilon = \frac{2}{\sqrt{3}} \cot \phi \qquad 1.8$$

where ε is the equivalent plastic strain and 2ϕ is the die angle (angle between the two intersecting channels). For a die angle of 90°, ε = 1.15 for one pass and for a die angle of 120°, ε = 0.67.

The cross-sectional area of the specimen does not change as a result of the extrusion. Therefore the specimen can be repeatedly deformed by extrusion through the same die leading to accumulation of large plastic strains. In addition, by changing the orientation of the specimen from one pass to the next and thereby changing the strain path to which the billet is subjected, different microstructures and mechanical properties can be produced. Coarse grained structures can easily be converted into a microstructure consisting of submicron sized grains/subgrains by this process. The level of grain refinement that is possible depends upon the die angle, alloy composition, total strain imparted, etc. The effect of equal channel angular extrusion on various aluminium alloys has been investigated (e.g. Horita *et al.*, 2001, Lee *et al.*, 2002). In spite of several inherent advantages, this process has still not entered commercial use. Issues relating to scaling up the process to handle industrial sized billets remain to be addressed.

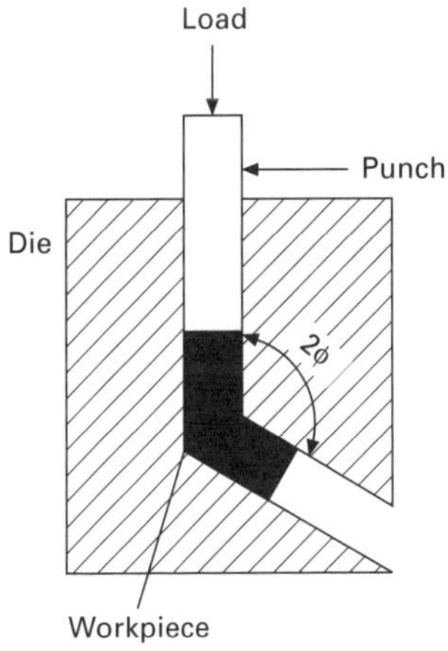

1.11 Schematic diagram illustrating the equal channel angular extrusion process.

1.6 Common problems and solutions

Even though aluminium alloys exhibit high levels of formability, some defects can occur during manufacturing as a result of improper choice of processing parameters. Some of these defects are briefly described below.

Variation in sheet thickness can sometimes result from elastic deflection of rolls due to the forces involved in rolling. This can usually be corrected by cambering the rolls so that the diameter of the rolls is slightly larger at the centre of the rolls compared to the ends. Inhomogeneous deformation during rolling can also lead to defects like edge cracking and alligatoring (which refers to the sheet splitting along the plane of the sheet). These defects can be avoided by using vertical edge rolls which keep the edges straight or by trimming the edges between passes (Dieter, 1986).

Sometimes, towards the end of extrusion of the billet, oxide particles from the surface of the billet can enter the extruded product. This is referred to as extrusion defect. A hollow region called pipe can also form at the back end of the extruded product towards the end of the extrusion of the billet. These defects can be avoided by stopping the extrusion process before the end of the billet is extruded and discarding the remainder of the billet. Surface cracking can also occur due to an excessive rise in the exit temperature of the extruded product. When this problem becomes severe and the surface cracks are closely spaced, this defect is referred to as a fir tree defect. Surface cracking can be avoided by reducing the ram speed thereby reducing the exit temperature. In some alloy grades, blisters can appear on the surface of the extruded product. Blistering is usually caused by expansion of entrapped hydrogen at the grain boundaries and is avoided by reducing hydrogen pick up by the alloy during solidification, pickling and cleaning, heat treatment, etc. (Saha, 2000). Another defect is variation in properties and structure between the front and back ends of the extrusion, especially in alloys that are press quenched and artificially aged. This is avoided by keeping the exit temperature constant. Detailed descriptions of these defects and their prevention have been given by Saha (2000) and Dieter (1986).

During the drawing of cups, a defect called earing can occur. Earing is a defect where the edges of the cup have a wavy appearance. This is avoided by using sheet with a proper crystallographic texture.

1.7 Conclusion

The deformation and processing of aluminium alloys has been covered in this chapter. A brief introduction to classification of aluminium alloys has been given followed by a description of microstructure evolution during cold and hot working operations. Some of the processing techniques commonly used in the forming of aluminium alloys have been discussed. The precipitation hardening behaviour of aluminium alloys has also been described in this chapter. Typical mechanical properties of some alloys have also been listed.

The technological importance of aluminium alloys is set to increase in the future. Developments in aluminium alloys are likely to be driven by the requirements of the automobile and aerospace industry. Much of the interest in aluminium alloys in the automobile industry is driven by the need to develop more lightweight and fuel-efficient vehicles. The higher specific strength of aluminium alloys, good formability and good corrosion resistance will ensure that the percentage of aluminium alloys used in automobiles will increase mainly at the expense of steel. Newer alloys are also being developed and investigated for use in the aerospace industry. Development of new alloys will also require innovations in forming and joining techniques.

1.8 Sources of further information

A significant amount of literature is available on aluminium alloys and their processing. References to relevant literature have been made at various points throughout this chapter. The various alloy types, their processing and typical mechanical properties have been described by Polmear (1989). Mechanical properties of all commercial grades of aluminium alloys has been compiled by Kaufman (1999). Constitutive equations for numerical modelling of deformation behaviour have been described by Kocks *et al.* (2003) and McQueen and Ryan (2002). Microstructure development during cold working has been described in detail by Kuhlmann-Wilsdorf (1989, 1999).

The processing of aluminium alloys by extrusion has been described in detail by Saha (2000). A detailed description of metal forming processes in general, including die design, theoretical analyses of forming loads, properties, etc., has been given by Lange (1985). The theory of precipitation hardening in various aluminium alloy systems has been covered by Porter *et al.* (1981). Processing maps showing mechanical properties and microstructure developing during the ageing of commercial alloy grades have been developed by Ber (2000a, 2000b). Some of the developments in the use of aluminium alloys in the automobile and aerospace industry have been described by Miller *et al.* (2000) and Heinz *et al.* (2000). Professional associations like the Aluminum Association, USA, Aluminum Extruders Council, USA, The Materials Society, USA, etc., also regularly bring out conference proceedings and publications on issues of interest to the aluminium industry.

1.9 References

Airod A., Vandekinderen H., Barros J., Colas R. and Houbaert Y. (2003), Constitutive equations for room temperature deformation of commercial purity aluminum, *J. Materials Processing Technology*, 134, 398–404.

Aluminum Association (1997), *Aluminum Standards and Data*, Washington D.C.

American Society for Testing and Materials (2000), E9-89a, *Standard Methods of Compression Testing of Metallic Materials at Room Temperature*, PA, USA, ASTM International.

Avner S.H. (1987), *Introduction to Physical Metallurgy*, New York, McGraw Hill.

Barker I., Hansen N. and Ralph B. (1989), The development of deformation substructures in face centred cubic metals, *Materials Science and Engineering A*, 113, 449–454.

Barlow C.Y. and Hansen N. (1989), Deformation structures in aluminium containing small particles, *Acta Metallurgica*, 37, 1313–1320.

Bay B., Hansen N. and Kuhlmann-Wilsdorf D. (1989), Deformation Structures in Lightly Rolled Pure Aluminium, *Materials Science and Engineering A*, 113, 385–397.

Ber L.B. (2000a), Accelerated artificial ageing regimes of commercial aluminum alloys, I – Al-Cu-Mg alloys, *Materials Science and Engineering A*, 280, 83–90.

Ber L.B. (2000b), Accelerated artificial ageing regimes of commercial aluminum alloys, II – Al-Cu, Al-Zn-Mg-(Cu), Al-Mg-Si-(Cu), *Materials Science and Engineering A*, 280, 91–96.

Blum W., Zhu Q., Merkel R. and H.J. McQueen (1996), Dynamic restoration mechanisms in hot torsion of Al-5Mg and Al, *Z. Metallkunde*, 87, 14–23.

Bray J.W. (1990), Aluminum mill and engineered wrought products, in *Metals Handbook*, 10th edn, volume 2, Materials Park, ASM international, 29–61.

Bryant J.D., Kung H. and Misra A. (1998), The effects of pre-ageing treatments on precipitation in Al-Mg-Si automotive body sheet alloys, in Das S.K., *Automotive Alloys II*, Warrendale, PA, The Materials Society, 3–18.

Chakkingal U., Suriadi A. and Thomson P.F. (1999), The development of microstructure and influence of processing route during equal channel angular drawing of pure aluminium, *Materials Science and Engineering A*, 266, 241–249.

Cooke M. and Larke E.C. (1945), Resistance of copper alloys to homogenous deformation in compression, *J. Institute of Metals*, 71, 371–390.

Davydov V.G., Ber L.B., Kaputkin E.Y., Komov V.I., Ukolova O.G. and Lukina E.A. (2000), TTP and TTT diagrams for quench sensitivity and ageing of 1424 alloy, *Materials Science and Engineering A*, 280, 76–82.

Deschamps A., Livet F. and Brechet Y. (1999), Influence of predeformation on ageing in an Al-Zn-Mg alloy – I, Microstructure evolution and mechanical properties, *Acta Materialia*, 47 (2), 281–292.

Dieter G.E. (1986), *Mechanical Metallurgy*, New York, McGraw Hill.

Dons A.L. (2001), The Alstruc homogenization model for aluminium alloys, *J. Light Metals*, 1, 133–149.

Dowling N.E. (1993), *Mechanical Behavior of Materials*, NJ, USA, Prentice-Hall.

Engler O. and Hirsch J. (2002), Texture control by thermo-mechanical processing of AA 6xxx Al-Mg-Si sheet alloys for automotive applications – a review, *Materials Science & Engineering A*, 336, 249–262.

Heinz A., Haszler A., Keidel C., Moldenhauer S., Benedictus R. and Miller W.S. (2000), Recent developments in aluminium alloys for aerospace applications, *Materials Science and Engineering A*, 280, 102–107.

Higgins R.A. (1993), *Engineering Metallurgy* (*Applied Physical Metallurgy*), London, Edward Arnold.

Horita Z., Fujinami T., Nemoto M. and Langdon T.G. (2001), Improvement of mechanical properties for Al alloys using equal channel angular pressing, *J. Materials Processing Technology*, 117, 288–292.

Hughes D.A. (1993), Microstructural evolution in a non-cell forming material: Al-Mg, *Acta Metallurgica*, 41 (5), 1421–1430.

Hughes D.A., Kassner M.E., Stout M.G. and Vetrano J.S. (1998), Metal forming at the Centre of Excellence for the Synthesis and Processing of Advanced Materials, *J. Materials*, 50(6), 16–21.

Kaufman J.G. (ed.) (1999), *Properties of Aluminum Alloys – Tensile, Creep and Fatigue Data at High and Low Temperatures,* Materials Park, OH, ASM International and Washington D.C., The Aluminum Association.

Khan Z.A., Chakkingal U. and Venugopal P. (2003), Analysis of forming loads, microstructure development and mechanical property evolution during equal channel angular extrusion of an aluminium alloy, *J. Materials Processing Technology*, 135, 59–67.

Kocks U.F. and Mecking H. (2003), Physics and phenomenology of strain hardening: the fcc case, *Progress in Materials Science*, 48, 171–273.

Korbel A. and Martin P. (1986a), Microscopic versus macroscopic aspect of shear band formation, *Acta Metallurgica*, 34 (10), 1905–1909.

Korbel A., Embury J.D., Hatherly M., Martin P.L. and Erbsloh H.W. (1986b), Microstructural aspects of strain localization in Al-Mg alloys, *Acta Metallurgica*, 34 (10), 1999–2008.

Kuhlmann-Wilsdorf D. (1982), Theory of dislocation cell sizes in deformed metals, *Materials Science and Engineering A*, 55, 79–83.

Kuhlmann-Wilsdorf D. (1989), Theory of plastic deformation: properties of low energy dislocation structures, *Materials Science and Engineering A*, 113, 1–41.

Kuhlmann-Wilsdorf D. (1999), Regular deformation bands and the LEDS hypothesis, *Acta Materialia*, 47 (6), 1697–1712.

Lange K. (1985), *Handbook of Metal Forming*, New York, USA, McGraw Hill.

Laue K. and Stenger H. (1981), *Extrusion*, Materials Park, OH, ASM International.

Lee S., Utsunomiya A., Akamatsu H., Neishi K., Furukawa M., Horita Z. and Langdon T.G. (2002), Influence of scandium and zirconium on grain stability and superplastic ductilities in ultrafine-grained Al-Mg alloys, *Acta Materialia*, 50 (3), 553–564.

Martin J.W. (1968), *Precipitation Hardening*, Oxford, UK, Pergamon.

McQueen H.J. and Ryum N. (1985), Hot working and subsequent static recrystallization of Al and Al-Mg alloys, *Scandinavian J. Metallurgy*, 14, 183–194.

McQueen H.J. (1988), Initiating nucleation of dynamic recrystallization, primarily in polycrystals, *Materials Science and Engineering A*, 101, 149–160.

McQueen H.J., Evangelista E. and Kassner M. (1991), The classification and determination of restoration mechanisms in the hot working of Al alloys, *Z. Metallkunde*, 82, 336–345.

McQueen H.J., Evangelista E., Jin N. and Kassner M.E. (1994), Dynamic recovery, dynamic recrystallization and energy dissipation efficiency in aluminum alloys, in Jonas J.J., Bieler T.R. and Bowman K.J., *Advances in Hot Deformation Textures and Microstructures*, The Minerals, Metals & Materials Society, 251–265.

McQueen H.J., Ryan N.D., Konopleva E.V. and Xia X. (1995), Formation and application of grain boundary serrations, *Canadian Metallurgical Quarterly*, 34 (3), 219–229.

McQueen H.J. and Celliers O.C. (1996), Application of hot workability studies to extrusion processing: Part II. Microstructural development and extrusion of Al, Al-Mg and Al-Mg-Mn alloys, *Canadian Metallurgical Quarterly*, 35 (4), 305–319.

McQueen H. and Ryan N.D. (2002), Constitutive analysis in hot working, *Materials Science and Engineering A*, 322, 43–63.

Miller W.S., Zhuang L., Bottema J., Wittebrood A.J., De Smet P., Haszler A. and Vieregge A. (2000), Recent developments in aluminium alloys for the automobile industry, *Materials Science and Engineering A*, 280, 37–49.

Perovic A., Perovic D.D., Weatherly G.C. and Lloyd D.J. (1999), Precipitation in aluminum alloys AA 6111 and AA 6016, *Scripta Materialia*, 41 (7), 703–708.

Polmear I.J. (1989), *Light Alloys (Metallurgy of the Light Metals)*, London, Edward Arnold.

Porter D.A. and Easterling K.E. (1981), *Phase Transformations in Metals and Alloys*, New York, Van Nostrand Reinhold.

Raj S.V. and Pharr G.M. (1986), A compilation and analysis of data for the stress dependence of the subgrain size, *Materials Science and Engineering A*, 81, 217–237.

Ravichandran N. and Prasad Y.V.R.K. (1991), Dynamic recrystallization during hot deformation of aluminum, A study using processing maps, *Metallurgical and Materials Transactions A*, 22A, 2339–2348.

Ren B. and Morris J.G. (1995), Microstructure and texture evolution of Al during hot and cold rolling, *Metallurgical and Materials Transactions A*, 26A, 31–40.

Richert M., Stuwe H.P., Richert J., Pippan R. and Motz Ch. (2001), Characteristic features of AlMg5 deformed to large plastic strains, *Materials Science and Engineering A*, 301, 237–243.

Saha P.K. (2000), *Aluminum Extrusion Technology*, Materials Park, OH, ASM International.

Segal V.M. (1999), Equal channel angular extrusion: from macro mechanics to structure formation, *Materials Science and Engineering A*, 271, 322–333.

Segal V.M., Reznikov V.I., Drobyshevsky A.E. and Kopylov V.I. (1981), Plastic working of metals by simple shear, *Russian Metallurgy*, 1, 99–106.

Sellars C.M. and Tegart W.J. McG (1972), Hot workability, *International Metallurgical Reviews*, 17, 1–24.

Smolej J., Gnamus M. and Slacek E. (2001), The influence of thermo-mechanical processing and forming parameters on superplastic behaviour of the 7474 aluminum alloy, *J. Materials Processing Technology*, 118, 397–402.

Wei B.C., Chen C.Q., Huang Z. and Chang Y.G. (2000), Ageing behavior of Li containing Al-Zn-Mg-Cu alloys, *Materials Science and Engineering A*, 280, 161–167.

Young C.T., Headley T.J. and Lytton J.L. (1986), Dislocation substructures formed during the flow stress recovery of high purity aluminium, *Materials Science and Engineering A*, 60, 391–407.

Yun M., Lokyer S. and Hunt J.D. (2000), Twin roll casting of aluminum alloys, *Materials Science & Engineering A*, 280, 116–123.

2
High temperature deformation, alloys and processing of magnesium

A ARSLAN KAYA, Materials Institute, Turkey,
M PEKGÜLERYÜZ, McGill University, Canada and
D ELIEZER, Ben Gurion University of the Negev, Israel

2.1 Introduction

Creep properties of magnesium alloys are of great importance as they are used by the automotive industry for power train applications where service conditions impose temperatures that are high enough for these alloys to cause creep damage. Although the subject temperatures (~150 °C) may appear rather low, due to the low melting temperature of magnesium, and its hexagonal crystal structure that allows for relatively rapid diffusion rates, creep becomes a dominant design criterion especially for such applications as engine blocks and transmission cases. Currently, it appears that the expansion of the current use of Mg alloys hinges, to a large extent, on the development of a diecastable and economically competitive alloy with good creep strength up to 200 °C. Yet another incentive for the greater use of magnesium alloys, though not for high temperature applications, would be the production of magnesium sheet metal at an economically attractive cost that would also offer good engineering properties. For both criteria a non-conventional technique such as twin-roll casting is envisioned worldwide.

Despite the fact that magnesium alloys are generally recognized as materials with relatively low creep resistance, one magnesium alloy in the magnesium-thorium system, namely HZ22, shows the highest ratio of service temperature (350 °C) over melting temperature compared to any material except some superalloys.[1] Although this alloy system is no longer exploited due to the radioactivity of thorium, the example it sets may be taken to indicate the true potential of magnesium alloys for future applications if a proper alloy design is achieved.

2.2 General deformation behaviour of magnesium

Magnesium metal has a hexagonal-close-packed (hcp) crystal structure, which has only two independent slip systems compared to, for example, f.c.c. crystal with five independent slip systems. Due to its crystal structure, the ability of magnesium to deform, especially at low temperatures, is limited. The possible slip planes and corresponding slip directions, which are not collectively operative at the room temperature, are given in Table 2.1.[2] Room temperature deformation occurs mainly by slip on the basal planes. Activation of prismatic planes can be due to high temperatures or presence of certain alloying elements such as indium and lithium.[3,4] In addition, twinning on the pyramidal planes, to a diminishing degree as the temperature is raised, provides means of deformation.

Although it was proposed that a favorable c/a ratio of magnesium crystal facilitates the operation of non-basal slip systems more easily than those in other hcp metals with a higher ratio, this view was later disputed based on a more rigorous analysis of the energies of stacking faults created by dissociated screw dislocations on basal and prismatic planes.[4,5] The low basal stacking fault energy of about 10 mJm^{-2} being about seven times smaller than that of prismatic stacking fault energy accounts for the occurrence of cross slip in magnesium only at high temperatures (above 225 °C) or ease of it as compared to other hcp metals.[4]

The electron configuration of magnesium does not permit any covalent component in its interatomic bonding as is the case in many other metals. This feature is one of the reasons for the relatively lower strength and elasticity modulus of magnesium.[4] In addition, the relatively large interatomic distances in hcp magnesium lattice and the resultant low Pierls stresses between the slip planes and large Burgers vectors (see Fig. 2.1 showing thick dislocation lines due to large strain fields) together facilitate twinning instead of slip. Indeed, twinning is frequently observed at ambient temperatures diminishing as the grain size reduces or the temperature exceeds about 225 °C.[2]

Mechanical behaviour of magnesium is greatly affected by grain size like any other metal. The Hall-Petch relationship expresses the dependence of yield strength on the grain size:

$$\sigma_y = \sigma_0 + k_y \sqrt{d} \qquad 2.1$$

where σ_y is the yield strength of the material, σ_0 material constant, k_y a material coefficient and d the average grain size. The k_y coefficient in the

Table 2.1 The slip planes and directions of magnesium crystal. Note that not all slip systems are operative under the same conditions[2]

Planes	Directions	Type	Operation
{0001}	$<11\bar{2}0>$	Basal	Below 225 °C
$\{10\bar{1}0\}$	$<11\bar{2}0>$	Prismatic	Above 225 °C
$\{11\bar{2}1\}$	$<11\bar{2}0>$	Pyramidal	High temperatures
$\{11\bar{2}2\}$	$<11\bar{2}3>$	Pyramidal	High temperatures
$\{10\bar{1}2\}$	$<10\bar{1}1>$	Pyramidal (Twinning)	Below 225 °C

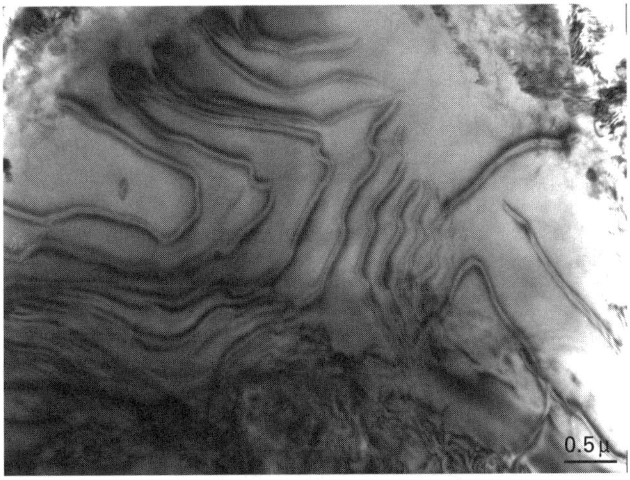

2.1 TEM micrograph showing some edge-dislocations in AZ91 matrix. Note that dislocations that would otherwise be seen as simple single dark lines in the case of, for example, aluminium, now appear as multiple lines, one of them being stronger than the other, possibly due to their large Burgers vectors and resulting large strain fields around each dislocation. (z: <$\bar{3}$21>).

Hall-Petch relationship is about four times greater for magnesium than, for example, that of aluminium.[6] Despite the implication of a high k_y towards greater strengthening effect by grain refinement, the ability of magnesium to deform increases as the grain size reduces, especially below 8μ where room temperature ductility is observed.[2] It is also known that sub-micron grain sizes, achievable with rapid solidification techniques, can create superplasticity in magnesium even at room temperature giving elongations up to 1,000%.[7] Although the mechanism of deformation in such fine grained magnesium is grain boundary sliding at least at high temperatures, plastic compatibility may also play a role in the increasing deformability as the grain size reduces. Plastic compatibility may be envisaged as the increased possibility for the stress acting on a particular glide plane in a grain to find a favourably oriented slip plane in a bordering grain and thus continue to cause dislocation movement.

Due to the low basal stacking fault energy the basal screw dislocations can readily dissociate, and create relatively large obstacles to hinder the movement of other dislocations. This, however, does not lead to a relatively high deformation hardening response in magnesium alloys as the dissociation does not create a sessile dislocation component.[4] Likewise, pile-up of more dislocations and large back-stresses at grain boundaries due to low inherent plastic compatibility because of hcp lattice, and the resultant lack of macroscopic yielding may also account for the relatively high k_y value in magnesium.

Magnesium is inherently prone to creep deformation due to high diffusion rates in its lattice ($D_{L,Mg}/D_{L,Al}$ and $D_{gb,Mg}/D_{gb,Al}$ ratios are greater than unity over a wide temperature range and the ratio of D_{gb}/D_L for magnesium is five times higher than that for aluminum at $0.67T_m$).[8] Furthermore, between approximately 200–250 °C depending on the alloy, additional pyramidal and prismatic slip planes become operative, lowering the creep resistance.

2.3 Creep behaviour of magnesium

Detailed studies exist on the creep behaviour of Mg and its alloys. Earlier interest on the creep behaviour of magnesium stemmed from its use in nuclear canning application.[9–11] The current interest, however, arises from the conceivable economic and environmental benefits of the use of magnesium alloys particularly in the automobile's power-train applications requiring creep resistance at temperatures up to 200 °C. Thus we must begin by revisiting our knowledge of general creep behaviour of metals considering magnesium specifically.

Creep is a time-dependent deformation process that manifests itself even when applied stress is constant and below the yield strength of material. While in an ordinary deformation the movement of dislocations, on an atomic scale, can cover large distances at each step, in thermally activated deformation this movement occurs due to diffusion of atoms and therefore is restricted to a few atomic distances at each step. In such a process, dislocations attempt to pass the obstacles in their path individually by mobilizing their pertinent segment locally. Hence, while the ordinary long-range deformation involves the movement of long segments of sequential dislocations, the thermally activated deformation component is associated with the movement of small sections of a dislocation, e.g., dislocation jogs, or the cutting of a dislocation by another. This nevertheless causes slow and permanent deformation called creep. Since atoms vibrate anywhere above absolute zero, thermal component always exists and creep is possible even at very low temperatures, given sufficient time.

Creep is a complex phenomenon in many engineering applications and it is perhaps more so in the case of magnesium alloys. It has been reported that creep resistance in interrupted creep tests is inferior and creep life is shorter compared to that in continuous creep indicating that the designs based on continuous creep test data might be misleading.[12] The service conditions exerted on a magnesium alloy, for example, in an engine block application, involve multiaxial and cyclic stress conditions, corrosive medium, inhomogenous and cyclic temperatures. Furthermore, especially with the currently popular high strength alloy AZ91 a severe microstructural instability of the cast structure also plays a major role in determining the creep response of magnesium alloys. Thus, the already questionable assumptions of conventional creep tests such as homogenous and constant stress-temperature distributions, uniaxially applied stress, non-corrosive test atmosphere may be even less representative of the conditions that apply to the case of magnesium applications due to the additional complexity brought about by the inherent microstructural instability and lack of a self-healing surface oxide in magnesium. This instability problem will be further discussed later.

When assessing magnesium and its alloys in terms of creep mechanisms, particular emphasis must be placed upon the creep modes corresponding to medium to high stresses, and low to moderate temperatures. This selective attention is mainly due to the service conditions imposed by the perceived use of magnesium alloys in automobile applications where such stress and temperature conditions prevail. However, it should be borne in mind that some magnesium alloys also serve in more severe conditions, at least in terms of temperature, such as nuclear canning applications.[9]

2.4 Modes of creep

Figure 2.2 shows a typical creep-strain versus time curve for a metal, where three distinct regions can be observed. After an initial and instantaneous restructuring of the metal, primary stage creep occurs with increasing creep rate. Here the work hardening is higher than the stress recovery. In the secondary creep stage (steady-state), the metal experiences a balance between work-hardening and recovery processes resulting in steady-state creep (constant creep rate or minimum creep rate). The slope of the plot in this region, i.e., steady-state creep rate is taken as an indication of the creep resistance of a material. The tertiary creep stage exhibits increasing creep rate due to necking resulting finally in fracture.

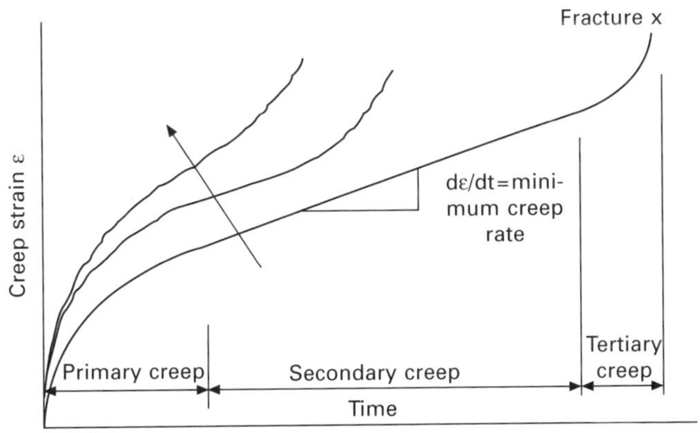

2.2 Typical creep-strain versus time curve showing the three stages of creep. The arrow indicates the trend with increasing load.

Normalized stress versus homologous temperature plots (Fig. 2.3), i.e., deformation mechanism maps, are helpful tools visually defining the conditions of temperature, stress and grain size, pertinent to a particular deformation mechanism.[13] The two main creep mechanisms, namely 'dislocation creep' and 'diffusional creep', appear as two distinct regions in these maps thus providing guidance in predicting the dominant deformation mechanism.

A strain-rate versus stress diagram at a constant temperature can also serve as a deformation mechanism map since the changes in the slope of the plots indicate the operation of a different dominant deformation mechanism (Fig. 2.4).[13] The first linear segment corresponding to the lower stress range

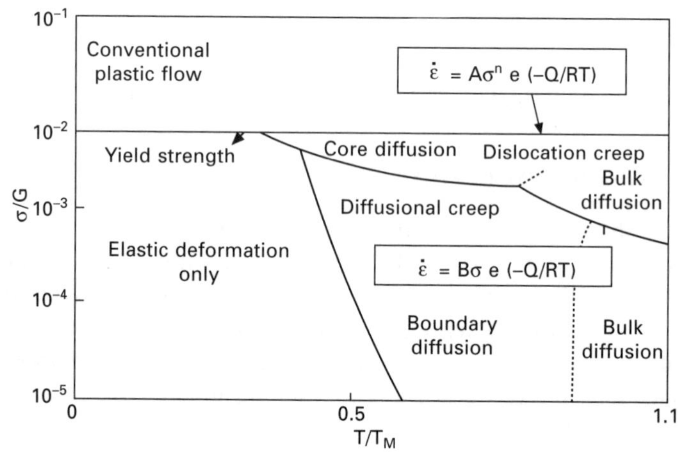

2.3 Normalised stress versus homologous temperature maps showing the various regions for different modes of deformation.[13]

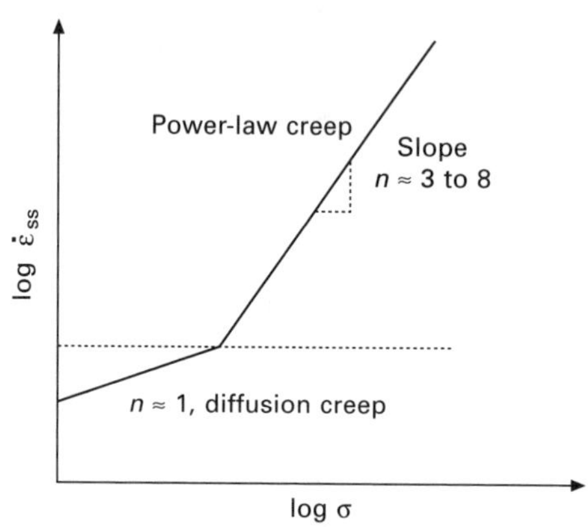

2.4 A change in slope in a strain-rate versus stress diagram at a constant temperature also indicates the operation of a different dominant deformation mechanism and thus can serve as a deformation mechanism map.[13]

is attributed to 'diffusional creep' and the second one with an increased slope at higher stresses to 'dislocation creep'. Upon moving towards even higher stress levels, the conditions of conventional plastic flow are encountered. The effect of grain size would manifest itself by shifting the plot to higher strain rates in the case of a fine-grained structure.

A deformation mechanism map for pure magnesium was first established by Frost and Ashby.[14] Watanabe *et al.* showed in their work on deformation mechanism maps of fine-grained Mg alloys the existence of a critical grain size above and below which lattice diffusion and grain boundary diffusion, respectively, govern grain boundary sliding (GBS).[15] Kim *et al.* further developed deformation mechanism maps for AZ61 and AZ31 alloys (Fig. 2.5) by normalizing the stress levels for high temperatures using the elastic modulus of pure magnesium calculated according to the equation:[8]

$$E\ (\text{MPa}) = 4.3 \times 10^4\ [1 - 5.3 \times 10^{-4}\ (T\text{–}300)] \qquad 2.2$$

The basic mathematical expression of the dominant mechanism at high stresses, i.e., 'dislocation creep', assumes the form of an Arrhenius type relationship in which strain-rate has a power law dependence on stress:

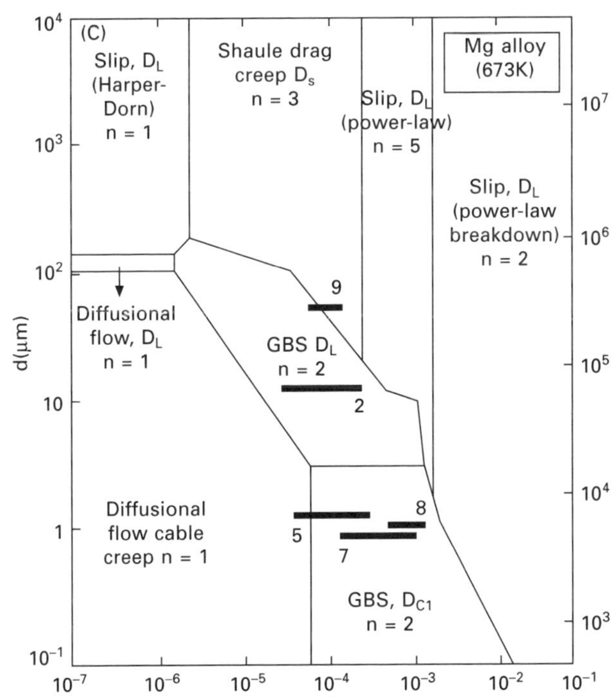

2.5 A deformation mechanism map of Mg alloy AZ61.[8]

$$\dot{\varepsilon}_{ss} = A\sigma^{n} \exp(-Q/RT) \qquad 2.3$$

'Diffusional creep', on the other hand, operates at lower stress levels and at high temperatures with a linear dependence of strain-rate on stress:

$$\dot{\varepsilon}_{ss} = B\sigma \exp(-Q/RT) \qquad 2.4$$

Since the diffusion is affected by grain size, a term, d, representing it may also be incorporated into this expression:

$$\dot{\varepsilon}_{ss} = C\frac{1}{d^2}\sigma \exp(-Q/RT) \qquad 2.5$$

In these equations A, B and C are material constants, and d grain size; R is the gas constant and T is the absolute temperature. Q represents the activation energy for creep, and n is known as the stress exponent whose values vary when examined over large ranges of stress and temperature, each abrupt change indicating the operation of a different rate-controlling mechanism of creep.[13] Although a clear-cut distinction between various modes of creep, and specifically among possible rate-controlling micro-mechanisms is often difficult, values of Q and n are being used to establish the operating creep mechanism and thus leading to contradicting interpretations among authors.

Some indications attributable to the operating creep mechanism may be found in the general shape of a strain versus time plot of a creep test, such as whether or not an instantaneous strain upon loading exists or whether the primary stage is normal or inverted. The lack of significant instantaneous strain upon loading and an inverted primary stage were related to a viscous glide process of dislocations, while the opposite of these observations was attributed to dislocation climb or cross slip of dislocations.[11]

2.4.1 Low stress creep regimes

Three different creep mechanisms, although not specific to magnesium, may be encountered depending on the material and test conditions in the low stress region. These mechanisms are diffusion creep, Harper–Dorn creep, and grain boundary sliding.[16–20] Diffusion creep is further subdivided into Nabarro–Herring creep operating through bulk diffusion at relatively higher temperatures and Coble creep operating through boundary diffusion at relatively lower temperatures.[21–23] However, there is disagreement on the significance and the occurrence of these three processes as viable creep mechanisms.[24–28]

Vacancies diffuse from grain boundaries subjected to tensile stress to the boundaries subjected to compression with a corresponding counterflow of atoms. This mechanism, independent of the alloy type, forms the basis of diffusion creep leading to elongation of grains in the direction of the applied stress. When this flow occurs through the grain interiors, i.e., lattice diffusion, the case corresponds to Nabarro–Herring creep, and if through grain boundaries then to Coble creep. If diffusion creep is operating, little or no primary creep is observed, and this mode is not associated with metallographic evidence such as slip lines.[10]

Although the grains retain their relative positions during diffusional creep, it is recognised that some sliding movement is necessary to maintain the integrity of the polycrystalline structure, without an eventual increase in the number of grains along the applied stress direction.[29] This accommodation process was first observed by Lifshitz and later by others.[30–32] Based on these principles Langdon has suggested a method to distinguish between the creep modes of low stress region by also making use of marker lines inscribed on the pre-polished specimen surface prior to creep testing, changes in grain aspect ratios during creep test, and particle-denuded zones that form in certain alloys during creep.[29]

In certain alloy systems used for nuclear reactor components such as Mg-Zr (Mg-0.5wt%Zr alloy, ZA) particle denuded zones form along grain boundaries perpendicular to the stress axis during creep.[33–35] An example of these denuded zones which are conveniently revealed by the interruptions along the hydride stringers forming on free surfaces is given in Fig. 2.6. Original interpretation of the denuded zones was based on diffusional creep.[34–38] However, this interpretation is still being debated.[29,39–43] It has been shown that such denuded zones, though not so pronounced and characteristic in terms of their location as in creep, may also be observed after annealing.[40,44,45] Dissolution of precipitates at migrating grain boundaries via grain boundary sliding is an alternative explanation for the formation of these denuded zones.[39,41]

Harper-Dorn creep occurs at even higher temperatures, i.e., close to melting point, leading to much faster creep rates than those expected from true diffusion creep.[46–48] Therefore it is explained in terms of a viscous flow of dislocations. A change in the shape of thc grains is again inevitable in this mode of creep as well. However, the change in the shape of the grains is not accompanied by a displacement of grains.[29] This mode of creep has not, to the best knowledge of the authors, been reported for magnesium alloys.

2.4.2 Creep via grain boundary sliding

Grain boundary sliding, on the other hand, is a major creep mechanism for magnesium alloys at the temperature and stress conditions they are currently being considered for, i.e., under-bonnet automotive applications. Under constant creep conditions the ratio of strain due to grain boundary sliding to the total strain remains constant, and this ratio increases with decreasing stress.[49] The process is attributed to a crystallographic deformation mechanism in the vicinity of the grain boundary rather than a viscous flow of grains over each

2.6 The appearance of a Mg–0.55% Zr alloy after creep testing at a low stress under conditions corresponding to diffusion creep, showing the development of denuded zones along the transverse grain boundaries and boundary offsets in the longitudinal hydride stringers (the tensile axis is vertical).[29] (courtesy of Langdon).

other. The movement of grains over each other leads to an increase in their number along the applied stress direction without a considerable elongation of the grains.[29] Such a displacement of grains is termed 'Rachinger sliding' and may be attributed to grain boundary sliding in the absence of a considerable change in the grain aspect ratio during creep.[50] This mechanism is akin to the superplastic behaviour of the fine-grained metals, which is well known in magnesium alloys.[29,51,52]

Even though finer grain sizes are desired for magnesium alloys to enhance many other mechanical properties, this leads to conditions that possibly facilitate grain boundary creep. Casting techniques such as high-pressure diecasting produce fine grain sizes. This process inherently produces bimodal grain size distributions as the solidification prematurely starts even before the liquid metal enters the mould and grain sizes as different as 0.5–50μ can exist in the same diecast structure.[53,54] Furthermore, as a result of rapid solidification rates metastable microstructures are produced. Due to this, together with the large freezing range of alloys like AZ91, microstructural instability becomes an inherent problem of diecast alloys complicating the creep behaviour of the material and its interpretation. Microstructural instability manifests itself as occurrence of a discontinuous reaction. Dargush *et al.* have suggested that the low resistance to grain boundary sliding is related to the discontinuous precipitation reaction in the Al-enriched peripheries of the grains.[54–56] Since this reaction requires the movement of high angle boundaries it is reasonable to assume that the reaction also facilitates grain sliding and migration during creep. It should also be borne in mind that, even at moderate temperatures, such regions of the microstructure are operating at relatively higher homologous temperatures compared to the grain interiors, thus leading to more diffusion in these regions. In support of this view, indeed, abundant occurrence of the discontinuous reaction in such regions has been observed after creep. Therefore its association with grain boundary sliding during creep should be regarded as highly possible.[57,58]

The major second phase in Mg-Al based alloys is the β-($Mg_{17}Al_{12}$) phase that exists in several distinct morphologies, namely grain boundary coarse particles and lamella of the colonies of discontinuous precipitation (Fig. 2.7) and two types of intragranular continuous precipitates, i.e., plates and plaques.[59] The role of β-$Mg_{17}Al_{12}$ in creep has been discussed since the 1950s.[60–62] It seems reasonable to assume that β-phase softens easily due to its low melting point (~460 °C), and therefore cannot serve to inhibit grain boundary sliding. Contrary to this view it was found that thinner sectioned die cast specimens containing finer and more of β-phase showed greater creep strength. Furthermore, heat treatment promoting β precipitation prior to creep seemed to improve creep strength.[56] Although it has been shown on bulk $Mg_{17}Al_{12}$ that β-phase, due to its strong covalent bonding, would not soften markedly below 260 °C, $Mg_{17}Al_{12}$ has very complex bonding showing a mixture of homopolar, heteropolar and normal metallic bonding where the softening of the compound cannot be completely excluded.[60,63–65] Hence precipitation of $Mg_{17}Al_{12}$ during creep and not its existence in the pre-creep structure seems to be the main factor in creep deformation of Mg-Al alloys.

2.4.3 Dislocation creep and its interpretation based on *Q* and *n*

It is known that at sufficiently high stresses dislocation glide takes place regardless of the temperature. However, stress levels that would not cause ordinary dislocation glide but are still higher than those required by the above-mentioned modes lead to a different dominant mechanism known as dislocation creep. At such stress levels, if temperatures are sufficiently high (at about 1/3 of T_m), dislocations are no longer confined to glide motion only in their original slip planes. Creep rates are essentially controlled by diffusion in this temperature range. Although the creep deformation essentially takes

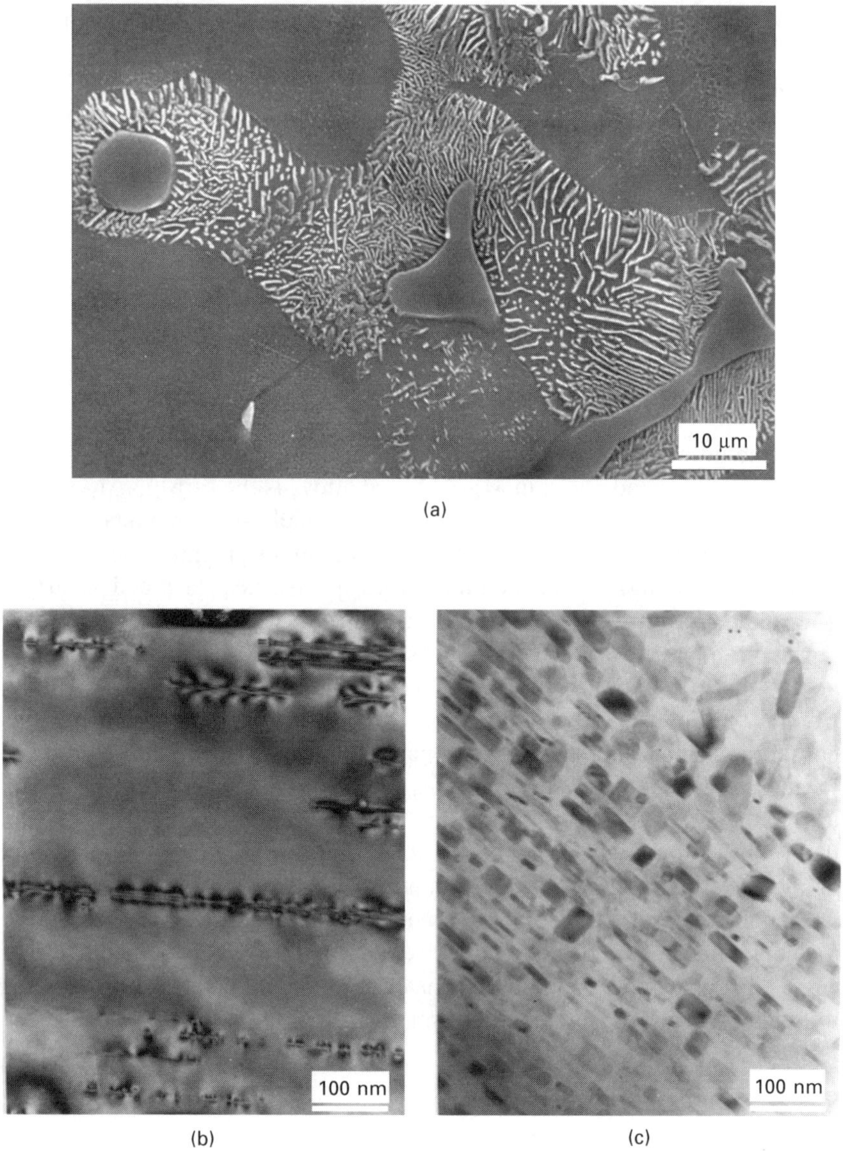

2.7 The major second phase in Mg-Al alloys, β-($Mg_{17}Al_{12}$), exists in several distinct morphologies in AZ91: (a) grain boundary coarse particles, the lamellar colonies of discontinuous precipitation and two types of intragranular continuous precipitates; (b) plates and (c) plaques.[59]

place through dislocation movement on basal or non-basal planes, this mode of creep involves different rate-controlling micro-mechanisms since the dislocations surmount the barriers on their path via short-range diffusion at each step. Due to thermal activation they can climb out of their slip planes via vacancy diffusion or cross slip, and thus overcome obstacles. Therefore dislocation creep may also be described as non-conservative motion of dislocations. The climb motion mentioned may be achieved via two different vacancy diffusion mechanisms at the same stress level, namely volume diffusion at higher temperatures and dislocation core diffusion at lower temperatures.[66] This allows the dissipation of interlocked dislocation networks as a recovery process. The deformation process is, in this case, a combination of climb and glide motions, glide being responsible for strain.[67] Although it is assumed that the contribution of climb to the steady state creep rate is zero, the deformation rate may not be solely due to glide as some energy may also be released by climb movement of dislocation segments.[68] It should further be pointed out that the steady state creep rate is affected by the stress through increasing dislocation velocity.[5,69,70]

In such successive processes, namely dislocation climb and viscous glide (also referred to as recovery), the slower one controls the steady state creep rate, giving rise to the different creep behaviour of pure metals and alloys.[10,11] In pure metals as a result of the recovery process an internal sub-grain structure forms. Alloys, on the other hand, are classified as Class-I and Class-II. The creep behaviour of the Class-I alloys is similar to that of pure metals. In Class-II alloys, however, some form of viscous drag process is assumed to operate on dislocations due to the presence of solute atoms with the result that the rate of glide motion is slower than the rate of climb. Glide motion therefore becomes the rate controlling mechanism with an activation energy equal to that of solute diffusion, and in such cases sub-grain formation is generally not observed.[71] The rate-controlling mechanism of dislocation creep is thus understood as the dragging of solute atmosphere around a dislocation in the lower stress range, and as the climb of dislocations in the higher stress range.[11,72]

Operation of a particular rate-controlling mechanism of dislocation creep can vary depending on the alloy, heat treatment, stress and temperature. Positive identification of a specific rate-controlling mechanism on the basis of Q and n is often difficult. The stress exponent, n, may be more indicative of a particular rate-controlling mechanism within the domain of dislocation creep.[58] Although there appears to be disagreement in the literature on the values of the activation energy and stress exponent these parameters are often taken to indicate a predominant rate-controlling creep mechanism.

Proliferation of Q and n values reported in the literature may be due to the sample or test conditions. Activation energy calculations can be complicated in the case of thermally unstable systems as is generally the case for magnesium alloys usually suffering from non-equilibrium solidification. It is conceivable that microstructural instabilities may induce or remove barriers to dislocation movement, e.g., formation or coarsening of precipitates, thus changing the creep strain rate.[73] At low stresses and high temperatures creep rates can decrease due to grain coarsening during the test, and the stress exponent may slightly increases due to the same reason. Some experimental scatter in the values of Q may also be attributable to concurrent grain growth.[10] Furthermore, there may exist decreasing levels of contribution to the total creep strain by grain boundary sliding with increasing stress levels in the dislocation creep region.[49]

Activation energy is a structure sensitive property and therefore it is assumed that its value for dilute alloys approaches that of elements. Furthermore, the activation energy is assumed to be insensitive to changes in temperature and stress within the stress-temperature domain of a particular micro mechanism controlling the creep rate.[49] However, when considering a wider span of stress and temperature, for a given temperature, the activation energy Q has a strong stress dependency, decreasing with increasing stress.[5,10,11] The general trend in hcp metals is that activation energy increases from values close to that of self-diffusion at ~0.5–0.6T_m to higher values at ~0.7–0.8T_m, T_m being the melting point in Kelvin.[10,74,75] For the same change in temperature range, hcp metals also show a decrease in the stress exponent.

Dislocation creep in pure magnesium

Since dislocation creep is a major creep mechanism under the service conditions (T, σ) of current magnesium applications this section focuses specifically on that mechanism. Pure magnesium shows little creep strain in the primary stage under low stress levels or under high temperatures.[10] The activation energy for the creep of pure magnesium was reported by Shi and Northwood to be 106 kJmol^{-1} in the temperature range 150–250 °C and under a stress range of 20–50 MPa.[76] Although this value is smaller than the activation energy for self-diffusion of magnesium, ~135 kJmol^{-1}, similar values of activation energies for the same temperature range were also reported earlier by others.[75]

Activation energies in the range of lattice self-diffusion of magnesium, i.e., 135 ± 10 kJmol^{-1}, have also been reported in the literature for the temperature range of ~200–477 °C.[10] Based on metallographic observations of straight slip lines across the grains it was suggested that this high stress-temperature region corresponded to a deformation mechanism via basal slip. Although basal slip, based on an activation energy value of ~135 kJmole^{-1} that was independent of stress, can be said to be controlled by lattice self-diffusion of magnesium, the micro-mechanism of creep still needs be determined. In this regard, the stress exponent n may be taken as an indication to distinguish between the two micro mechanisms, i.e., glide or climb controlled basal slip. Since the process of glide-controlled dislocation creep is associated with $n = 3$, the conclusion was made based on a higher stress exponent value of ~5, thus the specific rate-controlling process was identified as climb of dislocations, while Q is equal to the activation energy for solute diffusion.[10,71,75] This was further confirmed using the following theoretical steady-state creep rate equation developed by Weertman for dislocation climb:[10,77]

$$\dot{\varepsilon} = \alpha\left(\frac{D_t}{b^{3.5}M^{0.5}}\right)\left(\frac{G\Omega}{kT}\right)\left(\frac{\sigma}{G}\right)^{4.5} \qquad 2.5$$

where D_t is the coefficient for lattice self-diffusion, M is the number of dislocation sources per unit volume (equivalent to $0.27\rho^{1.5}$, ρ being the dislocation density), Ω is the atomic volume, and α is a constant in the range $0.015 < \alpha < 0.33$.

The stress exponent value was also assessed by Jones and Harris in terms of strain rate, and a lower value of 4.5 was determined for low strain rates ($\leq 10^{-6}$ sec^{-1}) where dislocation climb is the rate controlling mechanism.[78] For higher strain rates where non-basal slip predominates, n was found to increase with increasing temperature reaching a value as high as 8. At temperatures above about $0.6T_m$, however, creep behaviour changes and the rate controlling mechanism is understood to be the cross-slip of dislocations as the apparent activation energy of creep becomes larger than that for diffusion. At this high temperature range and under high stress levels an activation energy level of 220 ± 10 kJmol^{-1} was found corresponding to cross-slip of dislocations from basal to prismatic planes, that is, non-basal slip.[11,75] Thus it may be concluded for pure magnesium that, as compared to cross slip, climb occurs at lower temperatures and requires lower activation energies for a given stress level.

Dislocation creep in magnesium alloys

The overall creep behaviour of a dilute Mg alloy may be divided into low and high temperature ranges to facilitate understanding of the operating mechanisms.[11] In the low temperature range, two independent mechanisms have been shown to operate in an Mg-0.8Al alloy, the transition taking place between 327–477 °C and at lower temperatures as the stress increased.[11] The change of mechanism from glide to climb-controlled creep within the lower temperature range manifests itself with a change in the associated stress exponent from about $n = 3$ at lower stresses to $n = 6$ at higher stress.

At temperatures between 200–477 °C and at stresses below 10 MPa Class-II behaviour, i.e., uniform distribution of dislocations and absence of sub-grain formation associated with a viscous glide process, is observed. Extensive basal slip that may be observed metallographically, an activation energy of 140 kJmole^{-1} that matches well the value for interdiffusivity of aluminium (143 kJmole^{-1}) in magnesium, a low stress exponent of 3, lack of significant instantaneous strain upon loading, and a brief normal or inverted primary stage are all associated features of creep in this stress-temperature range. The experimental creep rates, however, were more than an order of magnitude slower than those predicted by the existing three theoretical models to calculate the creep rate based on a glide-controlled mechanism, i.e., dislocation-solute interaction.[79–81]

Above the stress levels associated with $n = 3$ the activation energy remains the same, while creep behaviour changes sharply. The unchanging value of activation energy while the creep mechanism is changing from viscous glide to climb is readily explained due to the small difference between the activation energies involved in each case, the interdiffusion of aluminium in magnesium (143 kJmole^{-1}) when glide is operative, and self-diffusion of Mg atoms (135 kJmole^{-1}) when climb is operative.[82,83]

A higher stress exponent, $n = 6$, is attributed to dislocation climb. This interpretation is supported by the observed creep rate, measurable instantaneous strain, extensive normal primary stage, formation of dislocation substructure and the agreement between the estimated and experimentally observed values of binding energy (~0.14 eV) between a dislocation and a solute atom.[11]

The high temperature behaviour in Mg-0.8Al alloy above 600–750 K is similar to pure magnesium. Extensive non-basal slip and an activation energy value that is significantly higher than those of lattice self-diffusion of Mg or interdiffusion of Al are observed depending on the applied stress level.[11]

As a rate controlling creep mechanism at high temperatures, glide on non-basal planes is excluded since the existing theoretical models for this mechanism that are based on a single or double kink formation does not produce results in agreement with experimental observations.[79,84] Whereas cross-slip of screw dislocations from basal to prismatic planes according to Friedel's model seems to be in excellent agreement with the experimental estimations in terms of the constriction energy which is equal to 160 kJmole^{-1}. A stress exponent of 4, measurable instantaneous strain, a normal primary stage, extensive non-basal slip and presence of free dislocations and some well-defined sub-boundaries altogether provided further evidence for the operation of a cross slip mechanism at high temperatures.[11] A similar trend also seems to be valid for non-dilute magnesium alloys. It was shown for Mg–Y–Nd–Zr and Mg–Y–Nd alloys that, as compared to cross slip, climb occurs at lower temperatures requiring lower activation energies at a given stress level.[73,85,86]

The effects of dispersoids on creep behaviour are also interesting and complex.[5] Oxide dispersions in Mg powder metallurgy samples exhibit self-diffusion controlled basal slip at relatively high stresses and low temperatures

(144–253 °C) as the rate-controlling mechanism. In that case, the stress exponent shows mainly dependence on temperature, with n ranging between 7.5–10. When the rate-controlling mechanism shifts to cross slip at higher temperatures n shows dependence on both stress and temperature while its value varies drastically, first decreasing with increasing stress from 15 to 7.5 and then increasing again to 18. This may be attributed to the fact that n has two contributing terms, one due to the activation energy Q which decreases with increasing stress, and a second one due to moving dislocation density which increases with increasing stress.[5]

2.5 Microstructure, properties and processing

In this section, the creep behaviour of magnesium alloys will be assessed on the basis of alloy systems. Effects of precipitate types and microstructure on creep of magnesium alloys will also be considered. Development of Mg alloys for high temperature applications requires considerations of several properties in conjunction with creep strength. Castability, especially diecastability, of a new alloy must be at an acceptable level, as this is the major processing route for production of parts made of Mg alloys, as well as corrosion and oxidation resistance. In addition, the alloys should not have an additional cost penalty by containing costly alloying elements such as the rare earths to be able to compete with relatively lower cost metals such as aluminum and steel.

AZ91, the most commonly used alloy, and likewise all the rest of the commercially exploited Mg-Al based alloys, utilizes solid solution and an intermetallic phase, $Mg_{17}Al_{12}$, for strengthening. The shortcoming in this alloy design is that the intermetallic phase β–$Mg_{17}Al_{12}$ (see Fig. 2.7), provides little or no benefit for creep strengthening. In an attempt to increase the number density of the plate-shaped intragranular precipitates toward better creep strength, Bettles *et al.* tried various alloying element additions in trace amounts with no apparent success.[87]

The creep resistance of AZ91 and AM series alloys may be acceptable up to 110–120 °C, reflecting the need for further improvements. The main development in creep resistant magnesium diecasting alloys has been achieved in two alloy systems in the past. One was the development of Mg-Al-Si diecasting alloys (AS21, AS41) in the 1970s. This alloy had, in addition to the β ($Mg_{17}Al_{12}$), another intermetallic, Mg_2Si phase with high temperature stability which was deemed to pin dislocations and the grain boundaries thereby imparting increased creep resistance to the alloy. However, the improvement in creep resistance was marginal and the alloys were more difficult to diecast. The other major development was the emergence of rare-earth (RE) containing alloys (AE41, AE42) which achieved major improvement in creep resistance due to the complete suppression of the formation of the β ($Mg_{17}Al_{12}$) phase and, instead, the presence of the Al_2RE intermetallic with excellent high temperature stability. RE addition was also found to be more effective in terms of preventing coarsening of β-particles and the grain size at 300 °C.[88] Both Si and RE additions, however, were observed to somewhat deteriorate the corrosion resistance of the alloy. Furthermore, the alloys containing RE additions are too costly for widespread automotive use and difficult to diecast with the additional drawback of low fatigue properties. Researchers working with rapidly solidified alloys have also explored different additions to the Mg-Al-based alloys to improve the creep resistance of alloys such as AZ91 with no apparent success. It was shown that the addition of silicon led to a smaller grain size and increased the hardness of the rapidly solidified alloy more when compared to the addition of similar amounts of cerium based misch-metal.[88]

Another attempt has been made by Mordike *et al.* to improve the creep resistance of AZ91 by reinforcing it with addition of short fibres of Al_2O_3 (Saffil) via squeeze casting.[85] Studies showed that creep fracture was matrix controlled rather than fibre cracking.[89–91] Molten magnesium may attack the reinforcement and be problematic, however, with careful control this reactivity has been shown to be useful in promoting good bonding with fibres and in turn to effective load transfer to the reinforcement.[92–94] Although it has resulted in some success, imparting about one order of magnitude longer creep life and less strain to fracture, this production method is costly, and furthermore, the gain in creep resistance seems to be lost when the fibre-strengthened alloys are extruded.[94,95]

The low creep resistance of magnesium is further worsened by other factors. When creep resistance is compared on the basis of fabrication technique, diecasting gives inferior results due to inherent features such as fine grain size and porosity that are imparted to the structure. Columnar or large dendritic microstructures that are beneficial for creep resistance are generally not achievable due to rapid solidification as well as breaking-up of dendrites during die-filling especially in high-pressure diecasting. Therefore, sand casting can produce better creep resistance in some magnesium alloys, such as ZRE1 (Mg-3w%RE-2.5w%Zn-0.6w%Zr), compared to the heat resistant aluminum alloys[96,97] although moderate tensile and fatigue properties in this case can be a limiting factor.

Another disadvantage in diecasting is imposed by the long freezing range of alloys like AZ91, which leads to additional shrinkage microporosity and in turn weakened creep resistance and pressure tightness.[98] Alloy constitutions bring about further inherent microstructural problems. A high cooling rate in castings leads to coring in Mg-Al based alloys, causing extensive formation of the divorced eutectic product (see Fig. 2.7), the coarse β ($Mg_{17}Al_{12}$) particles at grain boundaries, and microstructural instabilities that have been claimed to play an important role in weakening the creep resistance.[54]

For diecastings, any improvement of creep resistance in magnesium alloys needs to rely on the chemical constitution that would permit the formation of stable precipitates for strengthening at high temperatures without adversely affecting the diecastability, corrosion resistance, and the cost of the alloy as well as tensile and fatigue properties. Alloy systems that promise better creep resistance may not even be suitable for diecasting due to low fluidity or die-sticking problems. Mordike and Ebert have indicated the current potential of creep resistance in magnesium diecasting alloy systems such as Mg-Al-Si (max. 150 °C), Mg-Al-Re (max. 175 °C), Mg-Al-Ca-X and Mg-Re-Zn-Mn (max. 200 °C), and in alloys suitable for sandcasting such as Mg-Ag-Re-Zr (200–250 °C), Mg-Y-Re-Zr (250–300 °C), Mg-Sc-X-Y (>300 °C).[92] Developments in improving the creep strength of magnesium alloys will be discussed in detail below under specific headings.

2.5.1 Creep resistant magnesium alloys for high pressure die-casting

Mg-Al-Si alloys

The first development in creep resistant magnesium diecasting alloys for automotive applications has been the development of Mg-Al-Si diecasting alloys by Volkswagen in the 1970s. Silicon addition in AS alloys leads to the formation of the Mg_2Si precipitate that has low density, high hardness, low thermal expansion coefficient, and perhaps most importantly, unlike the β-$Mg_{17}Al_{12}$ phase, a high melting temperature (1,085 °C). Silicon, however, lacks the ability to tie up aluminum rendering it ineffective in hindering the formation of β-phase and discontinuous precipitation for a given level of aluminum. These alloys hence have two types of intermetallic, the β-($Mg_{17}Al_{12}$), and the Mg_2Si phase with high temperature stability, which is deemed to pin dislocations and the grain boundaries thereby imparting increased creep resistance to the alloy.

The Mg_2Si that has a face centred cubic crystal structure and is a metallurgically stable phase that forms in Chinese script morphology under slow cooling conditions, which would exhibit very low ductility.[99] In the diecast alloy, a fine distribution of the strengthening phase, Mg_2Si, is achieved with resultant high temperature resistance and a satisfactory level of toughness.[100] However, the creep resistance achieved in this system is moderate and the alloys are difficult to diecast.

The AS41 (Mg-4Al-1Si) and AS21 (Mg-2Al-1Si) alloys are the two standardized Mg-Al based heat-resistant alloys in use that display some improvement in creep properties in the range 130–150 °C. AS41 offers better heat and wear resistance than AZ91, albeit with reduced fluidity of its melt. The alloys of the AZ and AS series all exhibit the same relative decrease in strength as the temperature increases up to 200 °C, while AZ91 shows superior yield strength at all temperatures.[101]

AS41 has moderate castability, but due to lower Al content, castability in AS21 becomes an even bigger problem. Foerster suggested that silicon improves the fluidity of a magnesium alloy when 4% Al is present, making AS41 more castable than AS21.[102] Silicon addition must be accompanied by a reduction in aluminum content to lower the amount of $Mg_{17}Al_{12}$ in the microstructure in order to improve creep resistance. The improved creep resistance of AS21 over AZ91D is attributed to the smaller amount of the creep-induced precipitation and the grain boundary pinning provided by the Mg_2Si precipitates (Fig. 2.8).[54,55]

It is possible to obtain a compromise between castability and creep resistance in this alloy system by going to an AS31 (Mg-3Al-1Si) composition. Recently, a modified version of AS21X with small additions of rare earth elements has been developed. Rare earth additions at trace levels do not affect creep resistance but improve salt-spray corrosion resistance as well as castability.[103]

Mg-Al-RE alloys

A major development in creep resistant magnesium alloys has been the emergence of rare-earth (RE) containing alloys (AE41, AE42). This group of Mg-Al-RE alloys involves at least one, and in general a mixture, of the REs as precipitate-forming alloying additions in their compositions. Rare earth alloying additions are expensive and therefore a cheaper substitute known as mischmetal, a mixture of several rare earth elements that is enriched in one of the constituents is generally used. AE series constitutes the standardised alloys in the group of Mg-Al-RE alloys. This alloy system exhibits major

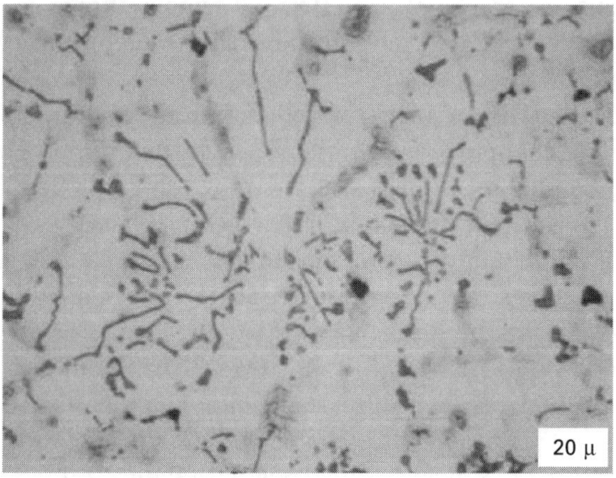

2.8 The appearance of Mg_2Si precipitates with Chinese script morphology in AS41 alloy. (X700).

improvement in creep resistance due to the complete suppression of the formation of the β-($Mg_{17}Al_{12}$) phase and the presence, instead, of the Al-RE containing intermetallics.[104] It has been shown that the ultimate tensile strength increases with the Al and RE content in the alloy, and that tensile yield strength primarily depends on the amount of RE and attains a limiting maximum level.[105] For a given rare earth content in the alloy the creep strength decreases with increasing aluminum content, while for a given aluminum content it increases with increasing amount of rare earth addition.[104]

There have been a number of studies aimed at identifying the second phases in the AE alloy family. Mg-Al alloys containing RE additions have been claimed to be suitable only for diecasting since slow cooling rates lead to the formation of coarse Al_2RE particles.[106] $Mg_{12}Ce$ type particles were also detected at the grain boundaries during creep. Both of these precipitates, unless they coarsen, are considered to inhibit grain boundary sliding and thus one beneficial for creep resistance.[106] Precipitation of Al_4RE compound in the alloy AE42 has also been reported.[54,96,107]

Depending on the composition, precipitates of other types of compound have also been mentioned in the literature. It has been suggested that when RE/Al weight ratio is above 1.4 all of the aluminum will be tied up as $Al_{11}RE_3$ in which case further precipitation of other phases such as another type of Al-Re phase or $Mg_{12}RE$ becomes possible.[96] In samples produced by cold chamber diecasting, $Al_{11}RE_3$ and $Al_{10}RE_2Mn_7$ formation at the grain boundaries was observed. Their TEM observation has revealed that the major phase was $Al_{11}RE_3$ (a body-centred orthorombic cell with lattice parameters a = 4.5 Å b =13.2 Å and c = 9.9 Å). No special orientation relationship was found between this phase and the Mg-matrix. EDS measurements gave an atomic ratio Al:RE = 3.3 ± 0.6 which is claimed to be in good agreement with the proposed stoichiometric ratio 11:3. The relative amounts of REs were reported to be (at %) $La_{27\pm2}Ce_{51\pm2}Pr_{6\pm1}Nd_{16\pm1}$. This phase was found to co-exist with a minor phase $Al_{10}RE_2Mn_7$. Lanthanum was said to go into $Al_{11}RE_3$ phase and Nd into $Al_{10}RE_2Mn_7$ phase. The creep resistance was attributed to precipitate morphology. Since the precipitates forming at the grain boundaries have a strongly branched morphology anchoring into several neighbouring grains together, this was deemed to contribute to grain boundary pinning.

Creep studies on AE42 alloy revealed that, just as in AZ91 and AS21 alloys, the activation energy was in the range of 30–40 kJ/mol, indicating the operation of the same creep mechanism of creep-induced-precipitation which results in grain-boundary migration.[54] $Mg_{17}Al_{12}$ was not observed but Al-supersaturation was noted. The difference between AZ and AS alloys and the AE was that the onset of the second creep mechanism at the high stress region at 150 °C was not seen in AE42 up to 100 MPa, and the creep strain was lower compared to those of AZ91 and AS21 (Fig. 2.9).

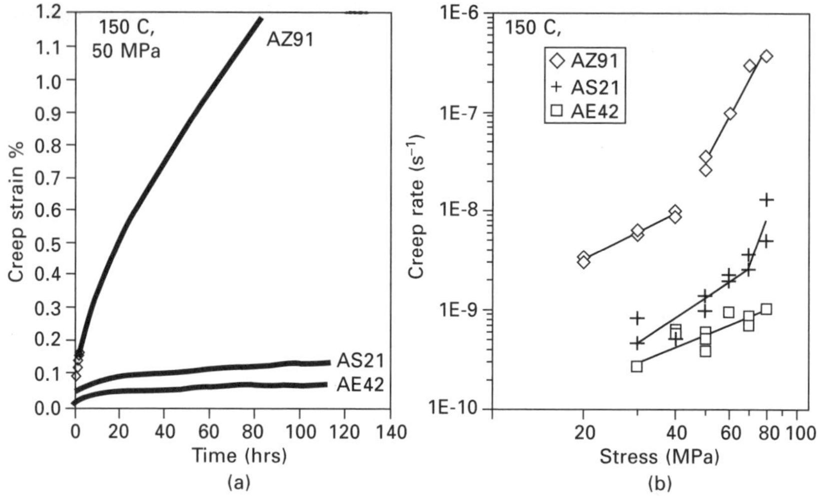

2.9 (a) Creep strain vs time and (b) creep rate vs stress in AE42, AS21 and AZ91.[54]

The microstructural evolution of AE42 alloy has been given in Fig. 2.10. Two different intermetallic phases, namely $Al_{11}RE$ and Al_2RE, and practically no $Mg_{17}Al_{12}$ are present in the diecast structure.[108] Creep testing at 150 °C and 175 °C showed a change in microstructure and changes in the relative amounts of these phases (Table 2.2). These changes are consistent with the reaction by which $Al_{11}RE_3$ decomposes to Al_2RE and frees Al, which then reacts with Mg to form $Mg_{17}Al_{12}$, which is accepted as being responsible for poor creep resistance in magnesium alloys.[65] Powell *et al.*, in order to see the role of stress on the decomposition of $Al_{11}RE_3$, exposed samples to the same thermal history but without the applied stress.[108] The authors observed the same microstructural changes and concluded that $Al_{11}RE_3$ suffered merely from thermal instability at 175 °C. A relationship was also discovered between

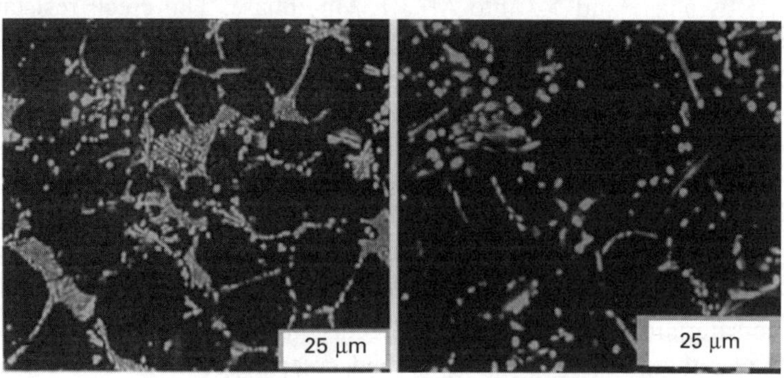

2.10 Microstructure of AE42 (a) diecast (b) after 175 °C exposure.[108]

Table 2.2 Change in the amounts of phases (wt %) with creep test in AE42 alloy[108]

Temp.	Mg	$Al_{11}RE_3$	Al_2RE	$Mg_{17}Al_{12}$
As-cast				
25 °C	97.5	1.8	0.8	0.0
150 °C	97.7	1.5	0.8	0.0
175 °C	97.0	1.2	1.3	0.6

the relative amounts of $Al_{11}RE_3$ vs. Al_2RE and the La:Nd. (below about 0.7 Al_2RE was seen and above 0.7 $Al_{11}RE_3$ seemed to form). They indicated that the formation of either $Al_{11}RE_3$ or Al_2RE is sensitive to individual rare earth elements, and that this might influence the stabilisation of the $Al_{11}RE_3$ phase in AE42 alloy.

AE42 alloy was perceived to have issues with cost, castability and fatigue strength and was not considered for widespread automotive use for a long time. In recent studies, a fresh look at the AE alloy family has been presented.[109] Different combinations of Al and RE are investigated namely Mg-3%-5%RE (AE35), and Mg-8Al-2RE (AE82). Future studies on AE42 should address thermal stability at 175 °C and conduct a thorough search in the ternary Mg-Al-RE alloy system to determine compositions that offer creep resistance, thermal stability and good castability. Phase diagram studies in the ternary system are also important.

Mg-Al-Ca alloys

Ca addition, as a cheaper and lighter alternative to RE elements, also contributes to high temperature properties.[110–111] When Ca is added to Mg-Al binary alloys, the type of precipitating compound is found to depend on the Ca/Al mass ratio. When this ratio is more than ~0.8 the presence of both Mg_2Ca and Al_2Ca was detected that resulted in a considerable increase in hardness, while below this ratio only Al_2Ca was observed to have formed (Figs 2.11 and 2.12). Both types of precipitate were observed to form along the grain boundaries with an increasing volume fraction as the Al and Ca content increased.[112]

Pekguleryuz *et al.* have shown that the creep resistance in Mg-Al-Ca ternary alloys (AX alloys) in the range of 2–6 wt% Al and 0.6–1.0wt% Ca show creep resistance at 150 °C similar to the AE42 alloy with the added benefit of good corrosion resistance.[111,113–115] The Al_2Ca compound has stability at high temperatures and the formation of the β-$Mg_{17}Al_{12}$ phase is suppressed. The beneficial effect of Ca addition was also proven in rapidly solidified AZ91 alloy. It has been shown that despite the smaller grain size

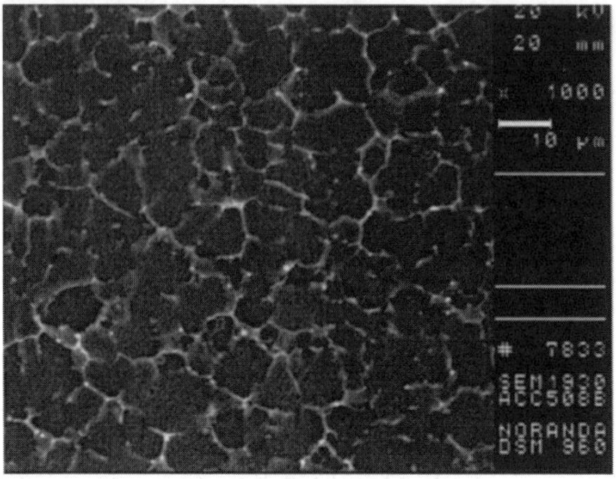

2.11 Diecast microstructure of Mg-5Al-0.8Ca.[111]

compared to conventional AZ91, a rapidly solidified structure involving fine Al_2Ca particles exhibited drastically smaller secondary creep rate.[114]

Although useful, Ca addition is not problem free. Calcium makes the alloy prone to hot cracking and die-soldering especially when the level is above 1%.[110,116] Subsequent addition of 8% Zn to the alloy has improved diecastability but the properties were found to vary over a wide range.[117] The second phase in this alloy was a Mg-Al-Ca-Zn containing complex phase. Presently, none of these experimental alloys with Ca addition has yet led to a commercially successful alloy. More work with the base alloy systems containing Ca is needed to optimise these alloys for diecasting. However, work has been carried out to use these alloys for semi-solid applications.[118] The Mg-Al-Ca system can be further investigated for optimum castability.

It has recently been shown that increasing the Ca to above 2% improves castability by eliminating hot-tearing.[119] Terada *et al.* studied creep mechanisms in Mg-5Al-1.7Ca alloy.[120] At 150 °C, the change in the stress exponent occurred at around 120 MPa, and in the low stress regime (100 MPa), in the 175–250 °C range, the activation energy was determined to be 142 kJ/mol. While the authors attributed this to dislocation climb, the mechanism may be better described by the mechanism of glide controlled by diffusion of solute atoms, i.e., that of aluminum since Q is identical to Q_{Al} 143 kJ/mole (diffusion of Al in Mg).

Mg-Al-Sr system

Mg-Al-Sr (AJ) alloys are a new addition to the creep-resistant magnesium alloys.[121,122] Various alloy compositions such as AJ51 (Mg-5Al-1Sr), AJ52 (Mg-5Al-2Sr) and AJ62 (Mg-6Al-2Sr) have been studied. AJ52 shows the

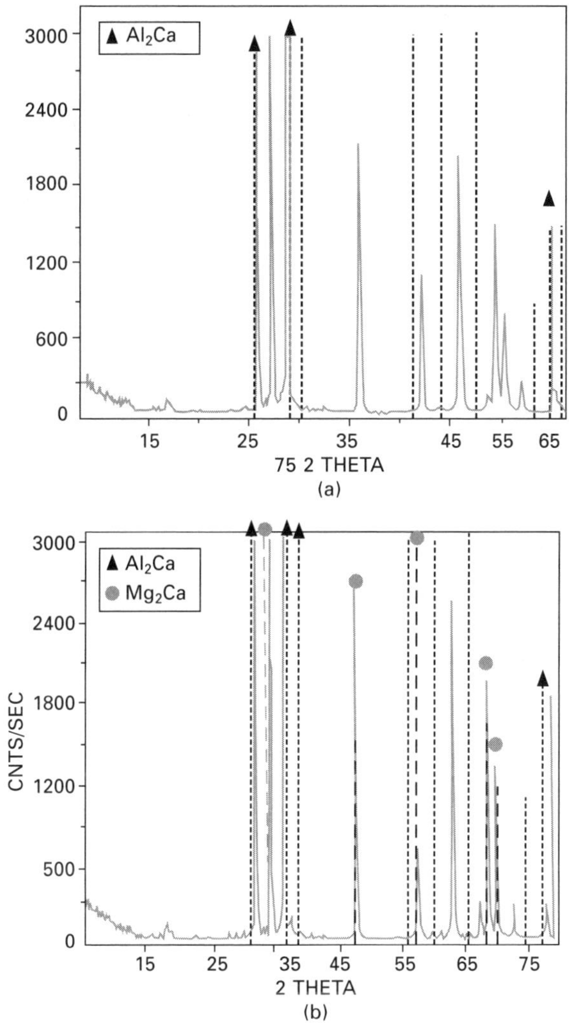

2.12 XRD from (a) Mg-5Al-0.6Ca alloy showing Al_2Ca. (b) Mg-5Al-0.8Ca showing the presence of Al_2Ca and Mg_2Ca.[111]

highest creep resistance, while AJ62x gives an excellent combination of creep performance and castability. The alloys have been shown to exhibit different microstructures based on the Sr/Al ratio.[123] For Sr/Al ratios below about 0.3, Al_4Sr intermetallic is only the second phase in the structure (Fig. 2.13). When the Sr/Al ratio is higher, a second intermetallic phase (a ternary Mg-Al-Sr phase) was also observed. Sr/Al controls the formation of $Mg_{17}Al_{12}$ as well. When the Sr/Al ratio is very low, there would be insufficient Sr to bind all Al and the excess Al would then form the $Mg_{17}Al_{12}$ phase. It was reported that the ternary Mg-Al-Sr phase was a new compound. The superior

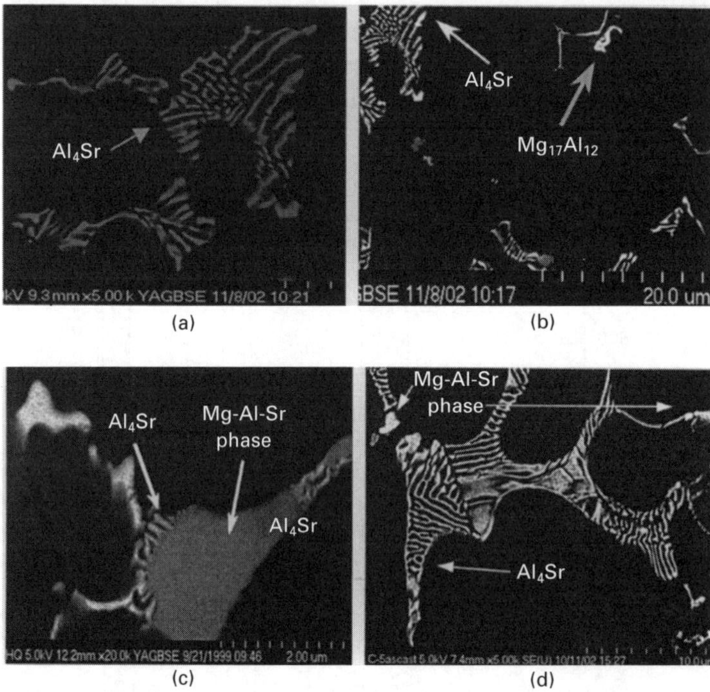

2.13 Diecast microstructures of (a) AJ51(Mg-5Al-1Sr) and (b) AJ62L (Mg-6Sr-1.6Sr) alloys with the Sr/Al ratio lower than 0.3, exhibiting Al_4Sr second phase. Microstructures of (c) AJ52 (Mg-5Al-2Sr) and (d) AJ62 (Mg-6Al-2Sr) alloys with the Sr/Al ratio higher than 0.3 exhibiting two types of second phases, Al_4Sr and a ternary Mg-Al-Sr compound.[123]

creep resistance of AJ52 and AJ62 alloys is attributed to the existence of this second phase in addition to Al_4Sr. Intermetallic Al_4Sr is a tetragonal phase and no phase coherency with the primary magnesium phase has yet been reported. Future work on this alloy system should focus on the determination of the ternary phase diagram and the identification of the ternary phases.

Magnesium diecasting alloys with combined additions of alkaline earth and/or rare earth elements

In recent years, many Al-containing Mg alloy systems have been developed with combined additions of rare-earth and alkaline earth elements. Patented alloys such as a Nissan alloy based on Mg-Al-Ca-RE system, and a Honda alloy ACM522 (Mg-5Al-2Ca-2RE) both claim improved creep resistance over AE42 alloy.[124,125] ACM522 alloy is based on specific AI/RE/Ca ratios to ensure AI-RE precipitates in addition to Al-Ca intermetallics. This alloy is

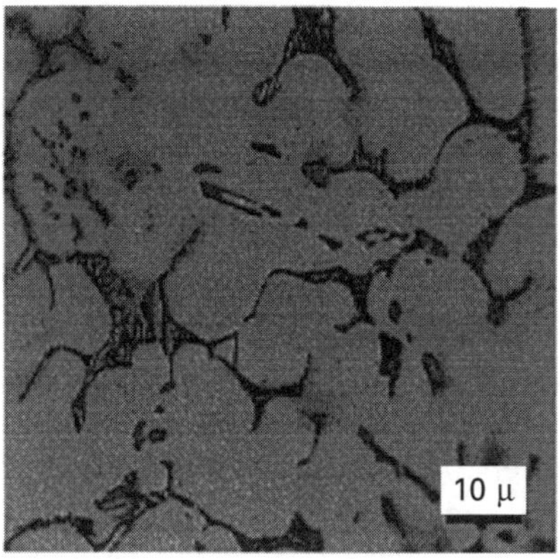

2.14 Microstructure of Mg-8Al-2.5RE-1.6Ca-1.3Mn (permanent-mould cast).[126]

used by Honda in oil pans for their hybrid car. The microstructures of these alloys exhibit a combination of Al-RE and Al-Ca intermetallic mixtures. The permanent-mould cast microstructure of Mg-8Al-2.5RE-1.6Ca-1.3Mn version of the alloy given in Fig. 2.14, shows Al_2Ca coupled eutectic phase as well as needle shaped Al-RE (Ca, Mn) and Al-Mn-RE intermetallics.[126]

Another alloy system based on Mg-Al-Ca-RE with optional additions of Sr and Zn is a patent by Dead Sea Magnesium and Volkswagen AG.[127] Two alloy formulations code-named MRI 153 and MRI 230 are reported to exhibit good elevated-temperature performance at 150 °C and 180 °C respectively.[128] Furthermore MRI 153 exhibits good diecastability similar to AZ91 alloy. The alloy system of this patent is quite complex, having a number of elements and would lend itself to further development to find optimum combinations of strength, creep resistance and castability.

Two separate alloy systems with combined additions of Sr and Ca but no rare earths have been developed by Noranda[129] and General Motors.[130] The Noranda alloy is a Mg-Al-Sr-Ca alloy with small amounts of Sr and Ca (AJX Alloy). Code-named Noranda N alloy, it has creep resistance comparable to AS41 at 150 °C but creep resistance improves at 175 °C being superior to AE42 alloy.[126] The GM version is a Mg-Al-Ca-Sr alloy with substantial Ca and a small amount of Sr (AXJ Alloy) and exhibits superior creep resistance.[131]

All these complex alloys offer great potential for high creep resistance and a major effort should be made to improve the diecastability of those alloy systems' compositions that offer the most creep resistance. This can be achieved by systematic and fundamental castability research or by trace element effects.

Comparative properties of magnesium diecasting alloys for powertrain applications

Formal comparative evaluation of the various alloys is in its initial stages. Important performance requirements for powertrain components, among others, are tensile and compressive creep resistance, bolt-load-retention, yield strength, fatigue resistance, corrosion resistance and diecastability as well as cost and availability. It is rather difficult to make a rigorous comparison of the alloys from published data due to differences in sample, test conditions and type of test. Creep tests are carried out at different stress-temperature conditions and results are reported at different creep times. An attempt is made in Tables 2.3 and 2.4 to summarise some of the data on various alloys and other data can be found in the literature.[119] Formal evaluation programs would be of great benefit to end users in identifying appropriate alloys for powertrain application development programs.

Diecastability is a special performance requirement of these alloys. The factors that contribute to good diecastability are low surface tension and relatively low liquidus for adequate fluidity, resistance to hot-tearing, optimum degree of wettability to avoid excessive reaction with steel dies, etc. Generally, increasing aluminum in Mg-Al based alloys improves diecastability by decreasing the liquidus and improving fluidity. The same can be achieved by increasing the diecasting temperature, within reasonable limits so as not to sacrifice die-life. A larger freezing range may also help diecastability by increasing the length of time the metal stays liquid and hence eliminate cold-shuts. Alloying elements that decrease the thermal conductivity or increase heat capacity are also beneficial for diecasting.[132] Each alloy would require a special set of diecasting parameters to fit the specific characteristics of the alloy.

2.5.2 Creep-resistant magnesium alloys for gravity or low pressure casting

Traditionally, automotive use of magnesium has been primarily in diecast parts to take advantage of the high productivity of the diecasting process that balanced the relatively high cost of magnesium alloys. With the price trends of magnesium since 2001, magnesium gravity cast or low-pressure parts are also becoming cost-effective.

The development of gravity cast alloys does not pose the additional challenge of meeting diecastability requirements. This would theoretically allow for a wider selection of alloy systems and compositions to be explored. Despite the low creep resistance of pure magnesium, it is known that alloying has been quite successful in the past in developing magnesium alloys for high temperature service for aerospace applications. This is evident in comparing

Table 2.3 Tensile properties of various die-casting Mg alloys and comparison with A380 Al alloy[119,123,128]

Alloy	20 °C			150 °C			175 °C		
	UTS, MPa	YS, MPa	%E	UTS, MPa	YS, MPa	%E	UTS, MPa	YS, MPa	%E
AZ91D	204 (240)	139 (160)	3.1 (3)	169 (157)	105 (108)	16 (23)	138	89	21
AS41	197 (215)	136 (140)	4 (6)	153 (155)	94 (95)	17 (24)	127 (127)	85 (85)	19 (25)
AS21x*	210	121	5.5	130	87	20	110	78	23
AE42	220 (230)	128 (145)	10 (11)	140 (157)	88 (100)	23 (20)	121 (130)	81 (91)	23 (23)
AX51	192	128	7	161	102	7	–	–	–
AX52	228	161	13	–	–	–	171	133	23
ZAX850	219	146	5	159	117	11			
AJ52x	212	134	6	163	110	12	141	100	18
AJ62x	240	143	7	166	116	27	143	103	19
AJ62Lx	276	153	12	163	108	19	143	109	27
AJX500	236	133	10	152	101	16	137	98	20
AXJ531	238	190	8	–	–	–	196	146	15
ACM522	200	158	4	175	138	7	152	132	9
MRI153*	197 (250)	157 (165)	2.2 (5)	–	118	–	139	113 (110)	3.4
MRI230D	235	180	5	205	150	16	–	–	–
A380	290 (290)	155 (180)	3.2 (3)	251 (257)	148 (159)	6.4 (7)	248 (228)	154 (148)	7.1 (10)

Table 2.4 Creep of die-cast Mg alloys and comparison with A380 Al alloy[119,123,128]

Alloy	Tensile creep, %						% Compressive creep 150 °C, 70 MPa, 200 hrs
	@ 50 MPa, 200 hrs		@ 50 MPa, 500 hrs		70 MPa, 100 hrs	70 MPa, 200 hrs (53)	
	150 °C	175 °C	150 °C	175 °C	175 °C	175 °C	
AZ91D	2.7	*	6.35	–	–	–	21
AS41	0.05	2.48	0.07	–	–	–	6.13
AS21x	0.19	1.27	–	–	–	8.95	3.97
AE42	0.06	0.33	0.08	0.44	–	0.18	2.16
AX52	–	–	–	–	0.06	–	–
AXJ531	–	–	–	–	0.06	–	–
AJ52x	0.04	0.05	0.03	0.09	–	0.14	0.24
AJ62x	0.05	0.05	–	–	–	–	1.73
AJ62Lx	0.13	0.29	–	–	–	–	–
MRI 153	0.18	–	–	–	–	–	–
MRI 230D	–	–	–	–	–	0.24	–
A380	0.08	0.04	0.10	0.05	–	0.22**	0.03

*failed after 80 hrs **A383 alloy

the 'maximum usefulness temperature' of various metals. A measure of maximum usefulness temperature is arbitrarily defined as the temperature at which the alloy can withstand a stress of ~69 MPa for 100 hours without fracturing.[133] Table 2.5 lists some of the metals (best possible alloys) with respect to this criterion. The criterion does not have practical significance but it may be used to evaluate the effectiveness of an alloying element on pure metals. For magnesium, it can be observed that the best alloys (mainly containing rare earth and in the past containing thorium) can compete very well with aluminum alloys.

Table 2.5 Best possible alloys of a base metal for elevated temperature service[133]

Base metal	Melting point (°C)	Temperature of useful strength for best alloy (°C)	% Absolute melting point
Mg	650	343	67
Al	660	288	60
Ti	1704	649	46
Fe (martensitic)	1538	732	56
Fe (austenitic)	1538	871	63
Ni	1455	1071	78

Yttrium-containing alloys

Low-pressure and gravity casting alloys of magnesium have been traditionally used in aerospace applications. There are well-known creep-resistant alloy systems based on rare earth elements and yttrium such as WE43, WE54. These alloys do not contain aluminum and are grain-refined effectively with zirconium. Mg-Y-RE (Nd) alloys are amenable to precipitation hardening. The microstructures of WE54 and WE43 contain a eutectic phase at the grain boundaries in the as-cast state that has been characterised as β-phase, $Mg_{14}Nd_2Y$, with a face centred cubic crystal structure.[134] The presence of another phase, Mg_9Nd with a face-centred tetragonal structure after homogenisation at 525 °C has also been reported.[135,136] Excellent room and elevated temperature tensile and creep properties of the alloy WE54 in the solution treated and aged condition are attributed to the formation of precipitates, namely β″, β′ ($Mg_{12}NdY$), and β($Mg_{14}Nd_2Y$).[134,137]

In addition to the standardised alloys in the Mg-Y-RE system, some experimental compositions have also been studied. Kamado *et al.* reported the ageing characteristics and tensile properties of some Mg-Gd-Y-Zr alloys.[138] The increasing age hardening with decreasing Y/Gd ratio was again attributed to the formation of β″ and β′ precipitates. Another important example is the positive contribution of zinc to creep resistance in experimental quaternary alloys in the Mg-Y-Zn-Zr system as reported by Morgan and Mordike.[139] An optimum amount of about 2%Zn was shown to result in better steady state creep rates for the Mg-Y-Zr alloys than those without zinc. The best creep resistance in Mg-Y-Zr alloys was obtained with 2% Zn and 2% Nd addition in the as-cast condition.[139]

Silver-containing alloys

Silver as an alloying element increases both strength and ductility in magnesium up to approximately 4% Ag.[60] A major casting alloy family is a quaternary system based on Mg-Zn-RE-Ag. These alloys were developed as a result of Ag replacement of Zn in Mg-Zn-RE alloys in order to improve the tensile properties.[106,140,141] The RE elements used in these alloys are the more soluble Nd-rich didymium misch metal (80% Nd, 16% Pr, 2% gd, 2% others) rather than the normal cerium-rich misch metal (Ce/La). Several casting alloy combinations developed have tensile properties comparable to aluminum casting alloys up to 250 °C. If the Ag content is below 2%, the precipitates are of the Mg-Nd-containing type. For Ag above 2% the equilibrium precipitate is a ternary intermetallic of possible $Mg_{12}Nd_2Ag$ composition.

Precipitation in silver-containing magnesium alloys has been summarised by Lorimer.[142] X-ray diffraction studies on Mg-12.5% Ag alloys reveal the following precipitation sequence with two types of GP (Guinier-Preston) zones:

$$Mg_{ssss} \rightarrow \text{G.P.I (plates on } <10\,\bar{1}\,1>\text{Mg)} \rightarrow \text{G.P.II (spheroids)}$$
$$\rightarrow \gamma' \rightarrow \gamma \cdot (Mg_3Ag) \qquad \text{mode 1}$$

In Mg-Ag-Zr alloys such as QK21 (Mg-2.4Ag-0.16Zr) and QK91 (Mg-8.7Ag-0.5Zr) transmission electron microscopy studies showed another type of precipitation sequence. In high-silver-containing QK91 γ' precipitates were seen during 200 °C ageing. No GP zones were observed even at lower temperatures of 75 °C. The proposed precipitation sequence was:

$$Mg_{ssss} \rightarrow \gamma' \cdot \text{(hexagonal)} \rightarrow \gamma \cdot (Mg_3Ag)\text{(hexagonal)} \qquad \text{mode 2}$$

Quaternary alloys of Mg-Ag-RE-Zr such as QE22 (Mg-2.5Ag-2RE(Nd)-0.7Zr) exhibit outstanding ageing response and good tensile properties up to 200 °C. The combination of silver and rare earth and a high volume fraction of homogeneously distributed precipitates formed through two different precipitation sequences is shown below. Both sequences are observed at 250 °C ageing, with γ and β intermediate precipitates transforming into $Mg_{12}Nd_2Ag$ at the end. Maximum age-hardening is associated with the presence of γ and β precipitates. The presence of hydrogen in the atmosphere during ageing can cause heavy precipitation of rare earth and zirconium hydrides within the grains and grain boundaries devoid of $Mg_{12}Nd_2Ag$ compounds.[143]

Rod-shaped zones // $[0001]_{Mg}$ → Rod-shaped γ//$[0001]_{Mg}$ (hexagonal)
/ \
Mg_{ssss} $Mg_{12}Nd_2Ag$ // $[1120]_{Mg}$ (hexagonal)
Lath-like equilibrium precipitates
\ /
Ellipsoidal zones → Equiaxed β (hexagonal)
mode 3

In a Cu-added version of Mg-Ag-RE-Zr alloys precipitation is slightly modified. The presence of 0.005–0.15% Cu changes from mode 1 to mode 3 precipitation above 1% Ag rather than above 2% Ag.

The yield and tensile strength of fully heat-treated quaternary Mg-RE-Ag-Zr alloys increase progressively up to 2% Ag, beyond which the change is minimal. Optimal heat treatment is a solution heat treatment of 4–8 h at 525 °C, cold-water quenching followed by artificial ageing at 200 °C.[106] The properties are influenced greatly by the type of rare-earth element used; cerium (Ce) or misch metal (MM) being the least and neodymium (Nd) being the most effective strengthener.[144] An alloy with lower Ag content, EQ21, was also developed with similar mechanical properties but at a lower cost.

Tensile, fatigue and creep properties of QE22 and EQ21 are shown in Table 2.6.[145] EQ21 with the lower Ag has slightly higher creep properties than QE22. Fatigue limits of silver-containing alloys are high compared to other magnesium and aluminum alloys.[146] This property makes the silver-containing magnesium alloys excellent candidates for aircraft wheel applications. Other properties of QE22 such as low propensity to brittle fracture and good creep properties have led to its use in missile and rocket applications such as the front cabin pedestals, frame of the front undercarriage in addition to wheel applications.[146] Other applications of these alloys are for compressor casings in the Rolls Royce RB11 engine and gearboxes for racing cars.[145]

The castability of QE22 alloy is relatively good. The alloy shows some susceptibility towards microporosity, surface sinks and segregation under

Table 2.6 Mechanical properties of silver-containing magnesium alloys[145]

Alloy/ temper	Composition, %				Tensile properties			Fatigue endurance N/mm^2 (5×108 cycles)	Creep (N/mm^2)	
	Ag	RE	Cu	Zr	YS (N/mm^2)	UTS (N/mm^2)	E %		200 °C	250 °C
QE22A-T6	2.5	2.0	–	0.6	175	240	2	90–100	85	32
EQ21-T6	1.5	2.0	0.06	0.6	175	240	2	90–100	95	36

poor feeding conditions that can be eliminated in sqeeze casting (Fig. 2.15).[146] Less porosity is seen at lower RE levels. Some tendency towards oxide inclusions due to the rare earth content is also seen, which can be minimised by avoiding turbulent filling of the die cavity. The alloy is not susceptible to hot tearing and cracking which translates into good castability and excellent weldability.[143,147,148]

Alloys containing zinc and RE addition

Possibly the most important gravity cast magnesium alloy system for automotive applications to date has been the Mg-Zn-RE system (ZE and EZ alloys). An alloy belonging to this group is ZE41 that offers moderate strength up to 150 °C after ageing of the cast material. Better properties in terms of castability, freedom from microporosity, strength and creep resistance up to 250 °C can be obtained with EZ33 (2.5–4.0% RE, 2.0–3.1% Zn, 0.5–1.0 Zr).[106] The grain boundaries in these alloys following ageing are decorated with Mg-RE precipitates. It has been suggested that the eventual equilibrium precipitate is $Mg_{12}Ce$ or $Mg_{27}Ce_2$. The ageing sequence is assumed to be similar to that in

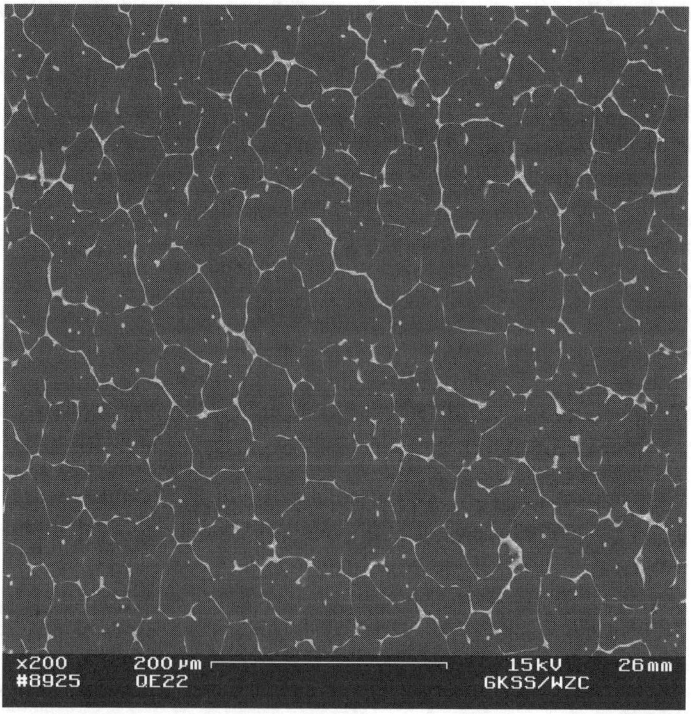

2.15 Microstructure of squeeze-cast QE22 in as-cast state.[146] (courtesy of C. Blawert *et al.*).

Mg-Nd alloys involving several stages including the formation of a coherent β″-phase that causes hardening.[106]

Since they do not contain aluminum, these alloys can be grain-refined by 0.6–0.7% zirconium addition. A study by Fisher represents a very different approach to modification of precipitates in this alloy system, and has led to a successful alloy used in the aircraft industry.[149] Hydrogenation of ZE62 at 500 °C for 24 hours followed by 64 h of ageing at 125 °C caused the intragranular precipitation of RE hydrides with a needle-like morphology resulting in remarkable properties.[106,149]

Alloys containing zinc and copper

Mg-Zn-Cu alloys were developed for a combination of good ductility and elevated temperature performance where Cu imparts a grain-refining effect and Zn good castability and ductility.[150,151] The microstructure is two-phase with α-Mg and a truly lamellar eutectic with the composition β-Mg $(Cu.Zn)_2$. Upon solution treatment and ageing two main precipitates are observed, β_1 needles and β_2 plates. Addition of Mn prevents overageing at 250 °C and improves overall yield strength. A practical casting alloy developed from this system is ZC63 (6Zn-2.5Cu-Mn) with good creep strength, good castability but low corrosion resistance. The alloy has been evaluated in select automotive applications.

Silicon-containing alloys

The commercial Mg-Al-Si diecasting alloys are hypoeutectic alloys with respect to the Mg-Mg_2Si pseudo-binary system. The eutectic point lies at 3.67% Mg_2Si in pseudobinary Mg-Mg_2Si. Any alloy to the hypereutectic side of this pseudo-binary system has not been considered commercially. It has been shown, however, that the hypereutectic alloys may be further strengthened by Al, Cu and RE additions via solid solution hardening.[152]

The use of Mg-Al-Si alloys in the sand and permanent-mould cast condition is difficult due to the existence of the Mg_2Si in a coarse Chinese script morphology, which lowers ductility extensively. Pekguleryuz *et al.* have shown that the addition of Ca in the range of 0.08–0.14% Ca modifies the Mg_2Si with a resultant increase in ductility and tensile strength. (Fig. 2.16).[100]

Recent alloy developments in magnesium gravity casting alloys for automotive powertrain applications

Two recent alloy development activities have the objective of developing cost-effective magnesium gravity-casting alloys with good creep resistance for automotive powertrains. One of these is the MEZ alloy in the Mg-Zn-RE

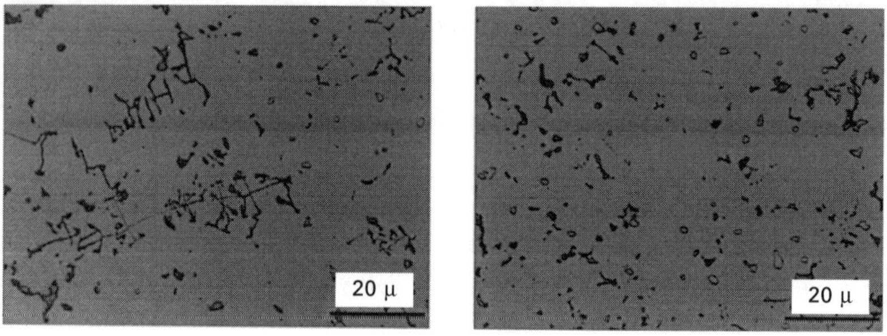

(a) (B)

2.16 (a) Mg_2Si Chinese-script morphology in AS41; (b) 0.1% Ca modified AS41 with refined Mg_2Si.[100]

system developed by MEL.[153] The alloy has 2.5% Zn and 0.5% RE. The permanent mould-cast microstructure comprises intermetallic particles of tetragonal $Mg_{12}(La_{0.43}Ce_{0.57})$ at the grain boundaries.[154] The alloys are age hardenable for improved strength. The second alloy development effort is from Dead Sea Magnesium and Volkswagen AG.[155] The alloys are coded MRI 201S and MRI 202S. The alloy MRI 201S in the T-6 condition with good mechanical properties and creep resistance and castability, is claimed to be a cost-effective substitution for WE43. The alloy MRI 202S with moderate strength and good creep performance is shown to be superior to the ZE41 alloy.

New gravity-casting alloy possibilities do exist with alkaline earth (Ca, Sr) additions as well. Very few gravity casting alloys have been explored in Mg-Al-(Alkaline Earth) and Mg-Zn-(Alkaline Earth) systems and these could provide a good starting point for new alloy development. In fact, without the constraints of diecastability, age-hardening and dispersion hardening can be effectively used to improve creep-resistance and elevated-temperature performance.

Properties of magnesium gravity casting alloys

Currently, it is difficult to make a comprehensive comparison of the various alloys from published data. A summary table is provided in Table 2.7.[156]

2.5.3 Processing microstructures and processability

Magnesium alloys are basically suitable for all casting techniques, and more notably so for high pressure diecasting (HPDC) which increases the chances for economic production due to various reasons such as lesser affinity of magnesium to react with die steels, its low heat content and good fluidity. Limitations, however, exist with certain alloys containing elements that have a tendency to react with steel and cause die-sticking problems.[157] In addition

Table 2.7 Properties of magnesium gravity casting alloys[156]

Properties	WE43-T6	ZE41-T5	ZC63-T6	ASX411-F	MEZ-T6	MRI 201S-T6	MRI 202S-T6
YS (MPa)							
20 °C	180	140	158	75	76	170	150
150 °C	175	120	134			170	145
UTS (MPa)							
20 °C	260	220	240	175	107	260	250
150 °C	210	170	180			245	220
Elongation (%)							
20 °C	6	5	4.5	7	2	6	7
150 °C	7	22	14			11	15
Stress to produce 0.2% creep-strain							
200 °C	160	50	63	–	–	160	100
250 °C	60	20	–	–	–	75	40
Corrosion ($mg/cm^2/day$)	0.10	3.1	*	–	–	0.10	0.12

* Better than AZ91C

to conventional casting methods, other relatively more advanced casting technologies, such as semi-solid casting (thixomoulding) and squeeze casting, are also suitable and give promising results. A comparison of the typical microstructures obtained via HPDC, squeeze-casting and thixocasting are given in Fig. 2.17.

These processes lead to more uniform grain size distributions and properties as compared to the bimodal grain structures obtained from HPDC. Furthermore, the undesirable microstructural defects such as porosity and cold-shuts that are almost inevitable in HPDC can be eliminated without sacrificing the fine grain sizes by employing squeeze-casting or thixo-casting. These improvements further impart the material very important abilities like weldability and heat-treatability.[158] Electromagnetic stirring is shown to have a significant role in breaking up the dendritic structure to a rosette structure during the solidification of melt to prepare the desired feedstock material for thixocasting (Fig. 2.18). Upon reheating to the semi-solid state, the rosette structure transforms to a globular structure, which enables an easy gliding of the solid alpha-phase particles in the liquid phase during the forming operation (Fig. 2.19) and hence, leads to a successful semi-solid die-casting and forging to the final component.

2.6 Conclusion

Magnesium diecasting alloys for elevated-temperature applications are coming of age. A number of aluminum-containing alloy systems with single or combined additions of rare-earth and/or alkaline-earth elements have been developed. Creep resistance in these alloys is attributed to grain boundary strengthening by intermetallic phases. Developing alloy compositions that avoid detrimental creep-induced precipitation, which contributes to grain boundary migration, is the key learning of the past decade. These new alloys are providing commercially viable alloys for automotive powertrain applications such as oilpans, transmission-case and engine components.

In the gravity and low-pressure casting area, the effort is still in its infancy and future R&D as well as interest on the part of car companies are expected to contribute to steady progress in this field. The role of REs and of yttrium in providing creep resistance is well known. New alternatives with Sr and Ca additions similar to diecasting versions can be explored in the future. Age-hardening and dispersion hardening can be effectively exploited for improving creep-resistance.

Needless to say, the search must continue for alloying additions to form useful intermetallic precipitates for better creep resistance. However, the limitations encountered in the utilisation of intermetallic precipitates may be taken to indicate that other types of precipitates, such as oxides, carbides and nitrides, need also be considered to obtain better strength, and creep resistance.

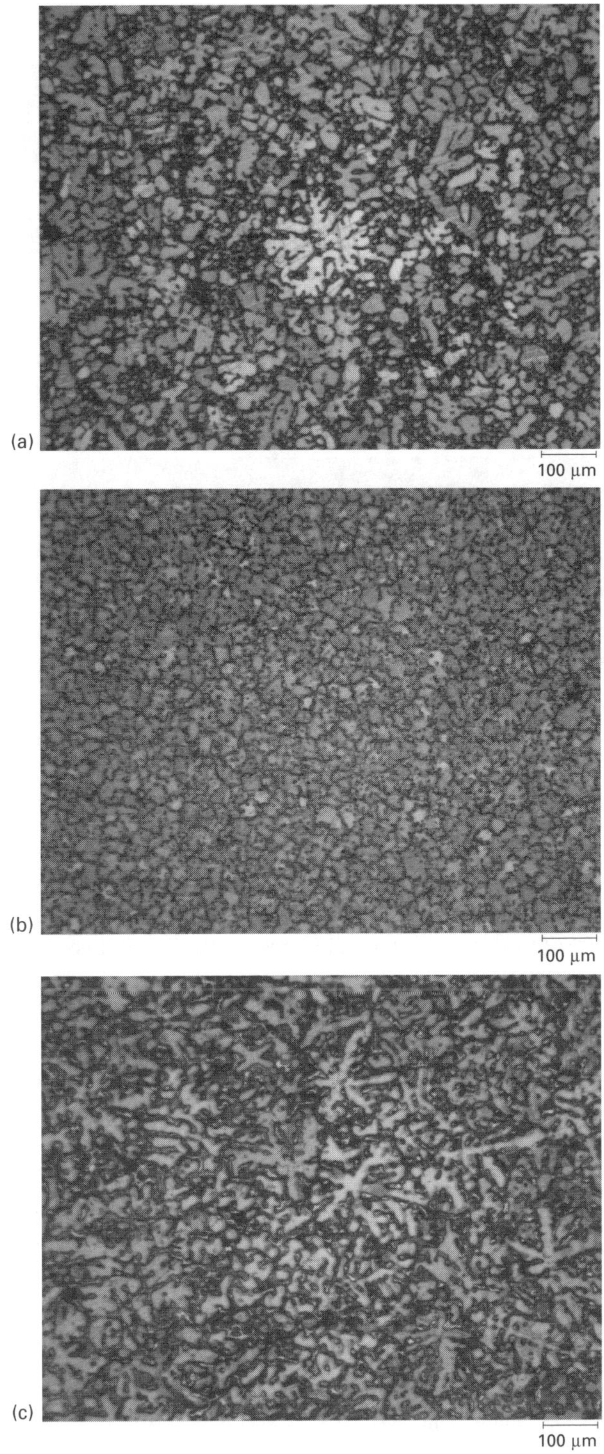

2.17 Comparison of microstructures of AZ91 (a) HPDC, (b) squeeze-cast, (c) thixomould.

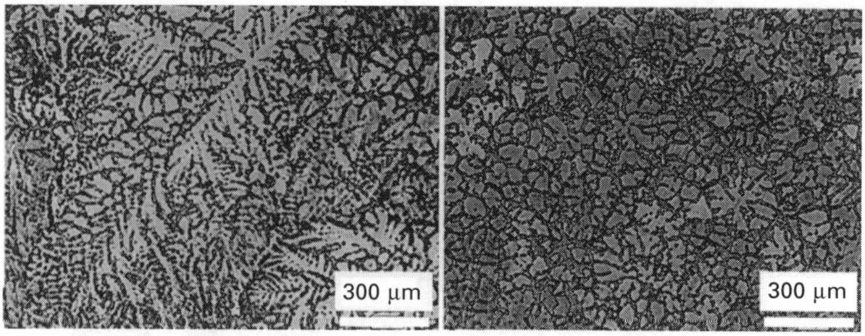

(a) (b)

2.18 Microstructure of laboratory as-cast thixotropic feedstock AZ91D (a) without electro-magnetic stirring (dendritic structure) and (b) with electro-magnetic stirring where dendrites are broken down to smaller rosettes. (Courtesy of Elhachimi *et al.*, CANMET).

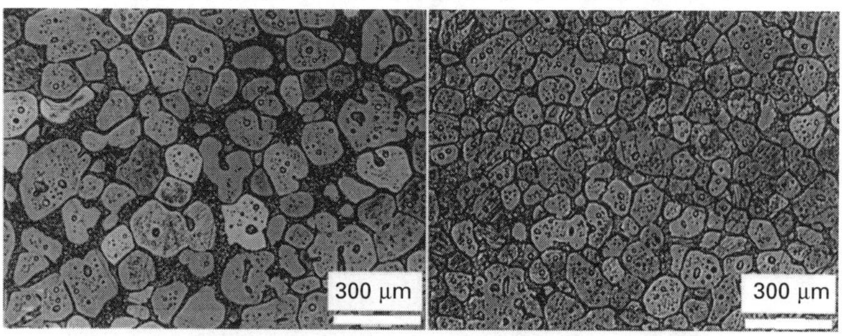

(a) (b)

2.19 Microstructure of laboratory prep d AZ91D (a) after heating to 580 °C and holding for 90 seconds and (b) after semi-solid die-casting into the final part. (The etchant is acetic acid-picral for 7–10 sec.). (Courtesy of Elhachimi *et al.*, CANMET).

Although incorporation of such particles may well dictate less conventional processing techniques such as spray forming, mechanical alloying, or powder metallurgy, opportunities may still exist at least for the *in-situ* formation of nitrides in casting processes. Therefore it may be worthwhile to look especially for nitride-forming compositions and/or conditions, and to understand their effects on the creep resistance of Mg alloys.

2.7 Sources of further information

Emley E.F. *Principles of Magnesium Technology*, Pergamon, Oxford, (1966).

Magnesium Alloys and Technologies, ed. K.U. Kainer, DGM, Wiley-VCH Publishing, (2003).

Raynor G.V. *The Physical Metallurgy of Mg and its Alloys*, Pergamon Press, Oxford, UK, p. 196, 1959.

Busk R.S. *Magnesium Product Design*, sponsored by the International Magnesium Association, Dekker Publishing, NY, USA, (1987).

Magnesium, edited by K.U. Kainer, Proceedings of the 6th International Conference on Magnesium Alloys and Their Applications, DGM, Wiley-VCH Publishing, Weinheim, Germany, (2004).

2.8 References

1. Mordike B.L. and Ebert T., *Materials Science and Engineering* A 302 (2001) 37–45.
2. Emley E.F., *Principles of Magnesium Technology*, Pergamon, Oxford, 1966.
3. Hauser F.E., Landon P.R. and Dorn J.E., *Trans. ASM* 50 (1958) 856.
4. Neite G., Kubota K., Higashi K. and Hehmann F., Ch. 4, *Magnesium-Based Alloys, vol. 8, Structure and Properties of Nonferrous Alloys*, volume ed. K.H. Matucha, Materials Science and Technology series, eds R.W. Cahn, P. Haasen and E.J. Kramer, Publisher VCH, p. 113–213 1996.
5. Milicka K., Cadek J. and Rys P., High Temperature Creep mechanisms in magnesium, *Acta Metal.*, v. 18 (1970) 1071–1082.
6. Courtney T.H. *Mechanical Behaviour of Materials*, NY, McGraw-Hill.
7. Lahaie D., Embury J.D., Chadwick M.M. and Giray G.T., *Scripta Metall. Mater.*, 27 (1992) 139–142.
8. Kim W.J., Chung S.W., Chung C.S. and Kum D., Superplasticity in Thin Magnesium Alloy Sheets and Deformation Mechanism Maps for Magnesium Alloys at Elevated Temperatures, *Acta Mater.* 49 (2001) 3337–3345.
9. Brown C., *Journal of Nuclear Materials* 12, 2 (1964) 243–247.
10. Vagarali S.S. and Langdon T., Deformation Mechanism in H.C.P. Metals at Elevated Temperatures-1 Creep Behaviour of Magnesium, *Acta Metal.*, v. 29 (1981) 1969–1982.
11. Vagarali S.S. and Langdon T.G., Deformation Mechanism in H.C.P. Metals at Elevated Temperatures-II Creep Behaviour of a Mg-0.8% Al Solid Solution Alloy, *Acta Metal.*, v. 30 (1982) 1157–1170.
12. Regev M., Botstein O., Bamberger M. and Rosen A., *Materials Science and Engineering* A302 (2001) 51–55.
13. Ashby M.F. and Jones D.R.H., *Engineering Materials 1*, Pergamon Press, Oxford, 1980.
14. Frost H.J. and Ashby M.F., '*Deformation Mechanism Maps*' Pergamon Press, Oxford, p. 44 1983.
15. Watanabe H., Mukai T., Kohzu M., Tanabe S. and Higashi K., *Acta. mater.* (1999) 47, 3753.
16. Harper J. and Dorn J.E., *Acta Metall.* 5 (1957) 654.
17. Luthy H., White R.A. and Sherby O.D., *Mater. Sci. Eng.* (1979) **39**, 211.
18. Ruano O.A. and Sherby O.D., *Mater. Sci. Eng.* (1981) **51**, 9.
19. Langdon T.G., Vastava R.B., in: Rohde R.W. and Swearengen J.C. (eds), *Mechanical Testing for Deformation Model Development*, ASTM STP 765, The American Society for Testing and Materials, Philadelphia, PA, (1982), p. 435.
20. Langdon T.G., in: Hondros E.D. and McLean M. (eds), *Structural Materials: Engineering Applications through Scientific Insight*, The Institute of Materials, London, p. 135 1996.
21. Nabarro F.R.N., in: *Report of a Conference on Strength of Solids*, The Physical Society of London, London, p. 75 1948.

22. Herring C., *J. Appl. Phys.* (1950) 21, 437.
23. Coble R.L., *J. Appl. Phys.* (1963) 34, 1679.
24. Ruano O.A., Wadsworth J., Wolfenstine J. and Sherby O.D., *Mater. Sci. Eng.* A165 (1993) 133.
25. Ruano O.A., Sherby O.D., Wadsworth J. and Wolfenstine J., *Mater. Sci. Eng.* A211 (1996) 66.
26. Burton B. and Reynolds G.L., *Mater. Sci. Eng.* A191 (1995) 135.
27. Greenwood G.W., in: Mishra R.S., Mukherjee A.K. and Murty K.L. (eds), *Creep Behavior of Advanced Material for the 21st Century*, The Minerals, Metals and Materials Society, Warrendale, PA (1999) 413.
28. Blum W. and Maier W., *Phys. Status Solidi* A 171 (1999) 467.
29. Langdon T.G., Identifying creep mechanisms at low stresses, *Materials Science and Engineering* A283 (2000) 266–273.
30. Lifshitz I.M., *Soviet Phys.* JETP 17 (1963) 909.
31. Raj R. and Ashby M.F., *Metall. Trans.* 2 (1971) 1113.
32. Cannon W.R., *Phil. Mag.* 25 (1972) 1489.
33. Squires R.L., Weiner R.T. and Phillips M., *J. Nucl. Mat.* 8 (1963) 77.
34. Harris J.E. and Jones R.B., *J. Nucl. Mat.* 10 (1963) 360.
35. Haddrell V.J., *J. Nucl. Mat.* 18 (1966) 231.
36. Nabarro F.R.N., *Proceedings of Symposium on Creep Behaviour of Advanced Materials for the 21st Century*, eds Mishra R.S., Mukherjee A.K., Murty K.L., Warrendale: T.M.S. (1999) 391.
37. Harris, J.E., *Metal Sci. J.* 7 (1973) 1.
38. Gifkins R.C. and Langdon T.G., *Scripta Metall.* 4 (1970) 563.
39. Rarety L.E., *J. Nucl. Mat.* 20 (1966) 344.
40. Vickers W. and Greenfield P., Diffusion-Creep in Magnesium Alloys, *J. of Nuclear Materials* 24 (1967) 249–260.
41. Wadsworth J., Ruano O.A. and Sherby O.D., *Metall. Mater. Trans.* A (2002) 33: 219–229.
42. McNee K.R., Greenwood G.W. and Jones H., *Phil Mag* A (2002) 82: 2773.
43. Khosrhoshahi R.A., Pilkington R., Lorimer G.W., Lyon P. and Karimzadeh H., *Proceedings of Third International Magnesium Conference*, editor G.W. Lorimer, The Institute of Materials, London (1997) 241.
44. Aigeltinger E.H. and Gifkins R.C., *J. Mater. Sci.* 10 (1975) 1889.
45. Pickles B.W., *J. Inst. Metals* 95 (1967) 333.
46. Barrett C.R., Muehleisen E.C. and Nix W.D., *Mater. Sci. Eng.* 10 (1972) 33.
47. Mohamed F.A., Murty K.L. and Morris J.W., *Metall. Trans.* 4 (1973) 935.
48. Ardell A.J. and Lee S.S., *Acta Metall.* 34 (1986) 2411.
49. Dorn J.E., Some Fundamental Experiments on High Temperature Creep, *Journal of the Mechanics and Physics of Solids*, 8 (1954) 85–116.
50. Rachinger W.A., *J. Inst. Metals* 81 (1952) 33.
51. Mabuchi M., Iwasaki H., Yanase K. and Higashi K., Low temperature superplasticity in an AZ91 magnesium alloy processed by ECAE, *Scripta Materialia*, 36 (1997) 681–686.
52. Watanabe H., Mukai T., Ishikawa K. and Higashi K., Low temperature superplasticity of a fine-grained ZK60 magnesium alloy processed by equal-channel-angular extrusion, *Scripta Materialia*, 46 (2002) 851–856.
53. Sequeira W., Murray M.T. and Dunlop G.L., Effect of Section Thickness and Microstructure on the Mechanical Properties of High Pressure Die Cast AZ91D,

Proceedings of the Third International Magnesium Conference, London, Institute of Materials (1997) 63.

54. Dargush M.S., Dunlop G.L. and Pettersen K., *Magnesium Alloys and their Applications*, Proc. volume sponsored by Volkswagen AG, eds B.L. Mordike and K.U. Kainer, Werkstoff-Informationsgesellschaft, Frankfurt, Germany (1998) 277–282.
55. Dargush M.S. and Dunlop G.L., Elevated temperature creep and microstructure of die cast Mg alloys, *Proc. of Materials 98*, Institute of Materials Engineering, Australasia Ltd., ed. Michael Ferry, Wollongong, Australia (1998) 579–584.
56. Dargush M.S., Hisa M., Caceres C.H. and Dunlop G.L., *Proc. of the Third International Magnesium Conf.*, ed. G.W. Lorimer, 10–12 April, Manchester, UK, The Institute of Materials (1996) 153–165.
57. Roberts C.S., *Magnesium and its Alloys*, John Wiley and Sons, New York, 1960.
58. von Buch F. and Mordike B.L., *High-temperature properties of magnesium alloys, Magnesium Alloys and Technologies*, ed. K.U. Kainer, DGM, Wiley-VCH Publishing (2003) 106–129.
59. Kaya A.A., Uzan P., Eliezer D. and Aghion E., An Electron Microscopy Investigation on As-Cast AZ91D Alloy, *J. of Materials Science and Technology*, vol. 16 (2000) 1001–1011.
60. Raynor G.V., *The Physical Metallurgy of Mg and its Alloys*, Pergamon Press, Oxford, UK, 196 1959.
61. Waltrip J.S. in *Proc. 47th Annual World Magnesium Conference*, Cannes 29–31 May, International Magnesium Association, (1990) 124.
62. Polmear I.J., overview on Magnesium alloys and applications, *Mater. Sci. and Tech.*, 10 (1994) 1–16.
63. Fukuchi M. and Watanabe F., *Journal of Japan Institute of Metals*, 39, (5) (1975) 493–498.
64. Fukuchi M. and Watanabe F., *Journal of Japan Institute of Metals*, 39, (5) (1980) 253–257.
65. Luo A. and Pekguleryuz M.O., Cast Magnesium alloys for elevated temperature applications, *J. Materials Sci.* (1994) 29, 5259–5271.
66. Lupinc V., in *Creep and Fatigue in High Temperature Alloys*, ed. J. Bressers, Applied Science Publishers, London, 7–40 1981.
67. Lagneborg, R., in *Creep and Fatigue in High Temperature Alloys*, ed. J. Bressers, Applied Science Publishers, London, 41–71 1981.
68. Shi L. and Northwood D.O., A dislocation network model for creep, *Scripta Metall.*, 26 (1992) 777–780.
69. Li J.C.M., *Dislocation dynamics*, eds. A.R. Rosenfield, G.T. Hahn, A.L. Bement and R.I., Jaffee, Mc Graw Hill (1968) 87.
70. Hirth J.P. and Nix W.D., *Phys. Status Solidii*, 35 (1969) 177.
71. Sherby O.D. and Burke M., Mechanical behaviour of crystalline solids at elevated temperature. *Prog. Mater. Sci.* 13 (1966) 323–390.
72. Sato H. and Oikawa H., *Strength of Metals and Alloys*, ICSMA 9, Freund Publishing House London, Haifa Israel (1991) 463.
73. Mordike B.L., Creep-resistant magnesium alloys, *Mat. Sci. and Eng.* A324 (2002) 103–112.
74. Tegart W.J.G. and Sherby O.D., *Phil. Mag.* 3 (1958) 1287.
75. Tegart W.J.G., Activation energies for high temperature creep of polycrystalline magnesium, *Acta Met.*, 9 (1961) 614–617.

76. Shi L. and Northwood D.O., Strain-hardening and recovery during the creep of pure polycrystalline magnesium, *Acta Metall. Mater.*, 42 (1994) 871–877.
77. Weertman J., *Rate Processes in Plastic Deformation of Materials*, eds J.C.M. Li and A.K. Mukherjee, p. 315, ASM, Metals Park, Ohio, (1975).
78. Jones R.B. and Harris J.E., *Proc. of the Joint Int. Conf. on Creep*, v. 1, 1–110, Inst. Mech. Eng. Proc., London, 1963.
79. Weertman J., *J. Appl. Phys.*, 28 (1957) 1185.
80. Friedel J., *Dislocations*, Pergamon Press, Oxford, 1964.
81. Takeuchi S. and Argon A.S., *Acta Metall.*, 24 (1976) 883.
82. Moreau G., Cornet J.A. and Calais D., *J. Nucl. Mater.*, 38 (1971) 197.
83. Shewmon P.G. and Rhines F.N., *Trans. Am. Inst. Min. Engrs.* 200 (1954) 1021.
84. Gilman J.J., *J. Appl. Phys.*, 36 (1965) 3195.
85. Mordike B.L. and Lukac P. in: Lorimer G.W., (ed.), *3rd Int. Magnesium Conf.*, 10–12 April, Manchester, UK, Inst. of Materials, London (1997) 419–429.
86. Mordike B.L. and Stulikova I., *Proceedings of the International Conference on Metallic Light Alloys*, Institution of Metallurgists, London (1983) 146–153.
87. Bettles C.J., Humble P. and Nie J.F., The effect of trace additions on the ageing behaviour of AZ91E, *Proc. of the 3rd Int. Conf.*, ed. G.W. Lorimer, Manchester, UK, April (1996) 403–418.
88. Gjestland H., Nussbaum G. and Regazzoni G., *Light-Weight Alloys for Aerospace Applications*, eds E.W. Lee, E.H. China and N.J. Kim, The Minerals, Metals & Materials Society, (1989) 139.
89. Pahutova M., Brezina J., Kucharova K., Sklenicka V. and Langdon T.G., Metallographic investigation of reinforcement damage in creep of an AZ91 matrix composite, *Materials Letters* 39 (1999) 179–183.
90. Sklenicka V., Pahutova M., Kucharova K., Svoboda M. and Langdon T.G., *Key Eng. Mater.* 171–174 (2000) 593.
91. Pahutova M., Sklenicka V., Kucharova K., Brezina J., Svoboda M. and Langdon T.G., *Proc. Second Israeli Int. Conference on Magnesium Science and Technology*, eds E. Aghion and D. Eliezer, Magnesium Research Inst. Israel (2000), 285.
92. Mordike B.L. and Ebert T., Magnesium, Properties-applications-potential, *Materials Science and Engineering* A302 (2001) 37–45.
93. Sklenicka V., Svoboda M., Pahutova M., Kucharova K. and Langdon T.G., Microstructural processes in creep of an AZ91 magnesium-based composite and its matrix alloy, *Materials Science and Engineering*, A319–321 (2001) 741–745.
94. Svoboda M., Pahutova M., Kucharova K., Sklenicka V. and Langdon T.G., The role of matrix microstructure in the creep behaviour of discontinuous fiber-reinforced AZ 91 magnesium alloy, *Materials Science and Engineering* A324 (2002) 151–156.
95. Blum W., Watzinger B. and Weidinger P., in *Magnesium Alloys and their Applications*, Conf. Proc. volume sponsored by Volkswagen AG, eds B.L. Mordike and K.U. Kainer, Werkstoff-Informationsgesellschaft, Frankfurt, Germany, p. 49, 1998.
96. Holta O., Videm M., Westengen H. and Albright D.L., Conference on *Diecasting Innovation*, by North American Diecasting Association, Indianapolis T95-053, Indiana, October (1995) 177–182.
97. Unsworth W., *Int. J. of Mat. and Prod. Techn.*, no. 4, (1989) 41.
98. Pekgüleryüz M.Ö. and Avedesian M.M., *Magnesium Alloys and Their Applications*, DGM Conf. Proc., eds B.L. Mordike and F. Hehmann, April (1992) 213–220.

99. Beer S., Frommeyer G. and Schmid E., *Magnesium Alloys and Their Applications*, DGM Conf. Proc., B.L. Mordike and F. Hehmann, eds, April (1992) 317–324.
100. Pekgüleryüz M., Luo A. and Aliravci C., *Modification of Sand Cast AS41 Magnesium Alloy*, Proc. Int. Symp. Light Metals Processing and Applications, The Metallurgical Society of CIM, Quebec City, Canada, Aug–Sept. (1993) 409–416.
101. Kr. Aune T. and Ruden T.J., *SAE 920070*, Detroit, Michigan, February (1992).
102. Foerster G.S., Designing Diecasting Alloys, *Light Metal Age*, v. 30, 9–10 (1972) 11–13.
103. Pettersen K., Westengen H., Skar J.I., Videm M. and Wei, L-Y., Creep Resistant Mg alloy Development, *Magnesium Alloys and Their Applications*. K.U. Kainer, ed., DGM, Wiley-VCH, Weinheim, Germany (2000) 29–34.
104. Mercer II.W., SAE Technical Paper No 900788, Detroit, Michigan, February (1990).
105. Kr. Aune T. and Westengen H., SAE Technical Paper No 950424, Detroit, Michigan, February (1995).
106. Polmear I.J., *Light Alloys, Metallurgy of Light Metals*, 3rd edn, Arnold, London, 1995.
107. Wei L.Y. and Dunlop G.L., *J. Alloys and Compounds*, 232 (1996) 264.
108. Powell B.R., Rezhets V., Balogh M.P. and Waldo R.A., Microstructure and Creep Behavior in AE42 Magnesium Die-Casting Alloy, *Journal of Metals*, TMS, August (2002) 34–38.
109. Bakke P., Pettersen K. and Westengen H., Enhanced Ductility and Strength Through RE Addition to Magnesium Diecasting Alloys, *2003 Magnesium Technology*, TMS, San Diego, March (2003) 171–176.
110. Holbrigl-Rosta F., Just E., Kohler J. and Melzer H.J., *Light Metals Age*, August (1980) 22–29.
111. Pekguleryuz M.O. and Renaud J., Creep Resistance in Mg-Al-Ca Alloys, *2000 Magnesium Technology*, H. Kaplan, J. Hryn and B. Clow, eds TMS (2000) 279–284.
112. Ninomiya R., Ojiro T. and Kubota K., *Acta Metall.*, vol. 43, No. 2 (1995) 669–674.
113. Pekgüleryüz M.O. and Luo A., *Creep Resistant Magnesium Alloys for Diecasting*, PCT (Patent Cooperative Treaty) application No. PCT/CA96/00091, 22 (1996).
114. Gjestland H., Nussbaum G. and Regazzoni G., *Mat. Sci. And Eng.*, A134 (1991) 1197–1200.
115. Zilberov A., Goren-Muginstein G.R. and Bamberger M., High temperature behaviour of a new Mg-Ca-Zn alloy. *Proc. of the 6th International Conf. Magnesium Alloys and their Applications*, K.U. Kainer, ed., DQM, Wiley-VCH, Wolfburg, Germany (2003) 134–139.
116. Holbrigl-Rosta F., Just E., Kohler J. and Melzer H.J., *Proc. 37th Annual World Magnesium Conf.*, IMA, New York (1980) 38–45.
117. Luo A. and Shinoda T., *Magnesium Alloys Having Superior Elevated-Temperature Properties and Die Castability*, US Patent 5, 855, 997, January 5 (1999).
118. EP 0799901A1, Mazda (1997).
119. Luo A., Recent Magnesium Alloy Development for Automotive Powertrain Applications, *Magnesium Alloys 2003*, Materials Science Forum, Trans Tech, Switzerland, vols. 419–422 (2003) 57–65.
120. Terada Y., Ishimatsu N., Sota R., Sato T. and Ohori K., Creep Characteristics of Ca-Added Die-Cast AM50 Magnesium Alloys, *Magnesium Alloys 2003*, Materials Science Forum, Trans Tech, Switzerland, vols. 419–422 (2003) 459–464.

121. Pekguleryuz M. and Labelle P., *Magnesium-Based Casting Alloys Having Improved Elevated Temperature Performance* US Pat. No. 6,322,644, Nov 27th (2001) and International Patent PCT WO 01/44529.
122. Pekguleryuz M., Labelle P., Baril E. and Argo D., *Magnesium Diecasting Alloy AJ62x with Superior Creep Resistance, Ductility and Castability*, SAE Technical Paper 2003-01-190, Detroit, March (2003).
123. Pekguleryuz M., Labelle P., Baril E. and Argo D., *Magnesium Diecasting Alloy AJ62x with Superior Creep Resistance and Castability*, 2003 Magnesium Technology, TMS, San Diego, March (2003) 201–207.
124. Samato K., Yamamoto Y., Sakate N. and Hirabara S., *Heat-Resistant Magnesium Alloy Member*, EP 0 799 901 A1 August 10 (1997).
125. Koike S., Wasizu K., Tanaka S., Baba T. and Kikawa K., SAE Technical Paper 200-01-1117, Detroit, March (2000).
126. Pekguleryuz M., Creep Resistant Magnesium Diecasting Alloys Based on Mg-Al-(Alkaline Earth Element) Systems, Materials Transactions. Special issue on Platform Science and Technology for Advanced Magnesium Alloys, *Japan Inst. Metals*, Vol. 42, No. 7 (2001) 1258–1267.
127. Bronfin B., Aghion E., Schuman S., Bohling P. and Kainer K.U., *Magnesium Alloy for High Temperature Applications*, US Patent 6,139,651, Oct 31 (2000).
128. Aghion E., Bronfin B., Friedrich H., Schumann S. and von Buch F., New Magnesium Alloys For High Temperature Applications, *2003 Magnesium Technology*, TMS, San Diego, March (2003) 177–182.
129. Lefebvre M., Pekguleryuz M. and Labelle P., *Magnesium-Based Casting Alloys Having Improved Elevated Temperature Performance*, US Pat. No 6, 342,180, Jan 29 (2002) and International Patent PCT WO 02/099147.
130. Powell B.R., Rezhets V., Luo A.A., Bommarito J.J. and Tiwari B.L., *Creep Resistant Magnesium Alloy Die Casting*, US patent 6,264,763, July 24 (2001).
131. Luo A., Balough M. and Powell B.R., SAE Technical Paper 2001-01-0423, Detroit, Michigan (2001).
132. Argo D., Pekguleryuz M., Labelle P., Vermette P., Bouchard R. and Lefebvre M., Process Parameters And Diecasting of Noranda's AJ52 High Temperature Mg-Al-Sr Alloy, *2002 Magnesium Technology*, H. Kaplan, ed., TMS, Seattle, Feb (2002) 87–93.
133. Reed-Hill R.E., *Physical Metallurgy Principles*, 2nd edn, D. Van Nostrand, New York, 1973.
134. Ahmed M., Lorimer G.W., Lyon P. and Pilkington R., *Magnesium Alloys and Their Applications*, B.L. Mordike and F. Hehmann, eds, DGM Verlag, April (1992) 301–308.
135. Stulikova I., Vostry P., Smola B. and Mordike B.L., *Proc. of 5th Int. Conf. on Heat Treatment of Materials*, IFHT, Budapest, (1986) 1993.
136. Worrall J.M., M.Sc. Thesis Manchester University, 1986.
137. Ahmed M., Lorimer G.W., Lyon P. and Pilkington R., *Magnesium Alloys and Their Applications*, B.L. Mordike and F. Hehmann, eds, DGM Verlag, April (1992) 251–257.
138. Kamado S., Kojima Y., Taniike S. and Hama S., *Magnesium Alloys and Their Applications*, B.L. Mordike and F. Hehmann, eds, DGM Informationsgesellschaft Verlag, April (1992) 169–174.
139. Morgan J.E. and Mordike B.L., *Met. Trans.*, vol. 12A, (1981) 1581–1585.

140. Duffy L.B., The Development of Magnesium Sand Casting Alloys, *Foundry Trade Journal*, Vol. 165, No 3432 (1991) 319–321.
141. Payne R.J.M and Bailey N., Improvement of the Age-Hardening Properties of Magnesium-Rare Earth Alloys by Addition of Silver, *Journal of Institutes of Metals*, vol. 88 (1959-1960) 417–427.
142. Lorimer G.W., Structure-Property Relationships in Cast Magnesium Alloys, *Magnesium Technology* (1987) 47–53.
143. Whitehead D.J., Magnesium-Silver-Didymium-Zirconium Casting Alloy QE22A, *AFS Transactions*, 61–22 (1961) 442–456.
144. Unsworth W., Role of Rare Earth Elements in the Development of Magnesium base Alloys, *Int. Journal of Materials and Product Technology*, Vol. 4, no. 4 (1989) 359–378.
145. King J.F., New Advanced Magnesium Alloys, *Advanced Materials Technology International, Specialty Metals and Alloys*, 12–19 1990.
146. Blawert C., Morales E., Dieringa H., Hort N., Azeem M.A. and Kumar S., Corrosion and wear properties of short-fiber reinforced QE22 magnesium alloy. In: Degischer, H.P. (Hrsg.): *Verbundwerkstoffe*, 14. Symposium Verbundwerkstoffe und Werkstoffverbunde. Wien (A), 02.-04.07.2003 (2003) 159–164.
147. Merrien M., Recent Developments in Magnesium Casting Alloys for Aircraft Structures, *Modern Casting*, Dec (1967) 84–90.
148. Nelson K.E., Premium Quality Magnesium Sand Castings, *AFS Transactions*, Vol. 61–69 (1961) 756–765.
149. Fisher P.A. *et al.*, *Foundry*, 95(8), 68 (1967).
150. Unsworth W., A new magnesium alloy for automobile applications, Magnesium in the Auto Industry: Prospects for the Future, Proc. 44th An. World Magnesium Conf. IMA, *JLMA*, May (1987) 22–27.
151. Unsworth W., *New ZCM Magnesium Alloy*, SAE technical Paper 880512, 1988.
152. Beer S., Frommeyer G. and Schmid E., Magnesium Alloys and Their Applications, *DGM Conf. Proc.*, B.L. Mordike and F. Hehmann, eds, April (1992) 317–324.
153. King J.F., Lyon P. and Nuttall K., US Patent 6, 193, 817 (2001).
154. Bettles C.J., Venkatesan K. and Nie J.F., Microstructure and Mechanical Properties of MEZ Casting Alloy, *Magnesium Alloys 2003*, Materials Science Forum, Trans Tech, Switzerland, vols. 419–422 (2003) 273–277.
155. Aghion E., Bronfin B., Friedrich H., Schumann S. and von Buch F., New Magnesium Alloys For High Temperature Applications, *2003 Magnesium Technology*, TMS, San Diego, March (2003) 177–182.
156. Pekgüleryüz M. and Kaya A.A., Creep Resistant Magnesium Alloys for Powertrain Applications, *Advanced Engineering Materials*, No.12, 5 (2003) 866–878.
157. Fink R., *Die-casting magnesium, Magnesium Alloys and Technologies*, ed. K.U. Kainer, DGM, Wiley-VCH Publishing (2003) 23–44.
158. Kainer K.U. and Benzler T.U., Squeeze-casting and Thixo-casting of magnesium alloys, *Magnesium Alloys and Technologies*, ed. K.U. Kainer, DGM, Wiley-VCH Publishing (2003) 106–129.

3 Thermomechanical processing of ferrous alloys

P MANOHAR, Carnegie Mellon University, USA and
M FERRY, University of New South Wales, Australia

3.1 Introduction

The steel industry is currently facing a number of challenges e.g. to become more flexible and responsive, less capital intensive, more energy efficient and environmentally 'greener'. The competition from non-ferrous metals and non-metallic materials is intensifying. Relentless pressure is applied by the ever-increasing quality demands from the customer and the need for cost reduction is driving the steel industry towards increased automation to produce steels of higher quality at a reasonable cost. The primary objective of steel making and processing has been augmented by the need to produce steel products with a wider range of desired properties such as strength, toughness, weldability and formability at low overall cost. The steel industry has risen to meet these challenges by developing a variety of alternative steel processing routes, both primary and secondary, in combination with judicious selection of steel compositions. The synergistic research and development activities conducted at academia, industries and other research laboratories have resulted into the development of several innovative steel compositions, process routes and technologies. A description and discussion of the principles, processes and practice of these modern microalloyed steels and thermomechanical processing (TMP) routes are the key objectives of the subject matter being presented in this chapter.

We begin by tracing the evolution of major primary and secondary steel manufacturing process routes and subsequently highlight the critical aspects of major technological advancements to gain an appreciation of contemporary steel processing technologies. The important mechanisms for microstructure development at high temperature, their characteristics, kinetics and modelling are described. Finally, a case study focusing on an industrial-scale manufacturing technology for rod rolling is presented. The case study will demonstrate the influence of the complex interaction of steel composition and process parameters on microstructural evolution and its impact on the choice, design and optimization of secondary manufacturing processes. It should be noted here that the focus of the chapter is on the thermomechanical processing (TMP) of low carbon and microalloyed steels that are most commonly used for such structural applications as ship plates, offshore platforms, oil and gas pipelines, and general-purpose structural products (bar, rod, wire, plate, strip, etc.). While TMP is also applied effectively to improve the properties of other metallic materials such as stainless steels, aluminum and copper alloys, these will not be discussed in this chapter. Sources of further information are provided in section 3.9 on the applications of TMP for a broader range of metallic and non-metallic materials.

3.1.1 Major primary manufacturing routes

The ability to produce useful components from liquid has been known for many centuries. Mass production of steel products began in the 1780s and progressed rapidly during the industrial revolution. The key stages in the development of steel manufacturing technologies are presented in Fig. 3.1. In the early days, liquid steel was cast into cast iron moulds to produce steel

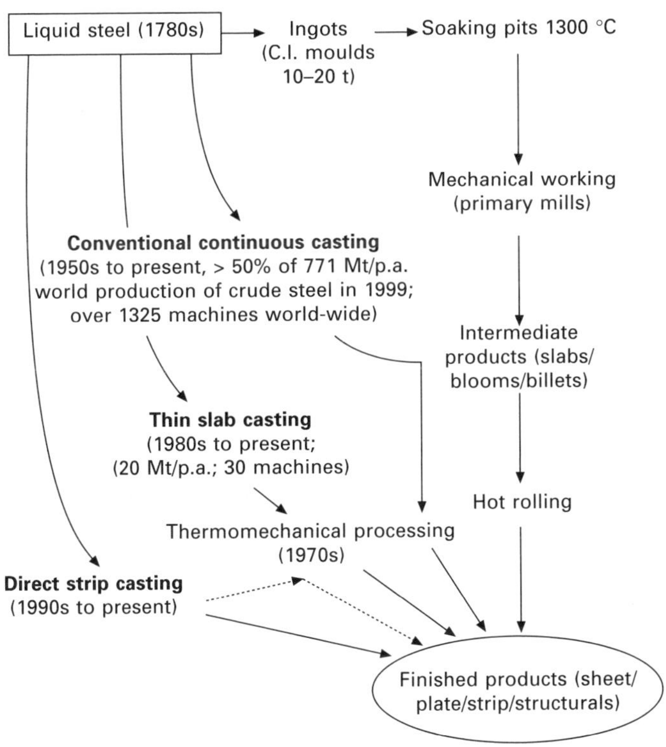

3.1 Key stages in the development of steel manufacturing technologies, after Manohar *et al.* (2000).

ingots weighing 10–20 tons. The ingots were reheated in soaking pits and then rolled in primary mills to make intermediate products such as slabs, blooms and billets. Continuous casting of steel (hereafter termed conventional continuous casting or CCC) first appeared in the 1950s and quickly gained ascendancy over ingot casting to develop into a mature industrial technology in which 70% of the world's crude steel production (~800 million tons (Mt) p.a.) is now obtained through this casting route. Following CCC, 200–250 mm thick cast slabs are subsequently hot rolled into plate, sheet, strip and various structural shapes.

In the last few decades, near net shape casting (NNSC) has attracted worldwide attention as these technologies have the capability of producing low-cost steel products with desirable mechanical and physical properties (Manohar *et al.*, 2000). NNSC involves the production of near-final-shape products directly from liquid steel with secondary processing such as hot rolling reduced to an absolute minimum. The development of thin slab casting (TSC,) which is an integral part of NNSC technology, began in the 1980s and today over thirty TSC machines are in operation around the world with an annual production of 20 m.t which is expected to rise to 40 m.t (Steffen and Tacke, 1999). Direct strip casting (DSC), a key NNSC technology, is now at the stage of full commercialization, where steel strips of ~1–2 mm thick are cast directly from liquid steel. This technology is expected to make a significant impact on steelmaking operations because it will open the flat product (sheet, plate, strip) market, which is traditionally the forte of integrated steel mills, to the minimills. NNSC technologies such as TSC and DSC have been shown to offer several advantages over CCC (Brimacombe and Samarasekera, 1994; Cramb, 1995; Manohar *et al.*, 2000).

In the context of secondary manufacturing processes, an alternative to conventional hot rolling (CHR) of steel, termed thermomechanical processing (TMP), was introduced in the 1970s and, since then, progress has been rapid with contemporary TMP routes now replacing, almost entirely, the CHR process. Some of these secondary manufacturing technologies will now be described.

3.1.2 Important secondary manufacturing processes

Thermomechanical processing is defined as a hot deformation schedule (rolling, forging, extrusion, etc.) designed to achieve a predetermined microstructure in austenite prior to its transformation to ferrite (DeArdo, 1995). The microstructure of austenite can be described by parameters such as grain size, composition, presence or absence of microalloy precipitates, degree of recrystallization and texture. Comparison of conventional and contemporary techniques for hot rolling of steel is shown schematically in Fig. 3.2. TMP has yielded a two- to threefold improvement in the mechanical properties of

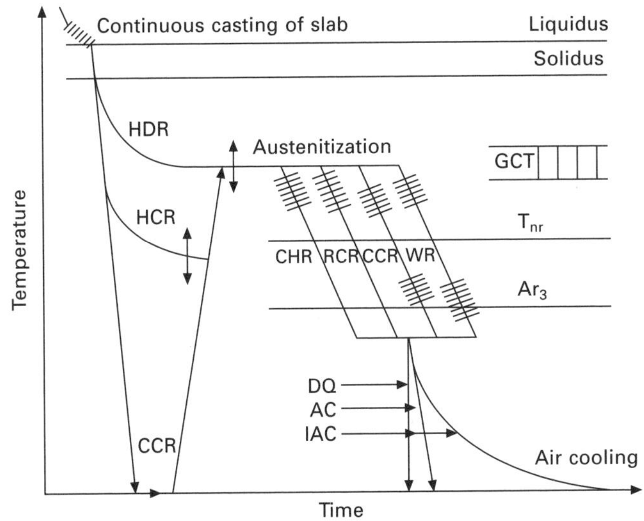

3.2 Schematic illustration of comparison of conventional and contemporary techniques for hot rolling of steels, after Manohar (1997).

steel products compared with CHR, especially when used in combination with the addition of microalloying elements (MAE) and accelerated cooling (AC). An example of the superior mechanical properties of steel products produced by TMP and AC is shown in Fig. 3.3.

Conventional hot rolling (CHR) involves reheating a continuously cast slab into the austenite phase field followed by soaking, continuous rolling to the desired thickness and subsequent air cooling to room temperature. In

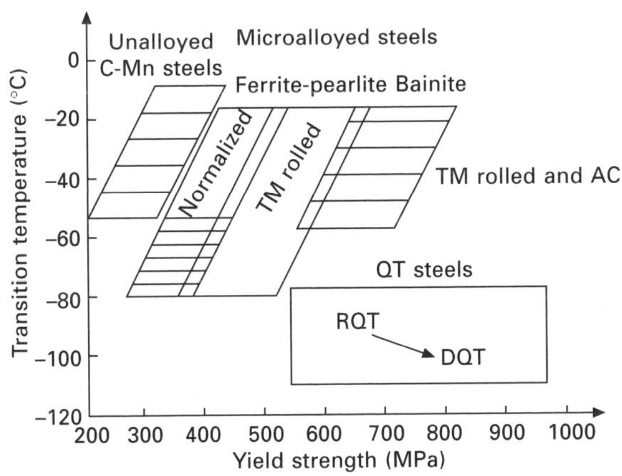

3.3 Comparison of properties of conventional C-Mn steels and thermomechanically processed microalloyed steels, after Mueschenborn *et al.* (1995).

conventional controlled rolling (CCR), the first two stages (reheating and soaking) are similar to CHR, however rolling involves two further stages, roughing and finishing. In the roughing stage, deformation is carried out above a certain critical temperature known as 'No-Recrystallization Temperature' (T_{nr}), and results in austenite grain refinement due to repeated static recrystallization cycles after each rolling pass. In the finishing stage, there is a significant time delay between the roughing and finishing passes to allow deformation of the steel to be carried out below T_{nr}. Austenite does not recrystallize which results in substantial grain flattening to produce a 'pancake' structure with the deformed grains containing a high internal dislocation density as well as deformation bands and twins (Kozasu *et al.*, 1975). A fine recrystallized austenite grain size after roughing combined with further deformation of the grains after finish rolling results in an increase in the surface area of austenite grain boundaries per unit volume (S_v). This provides substantial nucleation sites during the subsequent transformation to ferrite and results in considerable ferrite grain refinement.

A process termed recrystallization controlled rolling (RCR) is an alternative means of refining austenite and involves controlled reheating of the as-cast slab to produce a fine austenite grain size prior to roughing which is subsequently refined further by repeated recrystallization cycles during roughing above T_{nr} (Fig. 3.2). Some recent TMP technologies, such as hot direct rolling (HDR) and hot charge rolling (HCR), have been developed and these are also shown in Fig. 3.2. These processes involve the continuous cast slab being either rolled directly (HDR) or transferred to the hot rolling mill without cooling to room temperature (HCR) and offers several advantages over CHR or CCR, such as significant reduction in energy consumption, efficient furnace utilization, reductions in mill scale loss, improved delivery performance, reduced slab handling and surface and internal quality improvements (Peterson, 1994).

3.1.3 Post hot deformation cooling treatments

An area of ongoing interest is the influence of cooling after TMP. Figure 3.2 shows at least four possible options: air cooling, accelerated cooling (AC), interrupted accelerated cooling (IAC), and direct quenching (DQ). AC was first developed in 1965 for the production of strip and plate steel products with the objective of achieving a low coiling temperature (~ 600 °C) in a hot strip mill to minimize strength variations throughout the coil (Korchynsky, 1987). The use of AC was also shown to improve yield and tensile strengths and low temperature Charpy impact energy, without seriously affecting fracture appearance transition temperature (Tamehiro *et al.*, 1987).

The improved mechanical properties of AC-produced steel have been attributed to the refinement of the ferrite grain size and the formation of acicular ferritic, bainitic or martensite-austenite constituent microstructures. The gain in strength via AC allows a reduction in carbon content in the steel, which improves notch toughness and weldability. Microalloyed steels produced by AC have generated properties comparable to and often superior to quench and tempered steels (Korchynsky, 1987; Tamehiro *et al.*, 1987). AC steels are also more cost-effective than quench and tempered steels since AC eliminates the need for a separate heat treatment process after rolling.

Interrupted accelerated cooling (IAC) has also been shown to improve significantly the ultimate tensile strength (UTS) of Nb and Nb-Ti microalloyed steel plates (lower IAC temperatures resulted in a higher UTS) (Bognin *et al.*, 1987). A further example of the application of IAC is in the production of low C-Mn-Nb-Ti steels where it was found that this cooling technique has a significant influence on the type of microstructure (ferrite or bainite) and substantial ferrite strengthening by precipitation when the cooling was interrupted between 700–640 °C (Pereloma and Boyd, 1996).

3.2 Hot deformation and annealing behaviour

3.2.1 Mechanisms of strain softening

An understanding of strain hardening and softening of ferrite and austenite in steel is important as these phases undergo a significant amount of plastic deformation during TMP. A fraction of the energy of deformation is stored within the metal as strain energy associated with the various types of lattice defects (e.g. dislocations) that are produced in the metal as deformation progresses. The main consequences of the increase in the number of lattice defects are the increase in strength of the metal, called work hardening, and a reduction in ductility. The ability of the metal to undergo further deformation then depends on its ability to relieve the internal stresses through several competing softening (restoration) processes. Some of the more important restoration mechanisms will now be described. For a more extensive treatment of deformation and annealing of iron and steels, the reader is referred to several excellent books and reviews (section 3.9).

The driving pressure for softening by recovery and recrystallization is the stored energy of deformation. This is of the order 10–200 J/mol and generally much lower than the energy associated with phase transformations (Humphreys and Hatherly, 2003). The process of recovery reduces the internal energy of the metal through mechanisms such as annihilation of dislocations with opposite signs and/or re-arrangement of dislocations into low angle grain boundaries, a process termed polygonization. When recovery occurs following deformation, it is termed static recovery (SRV) but, in certain instances, recovery may occur during deformation, a process termed dynamic recovery (DRV). The mechanisms of recovery are complex and significantly influenced by parameters such as stacking fault energy (SFE) and crystal structure of the material. In high SFE phases such as body centred cubic (bcc) ferrite, both static and dynamic recovery is rapid while phases that exhibit a low to moderate SFE (< 80 mJ/m^2) such as face centred cubic (fcc) austenite, extensive recovery does not occur (Palmiere *et al.*, 1996). There are two main reasons why recovery is predominant in ferrite: (i) dislocation climb occurs readily due to its high SFE, and (ii) for a given temperature, the diffusivity of iron atoms in ferrite is ~100 times greater than in austenite.

If a significant proportion of the stored energy is released by either static or dynamic recovery, then recrystallization may not occur. However, recrystallization is possible if the stored energy remains high and if there are localised orientation gradients in the microstructure generated by the deformation. Recrystallization is initiated by the formation of strain-free nuclei, which subsequently grow and consume the deformed regions of the microstructure. Recrystallization that occurs after deformation is completed is termed static recrystallization (SRX) while recrystallization commencing and completing during deformation is termed dynamic recrystallization (DRX). In some cases, DRX may not proceed to completion during the deformation stage but further growth of these dynamically nucleated grains may be possible after deformation: a process termed metadynamic (post-dynamic) recrystallization (MDRX).

There are a number of factors that influence the type and kinetics of the operative softening process in ferrite and austenite. Material factors include composition, initial grain size, initial texture and the presence or absence of second phase precipitates. The principal processing variables include the mode of deformation, strain, strain rate, temperature of deformation and interpass (i.e. inter-deformation) time. The following sections summarise the concepts involved, and the kinetics, characteristics and industrial applications of recrystallization as it occurs in ferrite and austenite as a function of complex combinations of several of these factors.

3.2.2 Flow curves, restoration processes and constitutive relations

The combination of processing parameters that leads to a given softening phenomenon in steels can be understood from the schematic stress-strain (σ-ε) diagrams shown in Fig. 3.4. Such high temperature flow-curves are readily obtained using several testing techniques such as uniaxial tension, uniaxial compression, torsion, and plane strain compression.

The flow curves given in Fig. 3.4 are those for deformed austenite and show that flow stress increases with strain through work hardening up to a critical strain, ε_c, beyond which DRX is initiated. The material work hardens as strain increases beyond ε_c, but at a reduced rate due to further DRX and

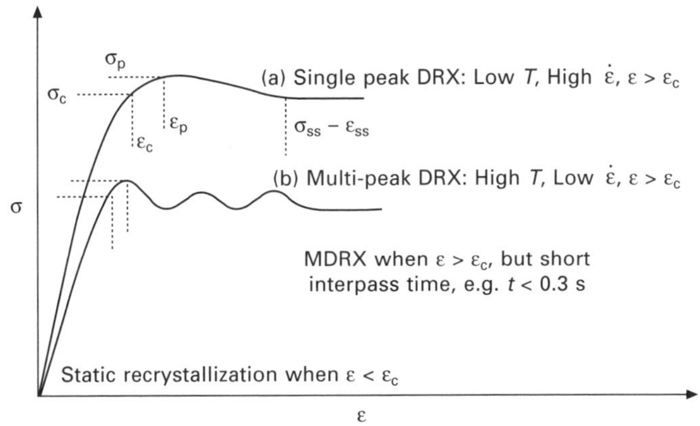

3.4 Schematic equivalent stress-strain diagrams showing various high temperature softening phenomena in steels as a function of process variables, after Manohar *et al.* (2003b).

DRV. The strain may reach another critical value called the peak strain (ε_p) with a corresponding peak stress (σ_p). Further straining beyond ε_p results in a decrease in flow stress since the rate of work hardening is now less than the softening caused by DRX. At even higher strains, deformation occurs under a steady state regime denoted by σ_{ss}-ε_{ss}. If DRX is not complete during a particular deformation pass, MDRX may occur during the interpass time. Finally, if the strain per pass is less than ε_c, DRX is not possible but SRX may occur in the time interval between passes. In contrast to austenite, warm ferrite deformation results in extensive DRV and leads to saturation in the flow stress with increasing strain. As a consequence, the retained strain in ferrite never approaches critical strain required for DRX.

Two types of flow curve are possible during DRX: single-peak and multiple-peak (curves (a) and (b) respectively in Fig. 3.4). Single-peak DRX, termed continuous DRX, occurs at high strain rates and low temperatures and in materials with a relative coarse grain size (Roucules *et al.*, 1994; Sakai, 1995). In contrast, multiple-peak DRX, termed cyclic DRX, occurs at low strain rates and high temperature and leads to grain coarsening (Sakai, 1995).

The Zener-Hollomon parameter, *Z*, (Zener and Hollomon, 1944) represents the combined influence of strain rate and deformation temperature on the flow stress in metals. For this reason, Z is often called the temperature-compensated strain rate and is given as:

$$Z = \dot{\varepsilon} \exp(Q_{def} / RT_{def}) \qquad 3.1$$

Where $\dot{\varepsilon}$ is the true strain rate, Q_{def} the apparent activation energy for deformation (J/mol), *R* the gas constant (8.318 J/mol.K) and T_{def} the deformation temperature (*K*).

The correlation between Z and the flow stress of steel (σ) was proposed by Sellars and McTegart (1966) according to:

$$Z = A(\sinh \alpha\sigma)^n \quad 3.2$$

Where A, α and n are temperature-independent constants.

Equation 3.2 can be written in terms of peak flow stress (σ_p) during hot deformation of steel (Medina and Hernandez, 1996):

$$\sigma_p = \frac{1}{\alpha} \ln[(Z/A)^{1/n} + \{(Z/A)^{2/n} + 1\}^{1/2}] \quad 3.3$$

Mathematical relations have been developed for the peak and critical strains and are typically of the following form (Lenard *et al.*, 1999):

$$\varepsilon_p = K_1 D_o^m Z^k \quad 3.4$$

$$\varepsilon_c = K_2 \varepsilon_p \quad 3.5$$

Where K_1 (0.0003–0.0007) and K_2 (0.7–0.8) are constants, m (0.3–0.7) and k (0.17) are exponents and D_o is the initial austenite grain size.

The determination of Q_{def} using mechanical test data and the Sellars-Tegart relation (eqn 3.2) can provide useful insight into the micromechanisms of deformation at elevated temperature (Dieter, 1986). However, calculated values are often not comparable with the activation energy of a particular atomic process, such as lattice, grain boundary or surface diffusion and problems often arise in its interpretation. Nevertheless, Q_{def} is often incorporated in hot working models (such as eqns 3.1–3.4 etc.) that are useful for furthering our understanding of the effect on processing parameters on the development of microstructure and properties of materials during hot deformation. Based on eqn 3.3, Table 3.1 gives values of Q_{def} for a range of steels and testing methods, which illustrates the considerable variability in this parameter. For a range of low carbon steels, however, Q_{def} is close to the activation energy for self-diffusion in pure iron (267 kJ/mol). Medina and Hernandez (1996) applied regression analysis to their steels in Table 3.1 and showed that as the composition of alloying elements approaches zero, Q_{def} approaches 267 kJ/mol. Based on this evidence, it was proposed that the principal mechanism for plastic flow in austenite is likely to be dislocation climb.

3.2.3 Industrial exploitation of softening processes

Conventional controlled rolling was the first widespread industrial exploitation of SRX (Guthrie and Jonas, 1990). In CCR, rolling is carried out in two stages: (i) roughing rolling at high austenitic temperatures (1150–1000 °C) which results in repeated SRX and refines the initial coarse austenite grain size to ~20 μm, and (ii) multi-pass finishing rolling below T_{nr} to the final product thickness with the temperature during the final pass decreasing to

Table 3.1 Values of Q_{def} for a range of steels and testing methods, as reported by Medina and Hernandez (1996)

Steel composition (mass %)	Testing method	Q_{def} (kJ/mol)	Steel composition (mass %)	Testing method	Q_{def} (kJ/mol)
Iron	Torsion	272	Fe-0.15C-0.21Si-0.74Mn	Torsion	276
Mild steel: 0.05C	Torsion	281	Fe-0.36C-0.2Si-0.82Mn	Torsion	272
Mild steel: 0.036C	Tension	309	Fe-0.53C-0.21Si-0.71Mn	Torsion	274
Steel: 0.25C	Torsion	304	Fe-0.11C-0.26Si-0.55Mn	Torsion	274
Fe-0.16C	Tension	286	Fe-0.11C-1.65Si-0.47Mn	Torsion	323
Fe-0.14C	Torsion	308	Fe-0.11C-0.26Si-0.68Mn	Torsion	278
Fe-0.85C	Torsion	339	Fe-0.15C-0.25Si-1.55Mn	Torsion	276
Fe-0.53C	Tension	270	Fe-0.44C-0.24Si-0.74Mn-0.26Mo	Torsion	285
Fe-1.2C	Torsion	389	Fe-0.44C-0.23Si-0.79Mn-0.38Mo	Torsion	287
Fe-1.09C	Tension	272	Fe-0.42C-0.27Si-0.74Mn-0.18Mo	Torsion	282
Fe-C-Mn low alloy steels	Torsion	312	Fe-0.15C-0.24Si-1.12Mn-0.021Ti-0.0105N	Torsion	285
Fe-18Cr-8Ni	Torsion	415	Fe-0.15C-0.27Si-1.25Mn-0.055Ti-0.01N	Torsion	294
Fe-10Cr-36Ni	Torsion	351	Fe-0.15C-0.26Si-1.1Mn-0.075Ti-0.0102N	Torsion	297
Fe-19Cr-14Ni	Torsion	508	Fe-0.11C-0.24Si-1.1Mn-0.043V-0.0105N	Torsion	275
Fe-0.53C-0.90Cr-0.1V	Compression	335	Fe-0.12C-0.24Si-1.1Mn-0.06V-0.0123N	Torsion	278
Fe-0.54C-1.68Si	Compression	331	Fe-0.11C-0.24Si-1.0Mn-0.093V-0.0144N	Torsion	279
Fe-0.46C-24.3Ni	Compression	507	Fe-0.11C-0.24Si-1.23Mn-0.093Nb-0.0112N	Torsion	288
Fe-0.12C-0.05Nb	Torsion	434	Fe-0.11C-0.24Si-1.32Mn-0.093Nb-0.0119N	Torsion	295

800–840 °C. Deformation below T_{nr} results in grain flattening (pancaking) along with internal deformation substructure and the increase in S_v raising the nucleation density of ferrite during its transformation from austenite. Final ferrite grain sizes in the range 5–10 μm are achievable by CCR, which produces high-strength steel plate with excellent toughness and weldability. CCR is used extensively in the production of steel plates for oil and gas pipelines and offshore platforms.

An alternative rolling process, termed recrystallization controlled rolling (RCR), has been developed. In RCR, the type and quantity of alloying elements in the steel are critically important (DeArdo, 1997; Siwecki and Engberg, 1996). Alloying additions are chosen to allow SRX to proceed relatively quickly between rolling passes and minimize grain growth by grain boundary pinning via the formation of a dispersion of fine particles. In RCR, recrystallization is allowed to occur between passes to produce fine-grained, recrystallized austenite, which subsequently transforms to ferrite to yield a final grain size in the range 8–10 μm. RCR generates hot-rolled products with a good combination of strength and toughness which are suitable for applications where heavy plate and thick-walled seamless tubes are required (Siwecki and Engberg, 1996).

In the high-speed production of strip, rod, tubular and bar products, multi-pass deformation of austenite at high strain rates often results in extremely short interpass times (0.03–0.7 s). In this case, SRX is not likely (see e.g. eqn 3.7 given in the following section) and the accumulation of strain per pass may exceed the critical strain, ε_c (Fig. 3.4), thereby resulting in DRX. Industrial processing by dynamic recrystallization controlled rolling (DRCR) can produce, by transformation to ferrite from dynamically recrystallized austenite, grain sizes in the range 3–8 μm (Samuel *et al.*, 1990). In the processing of some products, longer interpass times are possible (0.3–2.0 s) which may result in MDRX rather than DRX. This process is termed metadynamic recrystallization controlled rolling (MDRCR) (Roucoules *et al.*, 1994, 1995). It has been suggested that MDRCR enables the production of austenite and ferrite grain sizes comparable to those produced by DRCR but without the need for excessively high strain rates (Hodgson, 1997).

Steel strip produced by TMP may be further rolled in the ferrite phase field to produce thin gauge sheet for a range of applications. Ferrite rolling is not expected to result in DRX but substantial DRV will occur if deformation is carried out at an elevated temperature. In fact, warm rolling of interstitial free (IF) and ultra low carbon (ULC) steels is a recently developed processing strategy as it provides significant metallurgical and economic advantages (Bleck and Langner, 1997). Two important commercial processes for annealing cold/warm-rolled steel are batch annealing (BA-low heating rates) and continuous annealing (CA-high heating rates). The choice of annealing method is controlled by various productivity and quality factors (Humphreys and

Hatherly, 2003). For example, the major aim in cold rolling and annealing (CRA) of a range of carbon steel grades is to produce recrystallized strip with a particular preferred orientation of grains (recrystallization texture); an important variable that strongly influences the formability of these alloys (section 3.4.3)

3.3 Processing and microstructure

One of the major challenges in steel processing is the determination of the operative softening mechanisms under industrial processing conditions since it determines the microstructural evolution during processing. The type of the operating restoration process, in turn, is controlled by the steel composition and process route (section 3.2). For example, the strain rates and interpass times in plate mills are about 1–30 s^{-1} and 8–20 s, respectively, compared to the high strain rates (1–3000 s^{-1}) and short interpass times (0.015–1.0 s) in bar/wire/rod mills. Both material and processing variables discussed previously may result in situations where one or more softening mechanisms occur. An understanding of these complex interactions and utilizing them to improve properties of the steel products are key factors in the design of a successful industrial TMP sequence. A useful way of understanding these restoration mechanisms is in the study of the restoration kinetics.

3.3.1 Kinetics of microstructure evolution

The kinetics of SRX may be quantified using the classical nucleation and growth theory of phase transformations embodied in the form of the Johnson-Mehl-Avrami-Kolmogorov (JMAK) equation which relates fraction recrystallized (X_V) to annealing time (t):

$$X_V = 1 - \exp\,[-\,B(t - t_o)^n] \qquad 3.6$$

where t_o is the incubation period needed to form a nucleus, n the JMAK exponent and B is a function of nucleation rate, $\dot{N}$, and growth rate of nuclei, $\dot{G}$. As $\dot{N}$ and $\dot{G}$ in eqn 3.6 increase rapidly with temperature, the kinetics of recrystallization are expected to be a strong function of annealing temperature. Furthermore, both $\dot{N}$ and $\dot{G}$ are influenced by the stored energy of deformation and the rate of recrystallization is also expected to be dependent on deformation conditions and initial microstructure. Relationships of the form of eqn 3.6 are shown to reasonably predict the static recrystallization kinetics of both cold and hot deformed steel (Humphreys and Hatherly, 2003).

In general, the rate of recrystallization after hot working is difficult to predict using physically based models. Empirical models of recrystallization kinetics are frequently used which incorporate both microstructural and processing parameters and have the following form (Sellars and Whiteman, 1979):

$$t_x = A d_o^a \varepsilon^b Z^c \exp(Q_{rex} / R T_{rex}) \qquad 3.7$$

where t_x is the time for a certain volume fraction (x) to recrystallize during DRX, MDRX or SRX, d_o is the initial grain size, Q_{rex} an activation energy term, R the universal gas constant and T_{rex} the recrystallization temperature. The constants A, a, b and c are derived from experiment.

Empirical models may also be used to describe kinetics of precipitation. For example, the kinetics of strain-induced precipitation of Nb has been characterized by computing the time required for 5% precipitation, $t_{0.05p}$, according to the following relation (Dutta and Sellars, 1987):

$$t_{0.05p} = 3x10^{-6} \cdot \frac{1}{[Nb]} \cdot \frac{1}{\varepsilon} \cdot Z^{1/2} \exp\left(\frac{270000}{RT}\right) \exp\left(\frac{2.5x10^{10}}{T^3 (\ln K_s)^2}\right) \qquad 3.8$$

where

$$Z = \dot{\varepsilon} \exp\left(\frac{400000}{RT}\right)$$

and K_s is the supersaturation ratio given as follows:

$$K_s = \frac{[Nb]\cdot[C + 12/14N]}{10^{(2.26-6770/T)}}$$

The experimentally determined relations predicting the recrystallized grain size (d_{rex}) during or after hot working of ferrite and austenite are usually of the form:

$$d_{rex} = B d_o^e \varepsilon^f Z^g \qquad 3.9$$

where B, e, f and g are experimentally determined constants. In the case of DRX of austenite, there is only a small effect of both strain and initial grain size on d_{rex}, so the exponents e and f tend to zero. Empirical relationships such as eqns 3.2 to 3.9 are necessary components of models that describe industrial process routes for steel manufacturing. An example of the application of these models to industrial-scale manufacturing processes such as rod rolling is given in section 3.7.

3.3.2 Recrystallization-precipitation-time-temperature (RPTT) relations

The complex interaction of recrystallization and precipitation during TMP of microalloyed steels can be understood with the use of recrystallization-precipitation-time-temperature (RPTT) diagrams, as initially proposed by Hansen *et al.* (1980). Consider the RPTT diagram in Fig. 3.5 proposed by Kwon and DeArdo (1991). In this diagram, T_{sol} corresponds to the solution temperature above which Nb-rich precipitates are completely dissolved in austenite, T' is the temperature below which recrystallization and precipitation compete and T_R is the temperature below which substantial precipitation occurs prior to recrystallization. There are three distinct processing regimes as shown in Fig. 3.5.

Regime I. The steel is initially austenitized above T_{sol}, followed by deformation and holding between T_{sol} and T'. Recrystallization goes to completion (R_f) before the start of precipitation (P_s). This regime corresponds to the roughing stage during TMP.
Regime II. Precipitation occurs after the start of recrystallization (i.e. during the time interval between R_s and R_f) and also results in accelerated precipitation within the deformation substructure in the partially recrystallized austenite. Deformation within this regime should be avoided due to the undesirable properties associated with the development of a duplex microstructure (Cuddy *et al.*, 1980; Cuddy, 1981, 1982; Speer *et al.*, 1987). This regime should therefore correspond to the 'delay' period in the TMP schedule.

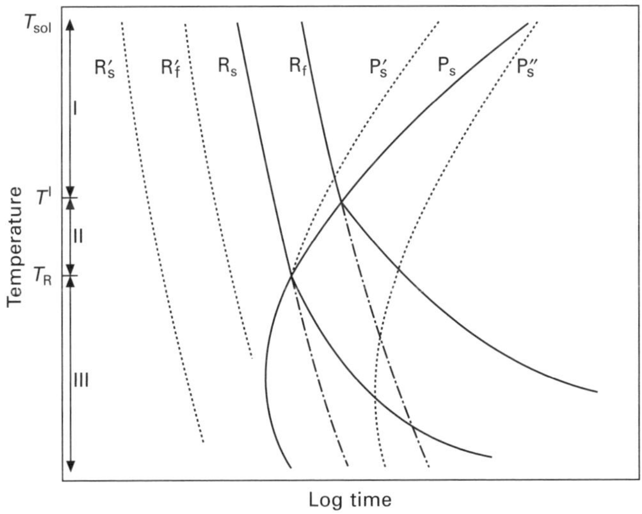

3.5 Schematic recrystallization–precipitation–time–temperature (RPTT) diagram showing interaction of precipitation and recrystallization, after Kwon and DeArdo (1991). R_s and R_f refer to the start and finish of recrystallization, respectively, in microalloyed steel; and R'_s and R'_f refer to the start and finish of recrystallization, respectively, in plain carbon steel; P'_s and P''_s refer to the hypothetical precipitation start times in deformed and undeformed austenite, respectively; and P_s is the actual precipitation start temperature. See text for more details.

Regime III. Precipitation commences prior to recrystallization and generates a fine particle dispersion with a large enough Zener pinning force (Zener, 1948) to impede recrystallization to such an extent that the deformation substructure may be fully retained after deformation (Palmiere, 1995). This regime is below T_{nr} (marked as T_R in Fig. 3.5) and therefore corresponds to the finishing stage during TMP.

It is clear that an understanding of T_{nr} is of critical importance for microalloyed steels in the design of suitable TMP schedules. Barbosa *et al.* (1987) used multi-pass torsion to determine T_{nr} of several steels containing different levels of Nb, V, Ti and Al with T_{nr} of another twenty steels reported in literature added to this data. Regression analysis was used to arrive at the following relation between T_{nr} and the composition:

$$T_{nr}\ (^{\circ}\mathrm{C}) = 887 + 464\ \mathrm{C} + (6445\ \mathrm{Nb} - 644\mathrm{Nb}^{1/2}) + (732\ \mathrm{V} - 230\mathrm{V}^{1/2}) + 890\ \mathrm{Ti} + 363\ \mathrm{Al} - 357\ \mathrm{Si}$$

Maccagno *et al.* (1994) have found this relationship to be satisfactory for calculating T_{nr} for Cu-bearing (0.4% Cu) and Nb/Nb-V steels. They compared this relationship with the experimentally determined T_{nr} and that calculated using the data from rolling mill logs and found good quantitative agreement between T_{nr} values.

3.3.3 Cold rolling and annealing – development of microstructure and texture

Cold deformation is generally restricted to low homologous temperatures ($T_{def} < 0.5T_m$) where T_m is the absolute melting temperature of the alloy. It is well known that microstructural evolution of ferritic, austenitic and dual-phase steels during deformation is dependent on many factors (Hutchinson, 1984; Ray *et al.*, 1994). During cold deformation of ferrite, there is a marked increase in dislocation density resulting in considerable work hardening and grains subdivide in a complex way to produce a myriad of features such as a cellular substructure, microbands, deformation bands and larger-scale heterogeneities such as shear bands (Humphreys and Hatherly, 2003).

In addition to this microstructural evolution, orientation changes take place during deformation and these are generally not random since deformation occurs on the most favourably orientated slip or twinning systems such that a preferred orientation or 'texture' is produced. The strength of the texture and the balance between the various texture components will depend on the texture of the starting material and the degree of deformation as the grains will align themselves according to crystal plasticity considerations. In addition to strain, a number of factors may affect the deformation texture, such as deformation geometry, temperature of deformation, crystal structure, SFE, initial grain size and the presence of a second-phase (Humphreys and Hatherly, 2003).

Deformation of ferrite

In bcc metals such as ferritic steels, slip is the principal deformation process during cold rolling and occurs in the close packed <111> directions by what is termed pencil glide but the slip plane may be any of the planes {110}, {112} or {123}. The choice of slip plane is influenced by factors such as the temperature of deformation (Hutchinson, 1984; Ray *et al.*, 1994; Humphreys and Hatherly, 2003).

The textures of rolled sheet metals are generally represented by the Miller indices {*hkl*}<*uvw*>, which signifies that {*hkl*} crystallographic planes lie parallel to the plane or the sheet, whereas their <*uvw*> directions lie parallel to the rolling direction (RD). More complex textures will consist of a number of components of different severities. It has been recognized that the best description of texture in a material is provided by the orientation distribution function (ODF), which specifies the frequency or occurrence of particular orientations in three-dimensional (Euler) orientation space (Bunge, 1982). Both cold rolling and annealing textures in ferritic steels are commonly described in terms of certain orientation fibres in Euler space. The orientation ranges of two of the more important fibres are: (i) α-fibre running from {001}<110> to {111}<110> along <110>//RD and (ii) γ-fibre running from {111}<110> to {111}<112> along <111>//ND. Further details of all the possible fibres in cold rolled low carbon steels may be found elsewhere (Ray *et al.*, 1994).

An ODF section (Bunge notation) at $\varphi_2 = 45°$ showing the location of the α and γ fibres is given in Fig. 3.6a and the texture of a typical cold rolled interstitial-free (IF) steel is shown in Fig 3.6b. Such rolling textures generally consist of two major orientation spreads: (i) an almost complete <111>//ND or γ-fibre texture with several prominent directions parallel to the rolling direction such as <110>, <112> and <123>, and (ii) a partial <110>//RD or α-fibre texture with prominent {001}, {112} and {111} rolling plane components. It is well known that the strength of the rolling texture of iron and carbon-base steels is affected by the hot band grain size and texture and the degree of cold rolling (Ray *et al.*, 1994). For example, texture strengthening during cold rolling is achieved by a decrease in initial grain size (for a given strain) and an increase in the degree of strain (for a given grain size). However, the rolling texture is largely independent of steel composition and processing variables (Ray *et al.*, 1994).

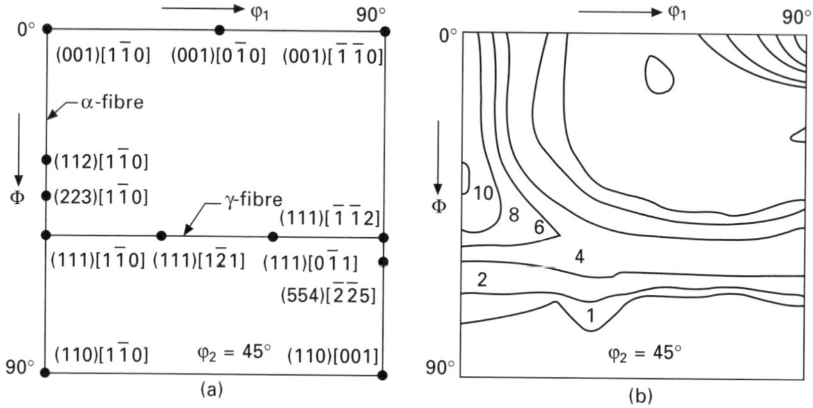

3.6 $\varphi_2 = 45°$ ODF section showing (a) the location of α and γ fibres and (b) the deformation texture of a typical cold rolled IF steel, after Hutchinson and Ryde (1997).

Annealing of cold deformed ferrite

In general, nucleation of recrystallization (SRX) is associated with microstructural inhomogeneities and the most significant of these are transition bands, shear bands, grain boundaries and the deformation zones around hard particles (Humphreys and Hatherly, 2003). In most carbon steels, the number of coarse carbides is low and particle stimulated nucleation is not likely to be significant. However, shear band nucleation is common in deformed coarse grained ferrite and in those steels containing high contents of interstitial elements. Shear bands are known to result in the creation of undesirable {110}<001> orientated nuclei, which indicate that a small grain size and low C and N contents are necessary (Ray *et al.*, 1994; Humphreys and Hatherly, 2003). The most desirable nucleation sites in ferritic steels are grain boundaries since the nuclei that form have orientations along <111>//ND or γ-fibre (Ray *et al.*, 1994). This is important since a strong <111>//ND texture with a balance of <112> and <110> parallel to RD in CRA steel results in superior sheet formability (section 3.4.3).

In contrast to rolling texture development, the annealing texture is affected by a host of parameters pertaining to steel composition and the entire processing history (Hutchinson, 1984; Ray *et al.*, 1994). The strength and distribution of this texture along the γ-fibre is affected by a myriad of processing parameters: degree of cold rolling, lubrication during rolling, annealing temperature and heating rate as well as material parameters: hot band microstructure and texture (dependent on hot deformation history); alloying additions and their distribution in the matrix and precipitation during annealing (Ray *et al.*, 1994). While the effect of each variable on texture development is well beyond the scope of this chapter, many detailed studies have been carried out over the years and these are documented in several reviews (section 3.9). Some of the more notable factors affecting the strength of the <111>//ND recrystallization texture in low carbon steels is summarised in Fig. 3.7. A useful example of the influence of rolling reduction and initial hot-band grain size the formation of the <111>//ND texture is given in Fig. 3.8. Here, a fine hot-band grain size is expected to provide copious sites for grain boundary nucleation thereby strengthening the <111>//ND texture.

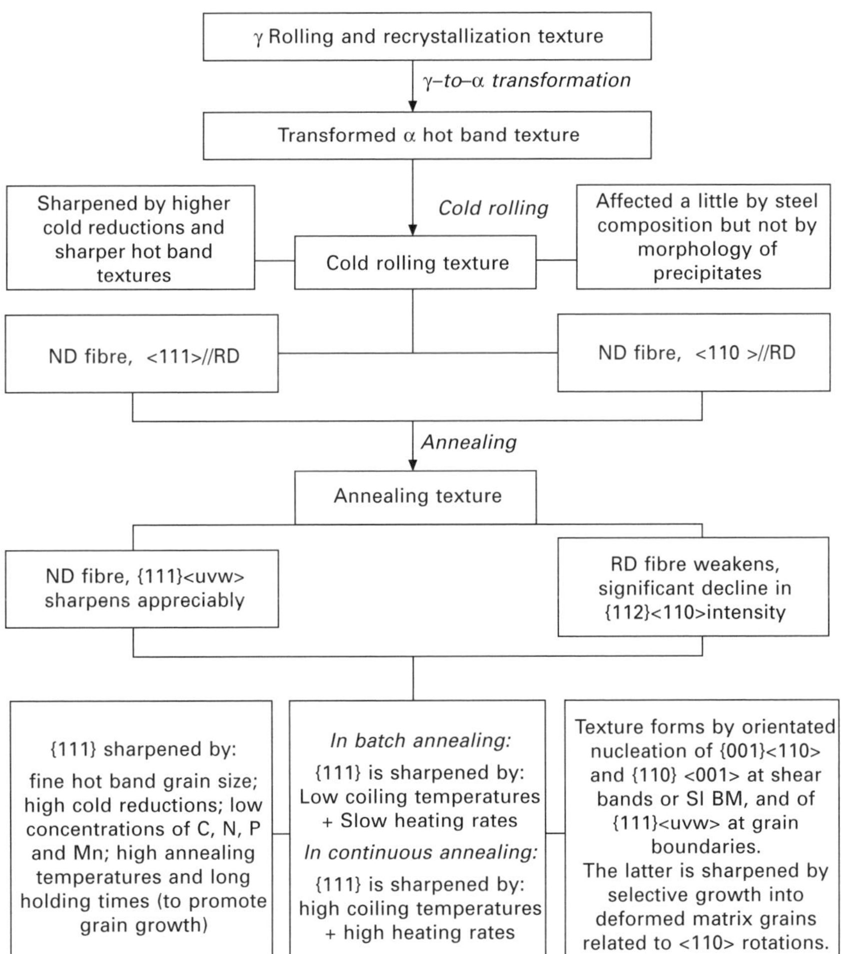

3.7 Factors affecting the strength of the <111>//ND recrystallization texture in low carbon steel, after Ray *et al.* (1994).

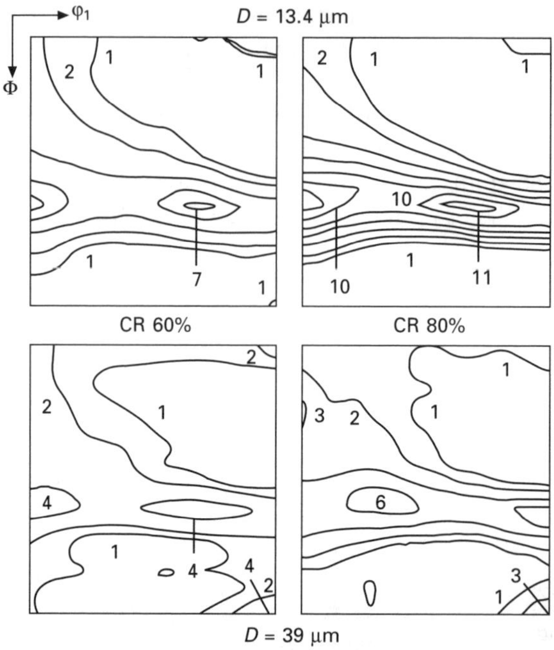

3.8 $\varphi_2 = 45°$ ODF sections showing the influence of cold rolling reduction and initial hot-band grain size on the strength of the <111>//ND texture and other texture components, after Hutchinson *et al.* (1990).

3.4 Microstructure and properties

3.4.1 Characterization and parameterization of microstructures

Microstructures must be characterized and described quantitatively for the purposes of design, specification, comparison, quality control and mathematical modelling of structure/property relationships. The size (e.g. nm, μm), shape (e.g. polygonal, acicular, globular, lamellar, dendritic), type (e.g. ferrite, pearlite, bainite, martensite, retained austenite), amount (i.e. volume fraction) and distribution (e.g. uniform, segregated, banded, clustered, along grain boundaries) of the main phases and the precipitates need to be measured for a detailed description of microstructure. A difficulty often encountered in the characterization of high-temperature microstructures of steels is that the prior austenite microstructure is lost at room temperature due to the $\gamma \rightarrow \alpha$ transformation during cooling. However, there are special techniques available to reveal the prior austenite microstructure as given in several references (Woodfine, 1953; Brownrigg *et al.*, 1975; Riedl, 1981; ASM Metals Handbook, 1985; Manohar *et al.*, 1996). The austenite microstructure is an important parameter affecting steel properties and several techniques are available to quantify the mean grain size, size distribution and morphology (Underwood, 1968; Leslie, 1981; Riedl, 1981; ASM Metals Handbook, 1985; ASTM E 112–96, 2002, E 930–99, 2002; E 1181–87, 2002; E 1382–97, 2002).

The types of microstructures observed after TMP and cooling in low carbon and microalloyed steels are quite varied and complex and therefore a clear understanding of phase identification, classification and nomenclature is of great importance. Particularly, bainitic microstructures in low carbon PC and MA steels differ in variety and form as compared to the classical bainitic microstructures. In this regard, the publications by Reynolds *et al.* (1991), Bainite Research Committee of ISIJ, e.g. Araki *et al.* (1991), Araki and Shibata (1995) and Bramfitt and Speer (1990) provide useful pointers in identifying the various bainitic microstructures in MA steels, along with their properties, conditions of formation and morphology.

3.4.2 Structure-property relationships at room temperature

The expression 'structure–property relationship' is used so often and in so many different contexts in the technical/scientific literature that it yields almost no intuitive understanding of its exact meaning. This is not to say that the expression is unimportant, on the contrary, such relationships are vital in product design, new product development and process control. After all, the thousands of steel grades available in the market today are designed, specified, manufactured, tested, bought and sold based on their structure and properties. Hence it is important to articulate as precise definitions of structure and property as possible.

The structure of steels, and indeed universally of all materials, exists at different levels of detail. For example, nano-scale structure means atomic bonding, crystal structure, solutes, precipitates and dislocations; microscopic structure implies different phases and inclusions; mesoscopic structure refers to individual crystals (grains), grain boundaries and interfaces; while macroscopic structure corresponds to the structure of bulk materials at polycrystalline level including texture. The word structure therefore is a sum total of all structural features mentioned here.

The word 'property' also means different things in different contexts. Properties may be broadly classified into three groups: intrinsic, functional and manufacturing. The first group (intrinsic properties) includes two sub-groups of properties: the first sub-group covers the mechanical properties such as yield and tensile strength, ductility (% elongation and % reduction in area), and toughness (?) while the second subgroup consists of physical properties, e.g., electrical, magnetic, thermal, optical, chemical and thermodynamic properties. The intrinsic properties are dependent upon and a consequence of the structural details at all levels. The question mark following toughness needs an explanation here. Toughness may be defined as the ability of a material to resist brittle fracture. This ability depends not only on the structure of material, but also on at least five other external parameters: the presence or absence of tri-axial state of stress, operating temperature, loading rate (impact loads), presence or absence of notches and their sharpness and section thickness. Thus, toughness is strictly not an intrinsic mechanical property. However, traditionally it has been included as a part of the mechanical property suite.

The second broad group of properties (functional properties) includes wear, corrosion, fatigue, and creep resistance. These properties are the most relevant properties in terms of the performance of a metal product in service. To some extent, these properties can be correlated to the structural details, however to a large extent they also depend on the external service conditions and geometry of a component. In this case, structure-property relationship is extremely difficult to model mathematically.

The last broad group of properties (manufacturing properties) expresses the relative ease with which a metallic product could be made. These properties include weldability, machinability, formability, forgeability and castability. Here again, these properties to some extent depend on the structure of the material, but they also depend significantly on the design of manufacturing tools and components and manufacturing process parameters. In addition, one should not forget the all-pervasive influence of material composition that impacts upon all structural features and all sets of properties.

From the foregoing discussion, it should be clear that it might not be possible to have a global model that can relate all structural details to all sets of properties through some mathematical functional relationship. With this in mind, it should be noted that what is about to be presented is limited to the models appropriate for correlating microstructural details to the mechanical properties in plain carbon steels or C-Mn steels that exhibit a predominantly ferrite + pearlite microstructure.

The yield (or 0.2% proof) strength of steels (σ) may be factorized into components consisting of intrinsic strength of pure iron, contributions by substitutional and interstitial solid solution hardening, dislocation and textural strengthening, strengthening due to dispersed particles or phase refinement strengthening and grain refinement strengthening as follows (Bhadeshia, 2001):

$$\sigma = \sigma_0 + \Delta\sigma_{comp} + \Delta\sigma_{tex,\rho} + \Delta\sigma_{micro} + \Delta\sigma_{gs} \qquad 3.10$$

where σ_o is the intrinsic strength of pure iron, $\Delta\sigma_{comp}$ is the contribution due to composition, $\Delta\sigma_{tex,\rho}$ is the strength differential due to texture and dislocation density, $\Delta\sigma_{micro}$ is the strengthening due to dispersion and refinement of microconstituents such as particles or inter-lamellar spacing of pearlite and $\Delta\sigma_{gs}$ is strength increase due grain refinement.

The first two factors in eqn 3.10 are given as follows (Rodriguez and Gutierrez, 2003):

$$\sigma_o + \Delta\sigma_{comp}\ (\text{MPa}) = 77 + 80\,[\%\text{Mn}] + 750\,[\%\text{P}] + 60\,[\%\text{Si}] + 80\,[\%\text{Cu}] + 45\,[\%\text{Ni}] + 60\,[\%\text{Cr}] + 11\,[\%\text{Mo}] + 5000\,[\%\text{C}_{ss}] + 5000\,[\%\text{N}_{ss}] \quad 3.11$$

where C_{ss} and N_{ss} are the mass percentages of C and N in solid solution, respectively. All other elemental compositions are also expressed in mass percent.

The contributions due to texture and dislocation density can be expressed according to the following relation (Gil-Sevillano, 1993):

$$\Delta\sigma_{tex,\rho} = \alpha M \mu b \rho^{1/2} \quad 3.12$$

where α is a constant (~ 0.33), M is the Taylor factor (average value 3) that correlates macroscopic flow stress to the critical resolved shear stress, μ is the shear modulus of ferrite (80 GPa), b is the Burgers vector (0.25 nm) and ρ is the dislocation density (numbers/m^2).

In eqn 3.12, the dislocation density is a function of strain during cold working: for annealed iron ρ~10^{12} m^{-2} with the maximum limit proposed to be ~10^{16} m^{-2} (Takaki, 2003) at true strains > 2 (equivalent to a cold reduction of ~90%). The relation between stored strain and dislocation density is highly non-linear in the true strain range 0–2.

The influence of a dispersion of second phase particles or interlamellar spacing of pearlite is accounted for according to the following model:

$$\Delta\sigma_{micro} = k' / \bar{L} \quad 3.13$$

where k' is a constant and $\bar{L}$ is the mean interparticle or interlamellar spacing.

For strengthening due to a given volume fraction of globular cementite particles (V_θ), the following constant has been proposed (Bhadeshia, 2001):

$$k' = 0.52 V_\theta \quad 3.14$$

Finally, the strengthening due to ferrite grain size is expressed by the famous Hall-Petch relation (Hall, 1951; Petch, 1953):

$$\Delta\sigma_{gs} = k d^{-1/2} \quad 3.15$$

where k is ~ 0.6 MPa·m$^{1/2}$ (Seto and Sakata, 2003) and d is the mean ferrite grain size.

The influence of compositional and structural parameters on the toughness of ferrite-pearlite steels is now considered. One way of quantifying the toughness of steels is to determine its ductile to brittle transition temperature (DBTT) by conducting Charpy impact tests over a range of temperatures. The lower the DBTT, the better the toughness of steel. The DBTT is modelled as follows (Pickering, 1978):

$$DBTT(°C) = -19 + 44[\%Si] + 700[\%N_{ss}] + 2.2[\%Pearlite] - 11.5d^{-1/2} \quad 3.16$$

Current literature on ultra-fine grained steels has reported a ferrite grain size of the order 1 μm is obtainable via a high strain TMP. Such a fine grain size results in a substantial and simultaneous increase of both strength and toughness of the steel. In fact, a very fine grain size was also found to be suitable for superplastic forming of high C steels at high temperatures where more than 500% uniform elongation was obtained (Serby *et al.*, 1975).

In summary, structure-property relationships are vital for the design, product development and control in steel processing. However, care must be exercised in defining the scope of structure-property modelling because both structure and properties may include a variety of multi-scale parameters. In the following section a functional property, formability, is discussed to gain an appreciation of the relationship between the structure and functional properties of steels.

3.4.3 Formability of sheet steel products

The demand for highly formable sheet steel for deep drawing applications has increased significantly in recent years. Conventionally, formable sheet steels have low alloying additions and carbon levels of typically 0.04 mass % C. These steels require full coil batch annealing in order to obtain the required mechanical properties and good drawability. However, this process is slow (heating rates of typically 0.33–0.83 °C/s), energy intensive and produces non-uniformity in properties in the sheet. More recently, steels containing about 0.003 mass % C (30 ppm) have been developed with improved forming properties through continuous annealing which is considerably faster (heating rates of 5–20 °C/s), and less energy intensive than batch annealing and produces steel with superior property uniformity.

The ability of sheet metals to be drawn into complex shapes is intrinsically associated with the texture of the grains that characterise the microstructure (section 3.6.2). The influence of preferred orientation is described, for practical purposes, by two parameters: normal plastic anisotropy (r-value) and planar anisotropy (Δr).

The normal plastic anisotropy is defined as the ratio of the true strains in the width and thickness direction:

$$r = \frac{\varepsilon_w}{\varepsilon_t} \quad 3.17$$

The planar anisotropy is defined as:

$$\Delta r = \frac{r_o + r_{90} - r_{45}}{2} \quad 3.18$$

where the subscripts 0, 90, 45 refer to the angle (°) that the longitudinal axis of the specimen subtends with the rolling direction. In the presence of planar anisotropy, the *r*-value will vary with direction in the sheet plane and an average value, $\bar{r}$, is often used:

$$\bar{r} = \frac{r_o + 2r_{45} + r_{90}}{4} \qquad 3.19$$

The *r*-value describes the rate of thinning of the sheet in comparison with the rate at which the sheet is being extended to form the desired shape. A random texture has a value of unity whereas a sample, which is highly resistant to thinning, has values greater than unity. Equation 3.19 therefore represents a convenient measure of normal anisotropy with high $\bar{r}$-values correlating with good drawability (Hutchinson, 1984; Ray *et al.*, 1994; Humphreys and Hatherly, 2003). It is useful to note that high normal anisotropy is not the complete requirement leading to superior formability. Variations in normal anisotropy in different directions of the sheet, Δr, leads to differences in flow behaviour in different directions which leads to the formation of ears (earing) when cylindrical cups are drawn or to wrinkling or localised thinning when more complex shapes are attempted. In general, high $\bar{r}$-values and low Δr are important for optimizing the drawability of sheet metals. It is important to note, however, that this is a simplified view of formability as other property requirements such as ductility and resistance to strain ageing must also be considered (Humphreys and Hatherly, 2003).

3.5 Processing and processability

Consider the evolution of production technologies presented in Fig. 3.1 with reference to the production of strip steels (e.g. up to 2 mm in thickness). When strip is produced from 200–250 mm thick slabs via CCC, TMP involves an overall reduction in the range of 125:1 to 100:1 in a continuous sequence of 20 to 30 reduction passes. On the other hand, TSC produces 50–70 mm thick slab and requires a reduction in the range 35:1 to 25:1 in 5 to 7 passes (usually in-line). In the extreme case, DSC produces 1–2 mm thick strip, which provides little scope for further reduction and results in a very restricted processing window for altering the as-cast microstructure. While it is critical to exploit fully any available hot deformation opportunity to the greatest extent, it is also complicated by the limitations imposed by load capability of the rolling mill, process logistics and the composition of a given steel grade as explained in the following section.

3.5.1 Steel composition and processability

The process of alloy design is ill structured, difficult to systematize and involves the heuristic knowledge of experts. In addition, relationships between compositional and processing parameters and product properties are non-linear and, thus, the assessment of processability is a challenging task. For example, an increase in the carbon and alloy contents of carbon steels will increase their strength as this property is proportional to carbon equivalent (CE = C + Mn/6). However, high CE also drastically reduces ductility, formability, toughness and weldability (Korchynsky, 2001). To counter the reduction in these properties, it may be necessary to use appropriate heat treatments such as normalizing and/or hardening and tempering which significantly add to the cost of the finished product. Since welding is a key and unavoidable method of fabrication for most engineering structures (bridges, pipelines, ships), an increase in CE is not a viable option for obtaining superior mechanical properties in steels. In this context, the intelligent design of composition and steel processing sequence provides an elegant solution to the problem.

The interaction of alloying, microalloying and TMP variables exerts a strong influence on metallurgical parameters such as grain size and shape, microstructure type (ferrite, bainite, martensite, etc.), dislocation density, texture, and size, shape, distribution and stability of non-metallic inclusions. These parameters in turn control the properties of the final steel product. For example, the $\gamma \rightarrow \alpha$ transformation begins by ferrite nucleation that occurs mainly at austenite grain boundaries and considerable ferrite refinement, and subsequent improvement in strength and toughness can be achieved by increasing the austenite grain boundary area by (i) refining the austenite (DRCR or MDRCR), (ii) preventing austenite grain growth during the time between passes (with the use of 0.005–0.007% Ti), (iii) accelerated cooling after hot rolling whereby substantial undercooling of austenite enhances the rate of ferrite nucleation and slows the rate of growth, and (iv) completely preventing austenite recrystallization to produce a pancake structure during deformation (CCR utilizing SRX). In the last case, further ferrite grain refinement is achieved by nucleation within the deformed austenite grains (intragranular nucleation).

An important class of steels, called microalloyed steels, contains 0.001–0.1 (mass) % microalloying elements (MAE) that significantly modify the mechanical properties (yield and tensile strength, toughness, formability, etc.) of the base steel. The various applications of these steels have been documented by Manohar (1997) and Davis (2001). Typical composition (mass %) of the base steel may contain 0.05–0.25% C and 1.0–2.0% Mn with optional additions of Mo (up to 1%), Ni (up to 3%) and Cr (up to 1%). Microalloying elements include Nb (0.03–0.1%), Ti (0.01–0.05%), B (0.0018–0.006%), V (0.06–0.15%) and Al (0.03–0.08%). MAE may be added either individually or in combination. Since MAE differs from one another in their physical and chemical features, they have different effects on precipitation (through reaction with C and N), recrystallization, phase transformation, and inclusion control (through reaction with S and O). It can be seen from Fig. 3.3 that the combination of microalloying, TMP and AC can increase the strength of the base C-Mn steel almost fourfold without sacrificing toughness. Major alloy systems that utilize different TMP for property optimization may contain V, V-N, Nb, Nb-Mo, V-Nb, Ti, V-Ti, Ti-Nb, Ti-Nb-Mo and Ti-Nb-Mn-Mo using not only grain refinement (Ti, Nb) and precipitation hardening (V, Nb) but also other strengthening mechanisms such as solid solution hardening (Mn) and the formation of acicular ferrite/bainite structures (Mn and Mo) via accelerated cooling.

Important microalloying elements in steel

Typically, 0.02–0.04% Niobium is added to steels to increase yield strength by precipitation hardening and grain refinement. Offshore platform steels (up to 75 mm thick plates) with yield strengths in the range 345–425 MPa is a typical application of Nb-based alloy systems. During hot rolling, strain-induced precipitation of Nb(C, N) is a highly effective method of retarding austenite recrystallization. Nb suppresses recrystallization by an order of magnitude when in solution (at $T > 1000$ °C) and almost by three orders of magnitude when precipitated (at $T < 950$ °C) (Kwon and DeArdo, 1987). As a consequence, the ability of Nb to form carbides will significantly increase T_{nr}. Dissolved Nb is also useful as it may be used to increase hardenability, which contributes to microstructure control when used in conjunction with accelerated cooling. Precipitation of Nb-base compounds in ferrite will further increase the strength of the final component.

Titanium reacts preferentially with N to form stable TiN precipitates which effectively control grain growth of austenite. Up to 0.025% Ti is most effective in preventing grain growth at temperatures up to 1300 °C and, hence, it is suitable for steels produced by DRCR and MDRCR. Dissolved or precipitated Ti has a similar but less pronounced effect as compared to Nb in retarding austenite recrystallization. The strong affinity of Ti for S may form hard carbosulfides, which is useful for preventing the undesirable effects exerted by elongated MnS particles. Since Ti also has a strong affinity for O, Ti addition is most effective in fully killed (e.g. using Al) steels. Addition at higher Ti levels is known to control the shape of sulphide particles through the formation of (Mn, Ti)S and globular $Ti_4C_2S_2$ along with the precipitation of TiC in ferrite, which further aids in strengthening of the final component. Ti-based microalloyed steels can be produced with yield strength in the range 345–550 MPa.

Steels containing up to 0.1% Vanadium are widely used in the hot-rolled, controlled-rolled, accelerated-cooled, normalized or quenched and tempered conditions for structural applications with a maximum yield strength of about 415 MPa achievable. Since V remains in solution during hot rolling, it does not strongly influence austenite recrystallization. However, precipitation of V-base compounds such as VN in ferrite during cooling results in substantial strengthening with an effectiveness that depends on cooling rate. A high cooling rate results in V remaining partially in solution whereas very slow cooling rates produce coarse (50–100 nm) VN particles. Optimum precipitation strengthening is obtained at a cooling rate of ~ 3 °C/s, which generates fine (4–10 nm) VN particles. The size of these particles is also dependent on the amount of N and Mn: a higher N content forms extremely fine (~2 nm) precipitates and a higher Mn content lowers the $\gamma \rightarrow \alpha$ transformation start temperature (Ar_3) which also produces fine particle dispersion.

In summary, steels based on V and V-Ti are the most suitable for DRCR while steels based on Nb and Nb-Ti are more suitable for CCR. The influences of TMP, AC and steel composition on the properties of the hot-rolled product are numerous and include improvements in strength, toughness and weldability. This is achieved through austenite grain refinement and ferrite strengthening by grain refinement, precipitation and solid solution hardening, as well as the development of multi-phase microstructures that are engineered to suit a desired application.

3.5.2 Processability and rolling load limitations

The main advantage of RCR is in the use of high rolling temperatures (> 1000 °C) which results in low rolling loads since restoration processes will occur easily (Fig. 3.9). However, RCR must be carried out between T_{nr}

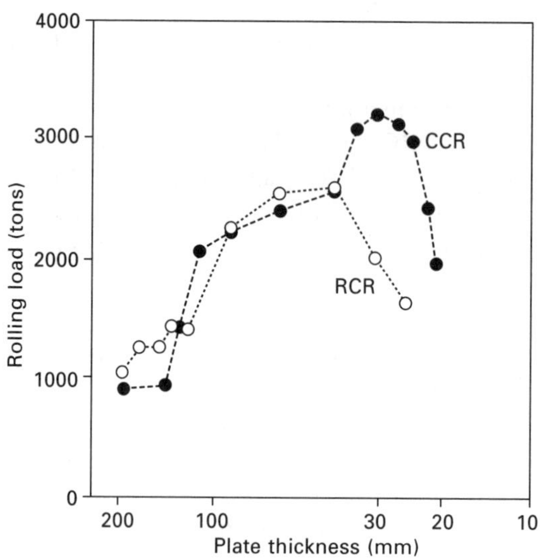

3.9 Rolling mill capability requirement for different TMP sequences, after Siwecki *et al.* (1995).

and GCT, which are influenced by steel composition, austenite grain size, prior TMP history and the precipitation sequence and kinetics. From a process logistics viewpoint, T_{nr} should be as low as possible to provide a large enough processing window to accommodate the necessary passes to obtain both the desired microstructure and final dimensions. Compared with RCR, CCR involves much lower finish rolling temperature (~800 °C) which generates higher rolling loads through deformation of unrecrystallized austenite (Fig. 3.9). Consequently, CCR is not suitable for mills that are limited by their rolling load capacity. In CCR, alloy composition should be such that T_{nr} is raised as high as possible to provide a sufficient gap between T_{nr} and the austenite transformation start temperature (Ar_3) to alleviate the rolling load problem. Higher Mn, Mo contents and a finer grain size lower Ar_3, while higher Nb content raises T_{nr}.

3.5.3 Secondary processing of strip-cast steels

One of the critical challenges for NNSC processes is the restricted TMP processing window available for modifying the as-cast microstructure. While CCC followed by CCR involves a large number of recrystallization cycles during hot rolling, TSC is restricted to a few hot rolling passes to adjust the austenite microstructure. In the extreme case of DSC, hot rolling is usually restricted to a single pass by in-line HDR in order to adjust the strip thickness and modify the as-cast microstructure. Since DSC of low carbon steel generates columnar austenite grains of 300–700 μm in length (parallel to ND) and 100–250 μm in width (Mukunthan *et al.*, 2000), complete austenite recrystallization is extremely difficult during in-line hot rolling. This is illustrated in Fig. 3.10, which shows the effect of austenite grain size on the rate of recrystallization at 1150 °C in low carbon steels. It is clear that the quantification of the influence of processing parameters, composition and austenite grain size on the recrystallization kinetics is a pre-requisite to optimize the limited downstream processing opportunity in DSC.

Perhaps the most critical challenge for DSC is the production of steel strip with mechanical properties similar to those obtained via conventional production routes. Compared to conventionally produced steel strip, strength is slightly higher and the elongation lower in strip-cast products. Nevertheless, it has been found that the lower elongation in strip-cast steel is not a major problem during forming and welding operations (Bledje *et al.*, 2000a, b).

3.6 Common problems and solutions

3.6.1 Abnormal grain growth

It is critically important to control austenite grain size during high temperature processing because it affects the kinetics of recrystallization and phase

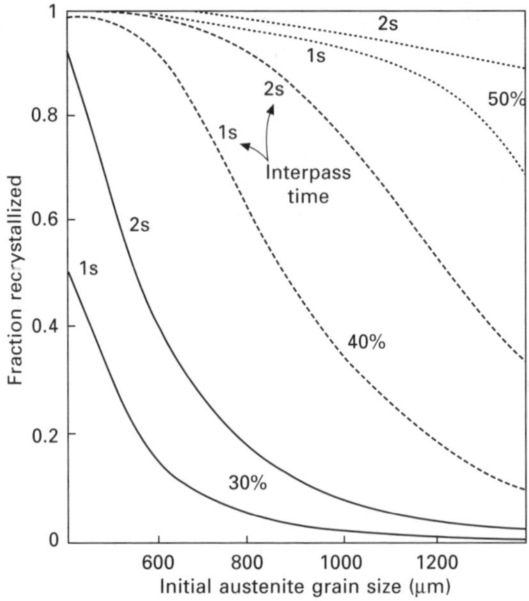

3.10 Recrystallization kinetics of coarse-grained austenite at 1150 °C in low carbon steels, after Korchynsky (1999).

transformation and therefore influences significantly the room temperature microstructure and properties of steels. It is well known that grain coarsening in polycrystalline materials occurs by two mechanisms: normal and abnormal grain growth (AGG). Normal grain growth consists of uniform and monotonous motion of all grains in grain size–time space and tends to achieve a quasi-steady state distribution of grain sizes (Hunderi and Ryum, 1982). In bulk materials, normal grain growth may be inhibited due to various factors (Manohar, 1997; Humphreys and Hatherly, 2003):

- narrow grain size distribution
- mean grain size that approaches the limiting grain size, as determined by the size and volume fraction of second phase particles
- existence of a strong texture
- solute atoms that cause drag on grain boundaries
- finite thickness of the material (e.g. in thin films).

If normal grain growth is inhibited, AGG may occur whereby a small number of grains grow preferentially to consume the surrounding microstructure. During AGG, the microstructure consists of a bimodal grain size distribution (Fig. 3.11) with the preferential growth of the large grains at the expense of the remainder. AGG continues until growing grains impinge upon each other producing a more uniform size distribution of extremely coarse grains (of

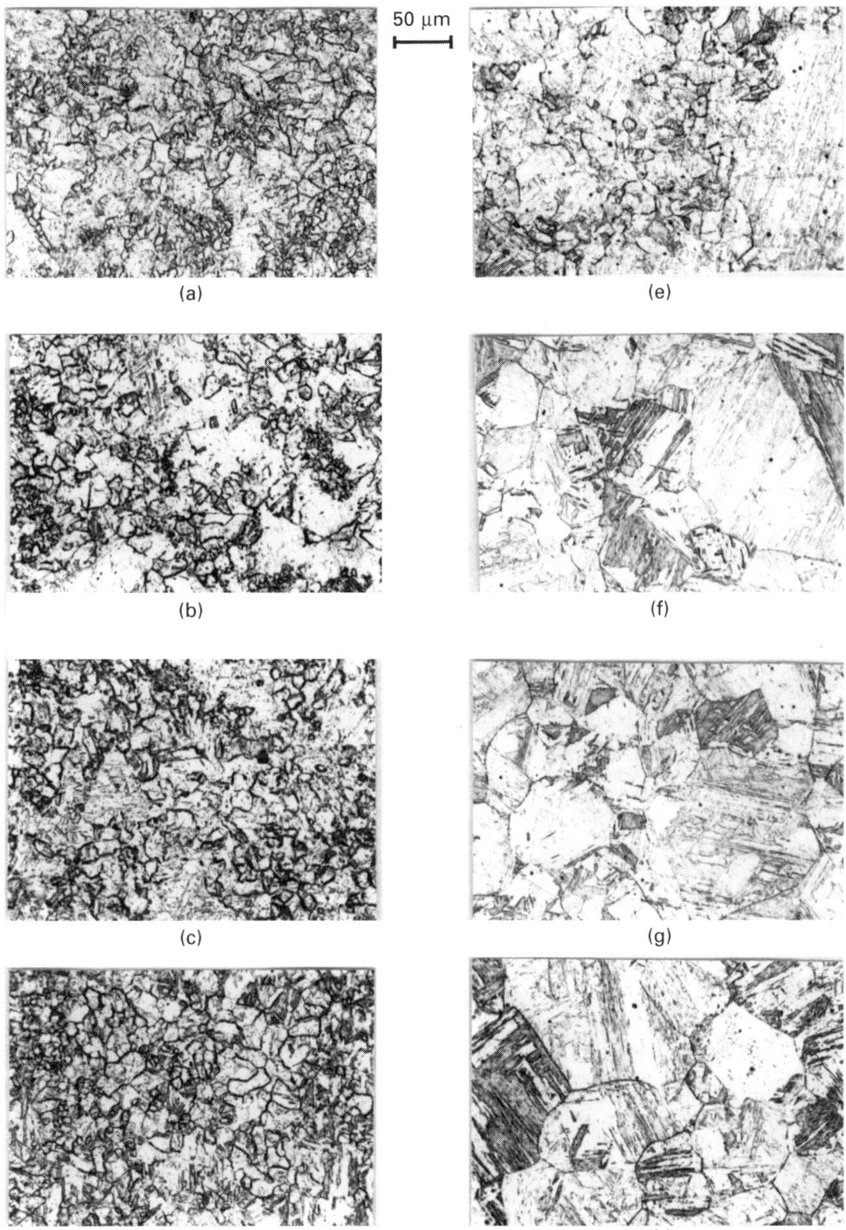

3.11 Representative microstructures of prior austenite during reheating of a Low Mn + High Mo, Nb, Ti microalloyed steel slab, holding time 1800 s at: (a) 900 °C, (b) 950 °C, (c) 1000 °C, (d) 1050 °C, (e) 1100 °C, (f) 1150 °C, (g) 1200 °C and (h) 1250 °C. Note the development of duplex microstructure in (e) due to AGG. Grain size is much coarsened, but more uniformly distributed following AGG as in (f), after Manohar (1997).

the order several hundred microns in MA steels to millimetre size in PC steels).

For a given material, AGG often occurs above a well-defined temperature termed the grain coarsening temperature (GCT). The concept of GCT is intimately coupled with the concepts of inhibition of normal grain growth by second phase particles and the subsequent unpinning of the grains due to progressive dissolution and coarsening of particles as the reheating temperature is increased. In addition, heterogeneity of the initial austenite microstructure has been found to affect GCT of Ti-Nb microalloyed steels (Fig. 3.12).

The grain size heterogeneity is quantified by determining a ratio (Z) of the mean size of the largest grain to mean grain size in a sample. It is clear from Fig. 3.12 that steels with high degree of grain size heterogeneity exhibit a lower GCT by as much as 50–100 °C compared to steels with a more uniform initial grain size distribution. In addition, considerable research on a range of microalloyed steels has confirmed that the type of precipitates {Nb(C, N), Ti(C, N) etc.} and their characteristics (size, volume fraction and distribution) are important factors in the determination of GCT for particular steel compositions (Webster and Allen, 1962; Erasmus, 1964a,b; Gladman and Pickering, 1967; George and Irani, 1968; Gladman, 1989, 1992; Palmiere *et al.*, 1994).

An understanding of GCT is extremely important in designing a viable hot rolling schedule because this temperature affects the first stage of TMP, namely, the reheating temperature. The reheating temperature affects the degree of dissolution of a given type of precipitate as well as the starting austenite grain size. During subsequent TMP, these factors significantly influence recrystallization, precipitation, austenite decomposition and, therefore, affect the final microstructure and properties. Reheating to temperatures below GCT generally results in a certain amount of undissolved microalloy precipitates. Therefore, microalloying elements are not available for precipitation hardening and/or hardenability control during further processing. Reheating to a temperature in the vicinity of GCT may produce a bimodal austenite grain size, which is difficult to eliminate during subsequent TMP, and generates inferior mechanical properties of the final product (Tanaka *et al.*, 1975; Cuddy, 1984). Reheating to a temperature much higher than GCT leads to AGG and a coarse austenite grain size, which is extremely difficult to refine by recrystallization during the successive deformation stages. It is clear that the most suitable reheating temperature is one that is high enough substantially to dissolve precipitates but low enough to restrict AGG.

3.6.2 Optimising the formability of sheet steel products

From a fundamental viewpoint, good formability of sheet steel requires a high $\bar{r}$-value (eqn 3.19); a parameter intimately associated with grains oriented

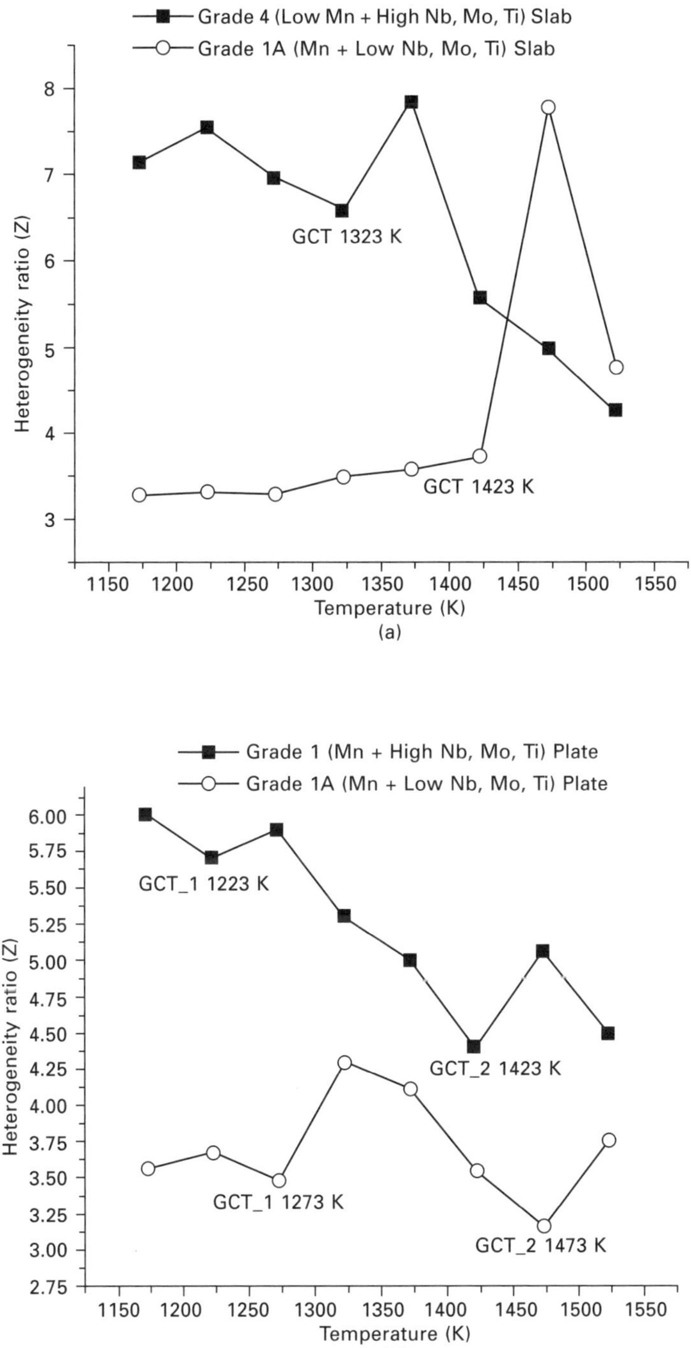

3.12 Influence of heterogeneity in austenite microstructure on GCT of Ti-Nb microalloyed steels: (a) slab steels, and (b) plate steels, after Manohar *et al.* (2003a).

such that <111> crystallographic directions are aligned in the normal direction to the sheet plane. However, low planar anisotropy is also important, which requires that the grains should also be randomly orientated parallel to the sheet plane. The preferred orientation of grains leading to high normal anisotropy is usually termed the <111>//ND or γ-fibre (section 3.3.3), and the production of this texture, and therefore highly formable sheet steels, is the major aim of industrial processing (Ray *et al.*, 1994). Since good drawability is associated with <111>//ND textures and poor drawability with <001>// ND textures (Hutchinson, 1984), the ratio of the intensities of the 222 and 200 X-ray reflections from a rolling plane specimen (I_{222}/I_{200}) is often used as a measure of formability and this is shown to correlate very well with $\bar{r}$ (Fig. 3.13)

As indicated in section 3.3.3, CRA textures that develop in carbon-base steels are affected by a host of parameters relating to steel composition and the entire processing history. It is the combination of these processing parameters, which ensures that any particular steel will possess the final microstructure and texture for superior formability. Figure 3.14a shows the strong effect of carbon on drawability which shows that $\bar{r}$ is dramatically increased at low carbon levels and that there is an optimum value associated with a carbon level near 3 ppm. As a further example, the influence of hot-band grain size and degree of cold rolling on formability is shown clearly in Fig. 3.14b, which correlates closely with the texture data given in Fig. 3.8. In recent years, the choice of alloying elements in steel and careful control

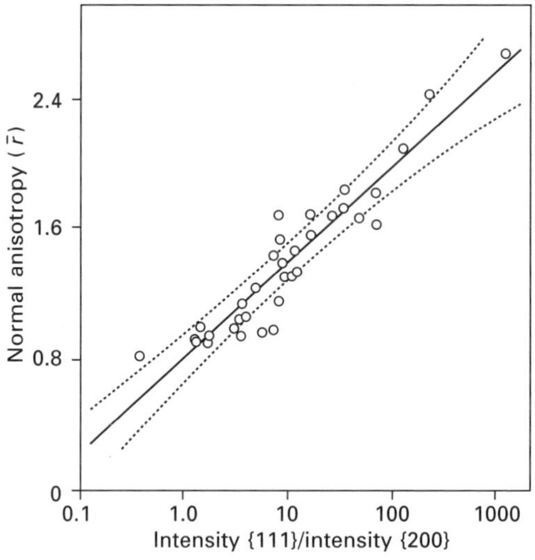

3.13 Relationship between plastic anisotropy, $\bar{r}$, and ratio of the intensities of the 222 and 200 X-ray reflections from a rolling plane specimen, I_{222}/I_{200}, after Held (1969).

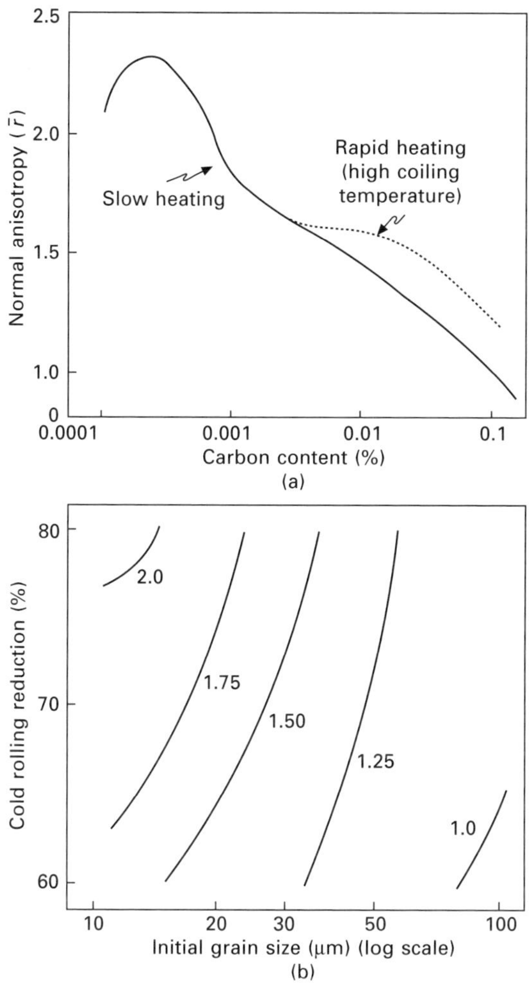

3.14 (a) Effect of percentage carbon on plastic anisotropy, $\bar{r}$, after Hutchinson (1984), and (b) Effect of hot-band grain size and degree of cold rolling on $\bar{r}$ (constant $\bar{r}$-values given as contours), after Hutchinson *et al.* (1990).

of the entire processing route has produced an excellent combination of very high $\bar{r}$-values (> 3), Δr-values close to unity and high ductility, thereby generating highly formable steel sheet products.

3.7 Case study: rod rolling

The importance of understanding the complex nature of microstructural evolution during an industrial-scale TMP process will be demonstrated here as a case study on rod rolling. The rod rolling process is characterized by continuous multi-pass deformation (up to 31 passes) at strain rates in the range 0.4–3000 s^{-1}, at temperatures in the range 900–1100 °C, and interpass times of 0.015–1.0 s. These are severe processing conditions and provide little scope for carrying out reliable laboratory-based or industrial-based experimental studies of microstructural evolution. In addition, microstructural evolution during rod rolling is complicated since several softening mechanisms may operate sequentially or concurrently depending on the choice of processing parameters.

To study continuous rod rolling, computer simulation has been used to gain an understanding of microstructural evolution and to evaluate the influence of austenite grain size and finish rolling temperature on final microstructure. An understanding of microstructural evolution is critical for both the optimization of the manufacturing schedule and adjustment of properties of the as-rolled product. For example, a fine austenite grain size is desirable at the end of rolling to decrease the hardenability. This helps to obtain a fine ferrite + pearlite structure via controlled rolling and cooling. Fine ferrite-pearlite structure is a basis for the elimination or reduction of the post-rolling heat treatments such as annealing or spheroidization and to improve the final mechanical properties. Furthermore, there is the need to reduce surface decarburization of high carbon spring steels and bearing steels by controlling the rolling parameters and by utilizing accelerated watercooling before the steel enters the air-cooling process. All of these requirements may be met by judicial choice of the TMP sequence as described in the following sections.

3.7.1 Process description

A typical continuous rod rolling process sequence is shown in Fig. 3.15. It consists of 31 passes, but as the last two passes act as sizing passes they are not included for rigorous analysis in the present study. It is evident from Fig. 3.15 that several delay zones exist in the TMP sequence: between passes #5 and #6; passes #11 and #12 (not shown in Fig. 3.15); passes #13 and #14 and passes #19 and #20 while two cooling zones are employed between passes #27 and #28 and final cooling subsequent to pass #31 where the temperature is reduced from the finish rolling temperature (FRT) of ~ 970 °C to ~830 °C using high-speed water cooling (~1000 °C/s). Further cooling of the rod occurs in a STELMOR™ air-cooling line at a much slower cooling rate of ~1.5 °C/s. During rolling, the speed of the rolling stock increases from 0.12 m/s at the first pass up to 110 m/s in the final pass as the 160 × 160 mm square billet is reduced to ~ 6.0 mm diameter round rod. Area reductions obtained in each pass vary in the range 17–27%, except for the final two passes where there is about 5% reduction per pass.

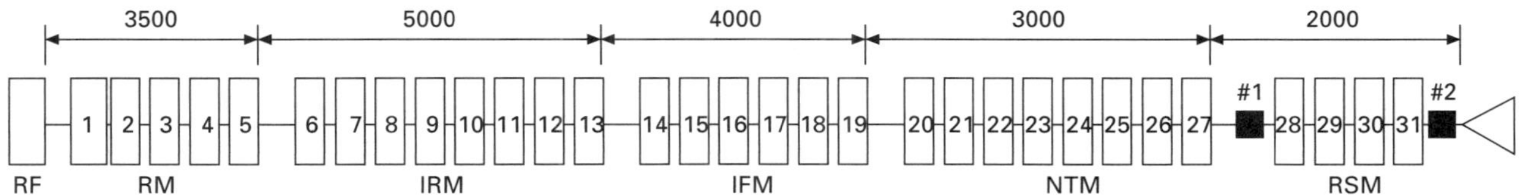

3.15 Schematic diagram showing a typical industrial-scale rod rolling sequence. (Legend: RF = Reheating Furnace; RM = Roughing Mill; IRM = Intermediate Roughing Mill; IFM = Intermediate Finishing Mill; NTM = Non-Twisting Mill; RSM = Reducing and Sizing Mill; #1 and #2 = Water Cooling Zones. Note: all dimensions are in mm), after Manohar *et al.* (2003b).

3.7.2 Process modelling

The rod rolling process involves a change in shape of rolling stock from a rectangular cross-section to a final circular cross-section via deformation through a series of grooved rolls. Since the roll grooves are non-rectangular (oval, round, diamond, square, etc.), the workpiece undergoes complex strain and stress paths that cannot be simplified as either plane strain or plane stress. The strain rate in the process changes at various stages within a given deformation step. For example, the strain rate is the maximum at the entrance to the roll (or in its vicinity) and decreases along the roll bite, finally becoming zero at the outlet. For this reason, it is necessary to use an average strain rate for a given pass, which is defined as the strain over a time interval taken for the workpiece to pass through the roll gap. The temperature of the workpiece during rolling depends on various factors such as rolling speed, initial temperature of the billet, the amount and nature of the plastic deformation, the cross-sectional shape of the workpiece during each pass, the cooling condition in the individual passes and distribution of cooling and equalization zone between stands (passes). While there is a significant temperature variation from the surface to centre of the workpiece, an average value has been used for modelling.

The average values of strain, strain rate and temperature for each deformation pass are given in Table 3.2. These parameters are obtained using both measured data on industrial processing lines and mathematical modelling as described in more detail elsewhere (Manohar *et al.*, 2003b). The initial (as-reheated) microstructure and rolling schedule are the basic inputs in the computer model. At each rolling stage, deformation conditions such as critical and peak strains and Zener-Hollomon parameter are subsequently computed on the basis of initial grain size, strain, strain rate and temperature. Relevant mathematical relations applicable to medium C-Mn steel are given in Table 3.3. These relations were used to compute the evolution of austenite grain size, fraction recrystallized and recrystallization type and its kinetics for each pass for a given deformation step.

3.7.3 Microstructural evolution in rod rolling

Detailed microstructural evolution, as computed by the simulation program, is given in Table 3.2. The data show that DRX followed by MDRX occurs in the roughing mill (passes #2 to #10), with the austenite grain size decreasing from 300 μm to ~17 μm. The remaining passes of the roughing mill (#11 to #13) show MDRX as the principal restoration mechanism with some austenite coarsening (~19 μm) due to (i) grain growth in the delay zone between passes #11 and #12 and (ii) in-process grain growth. The predominant restoration processes during the intermediate finishing mill passes (#14 to

Table 3.2 Simulated microstructure evolution in rod rolling

Pass#	T_{def} (K)	$\bar{\varepsilon}$	ε_c	ε_p	$\dot{\bar{\varepsilon}}$ (s^{-1})	ReX Type	Init_GS (μm)	ReX_GS (μm)	Final_GS (μm)
1	1273	0.347	0.329	0.405	0.4	MDRX(99)	300.0	41.0	43.0
2	1263	0.435	0.210	0.258	0.7	D(48)+MD	43.0	26.5	36.1
3	1253	0.461	0.212	0.261	0.8	D(53)+MD	36.1	23.4	31.2
4	1243	0.491	0.232	0.285	1.4	D(50)+MD	31.2	19.2	28.4
5	1233	0.320	0.232	0.285	1.3	D(14)+MD	28.4	21.9	32.5
6	1223	0.488	0.283	0.348	2.6	D(32)+MD	32.5	15.7	24.0
7	1213	0.354	0.271	0.333	2.7	D(11)+MD	24.0	16.1	22.2
8	1203	0.513	0.309	0.380	5.2	D(29)+MD	22.2	11.6	20.1
9	1193	0.398	0.315	0.387	5.4	MDRX(91)	20.1	11.7	18.6
10	1188	0.514	0.349	0.429	10.0	D(19)+MD	18.6	9.2	17.4
11	1183	0.437	0.364	0.448	12.8	MDRX(94)	17.4	8.8	19.2
12	1178	0.534	0.427	0.525	23.9	MDRX(92)	19.2	7.1	15.2
13	1173	0.465	0.410	0.502	25.0	MDRX(96)	15.2	6.9	19.1
14	1173	0.462	0.487	0.599	45.8	F/SRX(100)	19.1	6.1	13.9
15	1173	0.439	0.448	0.551	49.9	F/SRX(99)	13.9	5.1	10.2
16	1178	0.471	0.442	0.544	90.1	MDRX(99)	10.2	5.1	8.7
17	1178	0.449	0.427	0.525	96.9	MDRX(99)	8.7	5.0	8.1
18	1188	0.423	0.430	0.529	148.2	F/SRX(99)	8.1	3.6	7.2
19	1193	0.465	0.416	0.512	171.9	MDRX(97)	7.2	4.7	17.9
20	1198	0.474	0.597	0.735	325.4	P/SRX(88)	17.9	5.1	5.1
21	1203	0.424	0.402	0.494	331.5	MDRX(99)	5.1	4.3	4.7
22	1208	0.442	0.419	0.516	554.5	MDRX(99)	4.7	3.9	4.2
23	1213	0.438	0.403	0.496	596.6	MDRX(98)	4.2	3.9	4.2
24	1223	0.438	0.416	0.512	935.8	MDRX(99)	4.2	3.7	4.0
25	1233	0.483	0.412	0.507	1224.0	MDRX(95)	4.0	3.6	3.9
26	1243	0.456	0.425	0.523	1945.0	MDRX(99)	3.9	3.4	3.7
27	1253	0.481	0.412	0.507	2185.0	MDRX(95)	3.7	3.5	9.3
28	1233	0.335	0.584	0.719	2143.0	P/SRX(36)	9.3	5.0	5.0
29	1243	0.657	0.490	0.603	2876.0	D(13)+MD	5.0	3.0	3.3

Init_GS = initial austenite grain size prior to deformation; ReX_GS = recrystallized grain size; Final_GS = grain size after interpass grain growth. Note: numbers in parentheses in 'Rex Type' column indicate % volume recrystallized by the particular softening mechanism.

Table 3.3 Mathematical models employed in computing microstructure evolution in rod rolling

Parameter	DRX	MDRX	SRX
Fraction recrystallized (X_v)	$X_v = 1 - \exp[-0.8\,((\varepsilon-\varepsilon_c)/\varepsilon_p)^{1.4}]$ where $\varepsilon_p = 1.23\varepsilon_c$ (Hodgson, 1993)	$X_v = 1 - \exp[-0.693\,(t/t_{0.5})^{1.5}]$ (Hodgson and Gibbbs, 1992)	$X_v = 1 - \exp[-0.693(t/t_{0.5})]$ (Laasraoui and Jonas, 1991)
Time for 50% recrystallization ($t_{0.5}$)	Not applicable	$1.44 \cdot \dot{\varepsilon}^{-0.8}$ (Hodson, 1997)	$1.14 \times 10^{-13}\varepsilon^{-3.8}\,\dot{\varepsilon}^{-0.41}$ $\exp(252000/RT)$ (Laasraoui and Jonas, 1991)
Recrystallized grain size (D_{rex})	$3.9 \times 10^4 Z^{-0.27}$ (Hodgson, 1997)	$6.8 \times 10^4 Z^{-0.27}$ (Hodgson, 1997)	$0.5 \cdot D_0^{0.67}\,\varepsilon^{-0.67}$ (Laasraoui and Jonas, 1991)
Finding $\overline{D}$ when partial (Beynon and Sellars, 1992) and/or combined ReX (Lenard *et al.*, 1999) mechanisms operate	Partial SRX: DRX + MDRX:	$\overline{D} = F_X^{4/3} D_{rex} + (1 - F_X)^2 D_o$ $\overline{D} = D_{DRX} + (D_{MDRX} - D_{DRX}) \times F_{XMDRX}$	
Peak strain (ε_p) (Sellars, 1990)	$\varepsilon_p = 6.97 \times 10^{-4} D_0^3 \cdot Z^{0.17}$	Not applicable	Not applicable
Pass strain (ε) (Laasraoui and Jonas, 1991)	$\varepsilon = \varepsilon_{pass} + \Delta\varepsilon$ where $\Delta\varepsilon = K \cdot \varepsilon_{previous}(1 - F_X)$ and $K = 1$ if $F_x < 0.1$ and $K = 0.5$ if $F_x \geq 0.1$		
Zener–Hollomon parameter (Z) (Hodgson, 1997)		$Z = \dot{\varepsilon}\,\exp(300000/RT)$	
Austenite grain growth (Maccagno *et al.*, 1996)	When time for grain growth is ≥ 1s: SRX: MDRX: When time for grain growth is < 1s: SRX: MDRX:	 $D_f^7 - D_0^7 = 1.5 \times 10^{27}(t - 4.32t_{0.5})\exp(-400000/RT)$ $D_f^7 - D_0^7 = 8.2 \times 10^{25}(t - 2.65t_{0.5})\exp(-400000/RT)$ $D_f^2 - D_0^2 = 4.0 \times 10^7(t - 4.32t_{0.5})\exp(-113000/RT)$ $D_f^2 - D_0^2 = 1.2 \times 10^7(t - 2.65t_{0.5})\exp(-113000/RT)$	

#19) are either SRX or the MDRX. At this stage, the grain size is ~ 5 μm, but grain growth to ~18 μm occurs in the delay zone between passes #19 and #20. It is pertinent to note that MDRX dominates in almost all passes of the finishing mill (pass #20 to #27), except for pass #20 where, for the first time, a fraction of material remains unrecrystallized prior to the start of the next pass (partial SRX condition). Passes #28 and #29 are interesting as the strain rate is very high (> 2000 s^{-1}), the interpass time very short (~ 9–12 ms) and the deformation temperature low (~970 °C). Under these conditions, ε_c is too high to initiate DRX in pass #28 but is sufficient for SRX commence in the interpass time but since this time is short, only ~36% of the volume is recrystallized. This leads to a large amount of residual deformation that accumulates in pass #29 to produce an effective strain exceeding ε_c thereby resulting in DRX followed by MDRX to generate an austenite grain size of ~3.3 μm.

It is important to note that the austenite grain size increases from 3.3 μm to ~15 μm during rapid cooling from the FRT to the cooling stop temperature (CST), and during further slower cooling to Ar_3. Such grain growth is expected since medium C-Mn steels do not contain second phase particles or sufficient solute elements to limit grain boundary migration by particle pinning or solute drag. Furthermore, the grain size directly after rolling is very fine and will result in a high driving force for grain growth (Hillert, 1965). Hence, the advantage gained in the deformation stages by refining the austenite grain size through DRX, MDRX and pancaking (incomplete SRX) is partially lost during the cooling stages.

3.7.4 Process optimization

The insight provided by modelling of process and microstructural evolution can now be used to analyze the capabilities of rod rolling when critical processing variables are altered. The influence of austenite grain size just prior to $\gamma \rightarrow \alpha$ transformation on the final ferrite grain size is shown in Fig. 3.16a. This demonstrates that a 50% reduction in austenite grain size (obtained by post-deformation accelerated cooling) will generate a 20% refinement in ferrite grain size. As discussed in section 3.1.3, the ferrite grain size may be further refined if the cooling rate is increased through the $\gamma \rightarrow \alpha$ transformation range. On the other hand, a reduction in FRT for the stipulated cooling rates does not have a significant influence on austenite grain size prior to $\gamma \rightarrow \alpha$ transformation (Fig. 5.16b). The principal reason for this behaviour is the occurrence of rapid grain growth in the C-Mn steel during the cooling stages that clearly offsets the advantage of the reduction in FRT. To retard post-deformation grain growth, microalloying elements (e.g. Ti, Nb) should be added to the steel to take advantage of austenite grain refinement due to DRX and MDRX during the processing schedule.

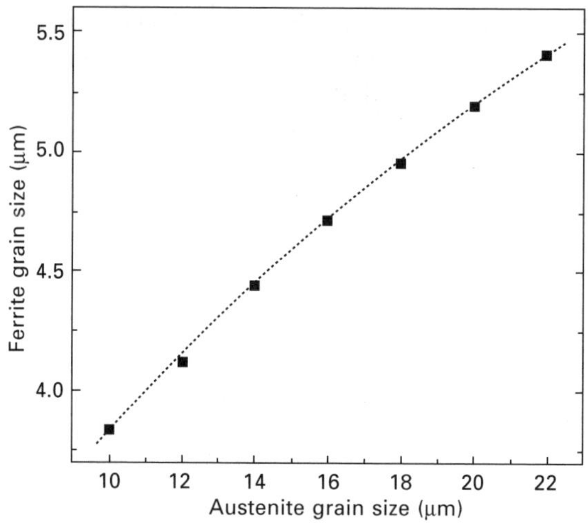

(a)

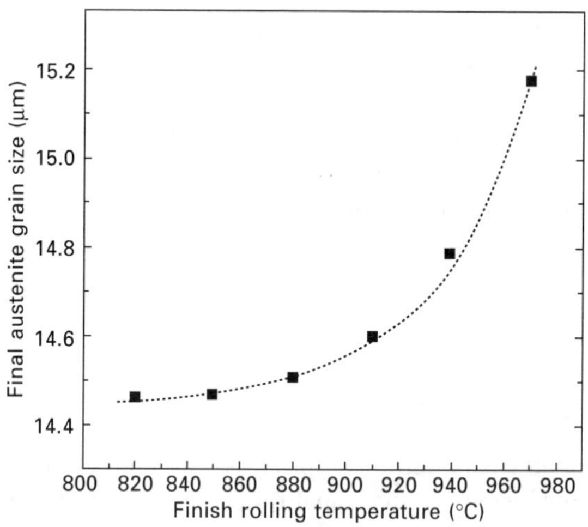

(b)

3.16 (a) Predicted ferrite grain refinement as a function of prior austenite grain size; and (b) austenite grain refinement as a function of finish rolling temperature, after Manohar *et al.* (2003b).

3.8 Conclusion

Greater demands placed on the quality, competitiveness and cost is driving innovative breakthroughs in contemporary steel making and processing technologies. Near net shape casting (NNSC) technologies including thin slab casting (TSC) and direct strip casting (DSC) are on the verge of full commercialization where steel products to near final shape are made directly from liquid steel. As a consequence, it has become imperative for downstream processing to utilize a range of thermomechanical processing (TMP) technologies to optimize steel properties. A combination of fundamental research and industrial trials focusing on mathematically modelling the complex problems of metal deformation and restoration processes at high temperature is facilitating further development. The insights gained through the synergistic application of research tools such as experimentation at laboratory and industrial scales, mathematical modelling and computer simulation are helping to develop and control new manufacturing technologies such as dynamic recrystallization controlled rolling (DRCR), metadynamic recrystallization controlled rolling (MDRCR) and warm rolling (WR).

The determination of the Zener-Hollomon parameter (Z, eqn 3.1) is an important building block in the mathematical theory of the high-temperature metal deformation process. This parameter accounts for the equivalence of the effect of lowering deformation temperature and increasing strain rate on flow strength of the metal. A hyperbolic sine relation between Z and flow strength allows the calculation and modelling of a high-temperature flow curve (eqns 3.2 and 3.3). Post-deformation softening behaviour depends on the austenite grain size, presence or absence of precipitates, strain, strain rate and interpass time. The kinetics of recrystallization is modelled in line with kinetics of phase transformations according to JMAK kinetics (eqn 3.6). However, more physically based models for recrystallization (eqn 3.7) and precipitation (eqn 3.8) articulated by empirical constants found through experimentation are used more commonly to solve practical problems.

As the degree of secondary processing available to adjust and control material properties decreases in step with emerging NNSC technologies, the design of alloy composition becomes even more critical. High temperature processability of a steel grade is related in a non-linear way to the compositional parameters. The addition of microalloying elements (up to 0.1 mass %) makes a significant difference to the mechanical behavior of steels. Niobium either in solid solution (at T_{def} >1000 °C) or precipitated (T_{def} < 950 °C) retards recrystallization kinetics by an order of magnitude compared to the recrystallization kinetics of steels with no Nb addition. The addition of Ti refines austenite grain size, modifies the shape of MnS inclusions, increases flow strength and retards recrystallization kinetics (although not as much as Nb). The room temperature strength of steels increases substantially with V addition. However, delaying or stopping recrystallization does have an undesirable side effect of increasing the flow strength of austenite so that greater rolling loads are required for continued deformation. This means that not all rolling mills are capable of delivering thermomechanical processing.

One of the important problems in the processing of steels is the occurrence of duplex microstructures due to abnormal grain growth (AGG). Progressive dissolution and coarsening of microalloy precipitates {Ti, Nb (C, N)} and heterogeneity in the prior microstructure leads to AGG in microalloyed steels. For any particular metal, there is usually a well-defined minimum temperature, which must be exceeded for AGG to occur. This is called the grain coarsening temperature (GCT). The design of a TMP schedule is done in such a way as to avoid the conditions that promote AGG. The TMP cycle may generate other problems such as the production of steel sheet products with inadequate formability. The final microstructure and texture of steel sheet is affected by composition and the entire processing history (section 3.3.3) and incorrect control of any one of these parameters will produce a product with inferior formability.

Finally, a full-scale industrial rod rolling process is described and mathematically modelled as a case study. The rod rolling process is characterized by multi-pass deformation sequences at widely varying strain rates (0.4–3000 s^{-1}), deformation geometry (square, round, oval) and interpass times (0.015–1.0 s). The importance of understanding the mechanisms and kinetics of competing softening processes such as static, dynamic and metadynamic recrystallization (SRX, DRX and MDRX) is demonstrated by detailed computation of microstructure in each deformation pass as shown in Table 3.2. It is paramount that appropriate mathematical models to compute recrystallization and precipitation kinetics, fraction recrystallized, austenite grain size and structure, and retained strain are developed and subsequently integrated to obtain a comprehensive understanding of the entire process. Such an approach yields insights into not only the optimization of existing process sequences but also the development new ones.

3.9 Sources of further information

3.9.1 Monographs

Byrne J.G. (1965), *Recovery, Recrystallization and Grain Growth*, McMillan, New York, USA.

Cotterill P. and Mould P.R. (1976), *Recrystallization and Grain Growth in Metals*, Surrey Univ. Press, London, UK.

Gladman T. (1997), *The Physical Metallurgy of Microalloyed Steels*, Institute of Materials, London, UK.

Honeycombe R.W.K. (1985), *The Plastic Deformation of Metals*. Edward Arnold, UK.

Honeycombe R.W.K. and Bhadeshia H.K.D.H. (1995), *Steels – Microstructure and Properties*, Edward Arnold, London, UK.

Humphreys F.J. and Hatherly M. (1996, 2003), *Recrystallization and Related Annealing Phenomena*. 1st and 2nd edns, Elsevier Science, Oxford, UK.

Lenard J.G., Pietrzyk M. and Cser L. (1999), *Mathematical and Physical Simulation of the Properties of Hot Rolled Products*, Elsevier Science, Oxford, UK.

Leslie W.C. (1981), *The Physical Metallurgy of Steels*, 1st edn, McGraw Hill, London, UK.

Sinha A.K. (1989), Ferrous Physical Metallurgy, Butterworths, Boston, USA.

3.9.2 Multi-author, edited compilations

Decomposition of Austenite by Diffusional Processes (1962), eds V.F. Zackay and H.I. Aaronson, TMS-AIME, USA.

The Hot Deformation of Austenite (1977), ed. J.B. Ballance, AIME, New York, USA.
International symposium on Hardenability Concepts with Applications to Steels (1978), eds D.V. Doane and J.S. Kirkaldy, Met. Soc. AIME.
Encyclopedia of Materials Science and Technology (2001), eds K.H.J. Buschow, R.W. Cahn, M.C. Flemings, B. Ilschner, E.J. Kramer and S. Mahajan, Elsevier Science, Oxford, UK.

3.9.3 Proceedings of conferences (chronological order) (1960–present)

Annual conference series on Mechanical Working and Steel Processing (MWSP) (1960–current), ISS, USA.
International Conference on Microalloying in Steels – Microalloying '75 (1975), Union Carbide Corporation, Washington D.C., USA.
International Conference on Strength of Metals and Alloys (ICSMA) (1967–2003), 13th conference in this series was organized at Budapest, Hungary, Aug. 25–30, 2003.
International Conference on Thermomechanical Processing of Microalloyed Austenite (1982), eds A.J. DeArdo *et al.*, TMS-AIME, USA
International Conference on HSLA Steels: Technology and Applications (1984), Philadelphia, USA, ASM.
International Conference on HSLA Steels (1984), eds T. Chandra and D.P. Dunne, Wollongong, Australia.
International Symposium on Accelerated Cooling of Steel (1986), ed. P.D. Southwick, Pittsburgh, TMS-AIME, USA.
International Conference on Processing, Microstructure and Properties of HSLA steels (1988), eds A.J. DeArdo *et al.*, TMS-AIME, USA.
THERMEC series of international conferences on Thermomechanical Processing of Steels and Other Materials (Tokyo 1988, Wollongong 1997, Las Vegas 2000, Madrid 2003), TMS, USA.
International conference series on Recrystallization and Related Phenomena (Wollongong, Australia 1990, San Sebastian, Spain 1993, Monterey, USA 1996, Tsukuba Science City, Japan 1999) in conjunction with ICGG conference series on Grain Growth (Rome, Italy 1991, Kitakyushu, Japan 1995, Pittsburgh, USA 1998), subsequently ReX and ICGG conference series merged to form a joint Recrystallization and Grain Growth Conference series (Aachen, Germany 2001, next: Annecy, France 2004).
International Conference on Mathematical Modelling of Hot Rolling of Steels (1990), ed. S. Yue, Hamilton, Canada.
International Conference on Processing, Microstructure and Properties of Microalloyed and Other Modern HSAL Steels (1992), eds A.J. DeArdo *et al.*, ISS-AIME, USA.
International Conference on Low Carbon Steels for the '90s (1993), eds R. Asfahani and G. Tither, TMS.
International Conference on Microalloying in Steels – Microalloying '95 (1995), eds M. Korchynsky *et al.*, ISS, Pittsburgh, USA.
International Conference on HSLA Steels – Chinese Society for Metals, Beijing, China: 1985 (HSLA Steels: Metallurgy and Applications), 1992 (HSLA Steels), 1995 (HSLA '95).
International Conference series (annual) on Metallurgy and Materials Science (1980–current), Riso National Laboratory, Roskilde, Denmark.

International Conference on Near-Net-Shape Casting in the Minimills (1995), Vancouver, Canada, eds J.K. Brimacombe I.V. and Samarasekera, Canadian Institute of Mining, Metallurgy and Petroleum, Canada.

International Conference on Thermomechanical Processing in Theory, Modelling and Practice (1996), eds W.B. Hutchinson *et al.*, The Swedish Society for Materials Technology, Stockholm, Sweden.

Deformation Processing of Metals (1999), Published in Phil. Trans. Royal Soc. 1441–1729, London, UK.

International Conference on Thermomechanical Processing of Steels (2000), Steel Division of Institute of Materials, London, UK.

International Conference on Thermomechanical Processing: Microstructure and Control, Sheffield (2002).

International Symposium on Microalloyed Steels (2002) (held in conjunction with the ASM Materials Solutions Conference), Columbus, OH, USA.

3.9.4 Review articles and books containing chapters

Doherty R.D., Hughes D.A., Humphreys F.J., Jonas J.J., Juul Jensen D., Kassner M.E., King W.E., McNelly T.R., McQueen H.J. and Rollett A.D. (1997), 'Current issues in recrystallization: a review', *Mater. Sci. Eng.*, A238, 219.

Hutchinson W.B. (1984), 'Development and control of annealing textures in low-carbon steels', *Int. Met. Rev.*, 29, 25.

Jonas J.J., Sellars C.M. and Tegart W.J. McG. (1969), 'Strength and structure under hot working conditions', *Met. Revs.* 130, 1.

Leslie W.C., Michalak J.T. and Aul F.W. (1963), 'The annealing of cold-worked iron', in *Iron and its Dilute Solid Solutions*, eds Spencer and Werner, Interscience, New York, 119.

Manohar P.A., Ferry M. and Hunter A. (2000), 'Direct strip casting of steel: historical perspective and future direction', *Mater. Forum*, 24, 15.

Manohar P.A., Ferry M. and Chandra T. (2001), 'Recrystallization of ferrite and austenite', *Encyclopedia of Materials Science & Technology*, eds. K.H. J. Buschow *et al.*, Elsevier Science, Oxford, 4, 3019.

Mishra S. and Därmann C. (1982), 'Role and control of texture in deep-drawing steels', *Int. Met. Rev.*, 27, 307.

Ray R.K. and Jonas J.J. (1990), 'Transformation textures in steels', *Int. Mater. Rev.*, 35, 129.

Ray R.K., Jonas J.J. and Hook R.E. (1994), 'Cold rolling and annealing textures in low and extra low low carbon steels', *Int. Mater. Rev.*, 39, 129.

Tanaka T. (1981), 'Controlled rolling of steel plate and strip', *Int. Met. Rev.*, 26, 185.

3.10 References

Araki T. and Shibata K. (1995), 'Microstructures and their characterization of modern very low carbon HSLA steels', *Proc. Int. Symp. HSLA '95*, eds L. Guoxun *et al.*, CSM, Beijing, China, pp. 13–21.

Araki T., Enomoto M. and Shibata K. (1991), 'Microstructural aspects of bianitic and bainite-like ferritic structures of continuously cooled low carbon (< 0.1%) HSLA steels', *Mater. Trans. JIM*, 32, pp. 729–736.

ASM Metals Handbook (1985), *Metallography and Microstructures*, 9th edn, ASM, 9, 170.

ASTM E 112-96 (2002), 'Standard test methods for determining average grain size', *ASTM Standards*, 03.01, pp. 251–274.

ASTM E 930-99 (2002), 'Standard test methods for estimating the largest grain observed in a metallographic sample (ALA grain size)', *ASTM Standards*, 03.01, pp. 718–724.

ASTM E 1181-87 (2002), 'Standard test methods for characterizing duplex grain size', *ASTM Standards*, 03.01, pp. 777–791.

ASTM E 1382-97 (2002), 'Standard test methods for determining average grain size using semiautomatic and automatic image analysis', *ASTM Standards*, 03.01, pp. 909–930.

Barbosa R., Boratto F., Yue S. and Jonas J.J. (1987), 'The influence of chemical composition on the recrystallization behaviour of microalloyed steels', *Proc. Int. Conf. on Processing, Microstructure and Properties of HSLA Steels*, ed. A.J. DeArdo, TMS, Pittsburgh, USA, pp. 51–61.

Bhadeshia H.K.D.H. (2001), *Bainite in Steels*, 2nd edn, IOM Communications Inc., London, UK, p. 286.

Beynon J.M. and Sellars C.M. (1992), 'Modelling microstructure and its effects during multipass hot rolling', *ISIJ International*, 32, pp. 359–367.

Bleck W. and Langner H. (1997), 'Softening behaviour of mild steels during hot deformation in austenite and ferrite range', *Proc. Int. Conf. Thermomechanical Processing of Steels and Other Materials – THERMEC 97*, eds T. Chandra and T. Sakai, TMS, Wollongong, Australia, pp. 611–619.

Bledje W., Mahapatra R. and Fukase H. (2000a), 'Development of low carbon thin strip production capability at project 'M'', *Iron and steelmaker*, 27(4), pp. 29–33.

Bledje W., Mahapatra R. and Fukase H. (2000b), 'Application of fundamental research at project 'M'', *Proc. Belton Memorial Symp.*, Sydney, Australia, ISS, pp. 253–261.

Bognin G., Contursi V.M., Tanzi G., Aprile A., DeFlorio G., Liguori A., Borsi P. and Ghersi M. (1987), 'Water blade jet for accelerated cooling of plates', *Proc. Int. Symp. on Accelerated Cooling of Steel*, ed. P.D. Southwick, TMS, Pittsburgh, USA, pp. 69–81.

Bramfitt B.L. and Speer J.G. (1990), 'A perspective on the morphology of bainite', *Met. Trans. ASM*, 21A, pp. 817–829.

Brimacombe J.K. and Samarasekera I.V. (1994), 'The challenge of thin slab casting', *Iron and Steelmaker*, 21(11), pp. 29–39.

Brownrigg A., Curcio P. and Boelen R. (1975), 'Etching of prior austenite grain boundaries in martensite', *Metallography*, 8, pp. 529–533.

Bunge H.J. (1982), *Texture Analysis in Materials Science*, Butterworths, London, UK.

Cramb A.W. (1995), 'Strip casting of steels: current status and fundamental aspects', *Proc. Int. Conf. Near-Net-Shape Casting in the Minimills*, Vancouver, Canada, eds J.K. Brimacombe and I.V. Samarasekera, Canadian Institute of Mining, Metallurgy and Petroleum, Canada, p. 355.

Cuddy L.J. (1981), 'Microstructures developed during thermomechanical treatment of HSLA steels', *Metall. Trans. ASM*, 12A, pp. 1313–1320.

Cuddy L.J. (1982), 'The effect of microalloy concentration on the recrystallization of austenite during hot deformation', *Proc. Int. Conf. on Thermomechanical Processing of Microalloyed Austenite*, eds A.J. DeArdo *et al.*, AIME, USA, pp. 129–139.

Cuddy L.J. (1984), 'Grain refinement of Nb steels by control of recrystallization during hot rolling', *Metall Trans. ASM*, 15A, pp. 87–98.

Cuddy L.J., Bauwin J.J. and Raley J.C. (1980), 'Recrystallization of austenite', *Metall. Trans. ASM*, 11A, pp. 381–385.

Davis J.R. (ed.) (2001), *Alloying: Understanding the Basics*, ASM International, USA.

DeArdo A.J. (1995), 'Modern thermomechanical processing of microalloyed steel: a physical metallurgy perspective', *Proc. Int. Conf. on Microalloying '95*, eds M. Korchynsky *et al.*, ISS, Pittsburgh, USA, pp. 15–33.

DeArdo A.J. (1997), 'The physical metallurgy of thermomechanical processing of microalloyed steels', *Proc. Int. Conf. Thermomechanical Processing of Steels and Other Materials – THERMEC 97,* eds T. Chandra and T. Sakai, TMS, Wollongong, Australia, pp. 13–29.

Dieter G.E. (1986), *Mechanical Metallurgy*, 3rd edn, McGraw Hill, p. 358.

Dutta B. and Sellars C.M. (1987), 'Effect of composition and process variables on Nb(C, N) precipitation in Nb microalloyed austenite', *Mater. Sci. Tech.*, 3, pp. 197–206.

Erasmus L.A. (1964a), 'Effect of aluminium additions on forgeability, austenite grain coarsening temperature, and impact properties of steel', *J. Iron Steel Inst.*, 202, pp. 32–41.

Erasmus L.A. (1964b), 'Effect of small additions of Vanadium on the austenitic grain size, forgeability, and impact properties of steel', *J. Iron Steel Inst.*, 202, pp. 128–134.

George T. and Irani J.J. (1968), 'Control of austenitic grain size by additions of Titanium', *J. Aust. Inst. Metals*, 13, pp. 94–106.

Gil-Sevillano J. (1993), *Materials Science and Technology*, eds R.W. Cahn *et al.*, Vol. 6, pp. 19–88.

Gladman T. (1989), 'Physical metallurgy of microalloyed medium carbon engineering steels', *Iron Making and Steel Making*, 16, pp. 241–245.

Gladman T. (1992), 'The theory and inhibition of abnormal grain growth in steels', *JOM*, 44, pp. 21–24.

Gladman T. and Pickering F.B. (1967), 'Grain-coarsening of austenite', *J. Iron Steel Inst.*, 205, pp. 653–664.

Guthrie R.I.L. and Jonas J.J. (1990), 'Steel processing technology', *Metals Handbook*, 10th edn, ASM, USA, 1: Properties and Selection: Irons, Steels and High Performance Alloys, pp. 107–125.

Hall E.O. (1951), 'The deformation and ageing of mild steel: III discussion of results', *Proc. Phys. Soc.*, series B, 64, pp. 747–753.

Hansen S.S., Van der Sande J.B. and Cohen M. (1980), 'Niobium carbonitride precipitation and austenite recrystallization in hot-rolled microalloyed steels', *Metall. Trans. ASM*, 11A, pp. 387–402.

Held J.F. (1969), 'Relationship between r-values and structure and properties of steel sheets', *in Mechanical Working and Steel Processing IV*, ed. D.A. Edgecombe, Gordon and Breach Science Publishers Inc., New York, USA, pp. 3–38.

Hillert M. (1965), 'On the theory on normal and abnormal grain growth', *Acta Metall.*, 13, pp. 227–238.

Hodgson P.D. (1993), 'Mathematical modelling of recrystallization processes during the hot rolling of steel', *Ph.D. Thesis*, University of Queensland, Australia.

Hodgson P.D. (1997), 'The metadynamic recrystallization of steels', *Proc. Int. Conf. Thermomechanical Processing of Steels and other Materials – THERMEC 97,* eds T. Chandra and T. Sakai, TMS, Wollongong, Australia, pp. 121–131.

Hodgson P.D. and Gibbs R.K. (1992), 'A mathematical model to predict the mechanical properties of hot rolled C-Mn and microalloyed steels', *ISIJ International*, 32, pp. 1329–1338.

Humphreys F.J. and Hatherly M. (2003), *Recrystallization and Related Annealing Phenomena*, 2nd edn, Elsevier Science, Oxford, UK.

Hunderi O. and Ryum N. (1982), 'On the stagnation of grain growth', *Acta Metall.*, 30, pp. 739–742.

Hutchinson W.B. (1984), 'Development and control of annealing textures in low-carbon steels', *Int. Met. Rev.*, 29, pp. 25–42.

Hutchinson W.B. and Ryde L. (1997), 'Principles and practice of texture control in cold rolled and annealed sheet steels', proc. international conference on *Thermomechanical Processing – Theory, Modeling and Practice [TMP]^2*, eds Hutchinson *et al.*, The Swedish Society for Materials Technology, Stockholm, Sweden, pp. 145–161.

Hutchinson W.B., Nilsson K-I. and Hirsch J. (1990), 'Annealing textures in ultra-low carbon steels', in *Metallurgy of Vacuum-Degassed Steel Products*, ed. R. Pradhan, TMS-AIME, Warrendale, USA, pp. 109–125.

Korchynsky M. (1987), 'Development of "controlled cooling" practice (a historical review)', *Proc. Int. Symp. on Accelerated Cooling of Steel*, ed. P.D. Southwick, Pittsburgh, USA, pp. 3–13.

Korchynsky M. (1999), 'New steels for new mills', *Scand. J. Metall.*, 28, pp. 40–45.

Korchynsky M. (2001), 'A new role for microalloyed steels: adding economic value', presented at Infacon 9 Conference, Quebec City, Canada.

Kozasu I., Ouchi C., Sampei T. and Okita T. (1975), 'Hot rolling as a high temperature thermo-mechanical process', *Proc. int. symp. Microalloying '75*, Union Carbide Corp., Washington D.C., pp. 120–135.

Kwon O. and DeArdo A.J. (1987), 'Suppression of static recrystallization by Nb (C, N) precipitation in HSLA steels', *Proc. Int. Conf. on Processing, Microstructure and Properties of HSLA Steels*, ed. A.J. DeArdo, TMS, Pittsburgh, USA, pp. 63–68.

Kwon O. and DeArdo A.J. (1991), 'Interactions between recrystallization and precipitation in hot-deformed microalloyed steels', *Acta Metall.*, 39, pp. 529–538.

Laasraoui A. and Jonas J.J. (1991), 'Prediction of temperature distribution, flow stress and microstructure during the multipass hot rolling of steel plate and strip', *ISIJ International*, 31, pp. 95–105.

Lenard J.G., Pietrzyk, M. and Cser L. (1999), *Mathematical and Physical Simulation of the Properties of Hot Rolled Products*. 1st edn, Elsevier Science, UK, p. 166.

Leslie W.C. (1981), *The Physical Metallurgy of Steels*, 1st edn, McGraw Hill, London, p. 251.

Maccagno T.M., Jonas J.J., Yue S., McCrady B.J., Slobodian R. and Deeks D. (1994), 'Determination of recrystallization stop temperature from rolling mill logs and comparison with laboratory simulation results', *ISIJ International*, 34, pp. 917–922.

Maccagno T.M., Jonas J.J. and Hodgson P.D. (1996), 'Spreadsheet modelling of grain size evolution during rod rolling', *ISIJ International*, 36, pp. 720–728.

Manohar P.A. (1997) 'Grain growth and continuous cooling transformation behaviour of microalloyed steels containing Ti, Nb, Mn and Mo', *Ph.D. Thesis*, University of Wollongong, Australia.

Manohar P.A., Dunne D.P., Chandra T. and Killmore C.R. (1996), 'Grain growth predictions in microalloyed steels', *ISIJ International*, 36, pp. 194–200.

Manohar P.A., Ferry M. and Hunter A. (2000), 'Direct strip casting of steel–historical perspective and future direction', *Mater. Forum*, 24, pp. 15–32.

Manohar P.A. and Chandra T. (2003a), 'The influence of $\alpha \rightarrow \gamma$ transformation on the grain coarsening behaviour of Ti–Nb microalloyed steels' *Proc. Mater. Sci. Tech. Conf. – MS&T '03*, TMS and ISS, Chicago.

Manohar P.A., Lim K.-H., Rollett A.D. and Lee Y. (2003b), 'Computational exploration of microstructural evolution in a medium C-Mn steel and applications to rod mill', *ISIJ International*, 43, pp. 1421–1430.

Medina S.F. and Hernandez C.A. (1996), 'General expression of the Zener Hollomon parameter as a function of the chemical composition of low alloy and microalloyed steels', *Acta Mater.*, 44, pp. 137–148.

Mueschenborn W., Imlau K.-P., Meyer L. and Schriver U. (1995), 'Recent developments in physical metallurgy and processing technology of microalloyed flat rolled steels', *Proc. Int. Conf. on Microalloying '95*, eds M. Korchynsky *et al.*, ISS, Pittsburgh, USA, pp. 35–48.

Mukunthan K., Strezov L., Mahapatra R. and Blejde W. (2000), 'Evolution of microstructures and product opportunities in low carbon steel strip casting', *J.K. Brimacombe Memorial Symp.*, ed. I. V. Samarasekera, Vancouver, Canada, p. 102.

Palmiere E.J. (1995), 'Precipitation phenomena in microalloyed steels', *Proc. Int. Conf. on Microalloying '95*, eds M. Korchynsky *et al.*, ISS, Pittsburgh, USA, pp. 307–320.

Palmiere E.J., Garcia C.I. and DeArdo A.J. (1994), 'Compositional and microstructural changes which attend reheating and grain coarsening in steels containing Nb', *Metall. Trans. ASM*, 25A, pp. 277–286.

Palmiere E.J., Garcia C.I. and DeArdo A.J. (1996), 'The influence of Niobium supersaturation in austenite on the static recrystallization behaviour of low carbon microalloyed steels', *Metall. Mater. Trans. ASM*, 27A, pp. 951–960.

Pereloma E.V. and Boyd J.D. (1996), 'Effects of simulated online accelerated cooling processing on transformation temperatures and microstructures in microalloyed steels. Part I. strip processing', *Mater. Sci. Tech.*, 12, pp. 808–817.

Petch N.J. (1953), 'The cleavage strength of polycrystals', *J. Iron Steel Inst.*, 174, pp. 25–28.

Peterson R.C. (1994), 'Titanium precipitation in hot charged and hot direct rolled microalloyed steels', *Ph.D. Thesis*, Monash University, Australia.

Pickering F.B. (1978), *Physical Metallurgy and the Design of Steels*, 1st edn, Applied Science, London, UK.

Ray R.K., Jonas J.J. and Hook R.E. (1994), 'Cold rolling and annealing textures in low carbon and extra low carbon steels', *Int. Mater. Rev.*, 39, pp. 129–172.

Reynolds W.T., Jr., Aaronson H.I. and Spanos G. (1991), 'A summary of the present diffusionist views on bianite', *Mater. Trans. JIM*, 32, pp. 737–746.

Riedl R. (1981), 'The determination of austenite grain size in ferrous metals', *Metallography*, 14, pp. 119–129.

Rodriguez R. and Gutierrez I. (2003), 'Unified formulation to predict the tensile curves of steels with different microstructures', *Mater. Sci. Forum*, 426–432, pp. 4525–4530.

Roucoules C., Hodgson P.D., Yue S. and Jonas J.J. (1994), 'Softening and microstructural change following the dynamic recrystallization in austenite', *Metall. Mater. Trans. ASM*, 25A, pp. 389–400.

Roucoules C., Yue S. and Jonas J.J. (1995), 'Effect of alloying elements on metadynamic recrystallization in HSLA steels', *Metall. Mater. Trans. ASM*, 26A, pp. 181–190.

Sakai T. (1995), 'Dynamic recrystallization microstructures under hot working conditions', *J. Mater. Proc. Tech.*, 53, pp. 349–361.

Samuel F.H., Yue S., Jonas J.J. and Barnes KR. (1990), 'Effect of dynamic recrystallization on microstructural evolution during strip rolling', *ISIJ International*, 30, pp. 216–225.

Sellars C.M. (1990), 'Modelling microstructural development during hot rolling', *Mater. Sci. Tech.*, 6, pp. 1072–1081.

Sellars C.M. and McTegart W.J. (1966), 'On the mechanism of hot deformation', *Acta Metall.*, 14, pp. 1136–1138.

Sellars C.M. and Whiteman J.A. (1979), 'Recrystallization and grain growth in hot rolling', *Met. Sci.*, 13, pp. 187–194.

Serby O.D., Walser B., Young C.M. and Cady E.M. (1975), 'Superplastic ultra-high-C steels', *Scripta Metall.*, Vol. 9, pp. 569–573.

Seto K. and Sakata K. (2003), 'Effect of grain refinement and its application to commercial steels', *Mater. Sci. Forum*, Vols. 426–432, pp. 1207–1212.

Siwecki T. and Engberg G. (1996), 'Recrystallization controlled rolling of steel', *Proc. Int. conf. on Thermomechanical Processing in Theory, Modelling and Practice*, eds W.B. Hutchinson *et al.*, The Swedish Society for Materials Technology, Stockholm, Sweden, pp. 121–144.

Siwecki T., Hutchinson W.B. and Zajac S. (1995), 'Recrystallization controlled rolling of HSLA steels', *Proc. Int. Conf. on Microalloying '95*, eds M. Korchynsky *et al.*, ISS, Pittsburgh, USA, pp. 197–211.

Speer J.G., Michael J.R. and Hansen S.S. (1987), 'Carbonitride precipitation in Nb/V microalloyed steels', *Metall. Trans. ASM*, 18A, pp. 211–222.

Steffen R. and Tacke K.-H. (1999), 'State of steel strip casting' (in German), *Stahl und Eisen*, 119(6/7), pp. 129, 236.

Takaki S. (2003), 'Limit of dislocation density and ultra-grain-refining on severe deformation in iron', *Mater. Sci. Forum*, 426–432, pp. 215–222.

Tamehiro H., Habu R., Yamada N., Matsuda H. and Nagumo M. (1987), 'Properties of large diameter linepipe steel produced by accelerated cooling after controlled rolling', *Proc. Int. Symp. on Accelerated Cooling of Steel*, ed. P.D. Southwick, Pittsburgh, TMS, USA, pp. 401–413.

Tanaka T., Funakoshi T., Ueda M., Tsuboi J., Yasuda T. and Utahashi C. (1975), 'Development of high-strength steel with good toughness at arctic temperatures for large-diameter linepipe', *Proc. Int. Symp. on Microalloying '75*, Union Carbide Corporation, Washington D.C., pp. 399–409.

Underwood E.E. (1968), *Quantitative Stereology*, Addison-Wesley, USA.

Webster D. and Allen G.B. (1962), 'Grain growth in Vanadium steels', *J. Iron Steel Inst.*, 200, pp. 520–526.

Woodfine B.C. (1953), 'Some aspects of temper-brittleness', *J. Iron Steel Inst.*, 173, pp. 240–255.

Zener C. (1948), referenced in: 'Grains, Phases and Interfaces: An interpretation of microstructure', by C.S. Smith, *Tran. AIME*, 175, pp. 15–49.

Zener C. and Hollomon J.H. (1944), 'Effect of strain rate upon plastic flow of steel', *J. Appl. Phys.*, 15, pp. 22–32.

4
Superalloys

S TIN, Cambridge University, UK

4.1 Introduction

As high-temperature structural materials, Ni-based alloys have a long and rich history. Compared to other metallic alloys, many commercially produced Ni-based alloys exhibit an unparalleled combination of high temperature strength, ductility and environmental resistance. The superior physical and mechanical properties of these Ni-based alloys at elevated temperatures justify classification of these truly unique materials as 'superalloys.'

Many of the initial advances in superalloy metallurgy have coincided with developments in turbine engine technology. Frank Whittle and Francis von Ohain working independently in the United Kingdom and Germany, respectively, simultaneously developed designs for jet engines in the 1940s. In these initial engine designs, components in the 'hot' section of the engine that converts the combustion gases into useful work energy were fabricated using early austenitic stainless steel alloys. Rex-78 (60Fe-18Ni-14Cr-4Cu-0.6Ti-4.0Mo-0.01C-0.015B wt.%) was used in many of Whittle's early designs, while the German manufactured turbine engines used Tinidur (53Fe-30Ni-15Cr-2Ti wt.%) for turbine blade components. Although these engine designs were demonstrated to work, the operational lifetimes were extremely short due to failure of these critical components. For these revolutionary engines to become commercially viable, engineers in both the UK and Germany set out in search of structural materials capable of surviving the extreme conditions in a turbine engine.

After World War II, strategic military and commercial applications of turbine engines were recognized and massive research efforts were initiated in the United States and England. Designs from Whittle's jet engine were exported to the United States where industrial machinery manufacturers General Electric and Pratt & Whitney began work on developing their own version of the turbine engine. Some of the earliest nickel-based superalloys to be developed specifically for these applications were Nimonic 80A and Inconel-X. Pratt & Whitney later developed their own superalloy variant,

Waspaloy, named after their famous Wasp engines[1]. The critical characteristic that distinguished these early Ni-base alloys was the strengthening effect associated with precipitation of intermetallic phases. As will be discussed in the following sections, the high-temperature structural properties associated with these multi-phase microstructures are truly unique as they are superior to either of the constituent phases considered individually.

4.2 Physical metallurgy

4.2.1 Superalloy chemistries

Many modern superalloys contain a multiplicity of 'major' alloying additions along with a number of 'minor' constituents as well[2,3] (Fig. 4.1). Each of the intentional elemental additions is intended to modify certain aspects of the microstructure or intrinsic property to yield a specific response. For example, additions of Al, Ta, Ti and Nb tend to promote the formation of stable intermetallic precipitates (Ni_3Al, Ti, Ta, Nb) based on the $L1_2$ crystal structure. These ordered precipitates, commonly referred to as the γ' phase, are considered to be the primary strengthening mechanism against creep deformation since they effectively inhibit dislocation motion. The mechanisms by which these precipitates prevent deformation will be described in greater detail in the following sections. In many of the early Ni-base superalloys, however, the levels of Al and Ti were limited since these reactive elements oxidized readily when melted in air. Consequently, metallurgists could not truly take advantage of the strengthening associated with γ' precipitation until commercialization of vacuum induction melting (VIM) equipment occurred in the 1950s.

A wide variety of elements are utilized to provide a high degree of solid-

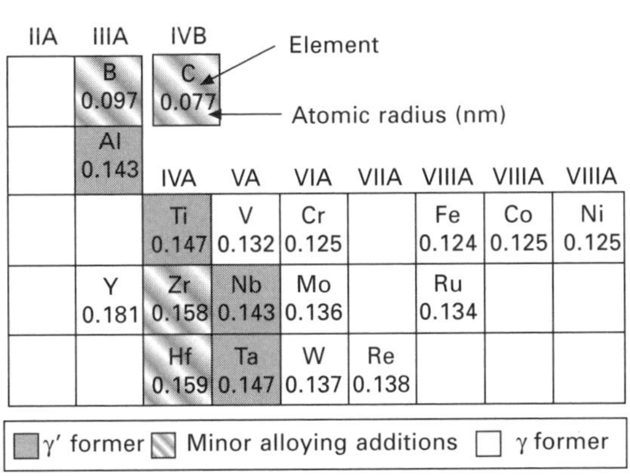

4.1 Alloying elements present in Ni-based superalloys (Adapted from ref. 2).

solutioning in Ni-based superalloys. Refractory additions of Co, Re, W, Mo, V, Cr and the platinum group metals are commonly added to provide solid solution strengthening of both the austenitic matrix and the γ' phase at elevated temperatures. Since the effectiveness of each elemental addition varies, many theories exist to describe the basis by which solid-solution strengthening occurs in Ni-base superalloys. The most compelling arguments suggest that the resulting increase in flow stress can be attributed to a combination of short-range ordering effects and local lattice and modulus distortions associated with the solute atoms in the austenitic matrix. Differences in solubility of the refractory elements results in a lesser degree of solid-solution strengthening in the ordered γ' phase and preferential chemical partitioning upon precipitation. In many instances, additions of certain elements can enhance more than one property of the alloy. For example, as well as strengthening the alloy, additions of Al and Cr also play a major role in determining the resistance of the alloy to oxidation and hot corrosion. Because of these subtle, yet important interactions, the physical metallurgy of Ni-base superalloys can be quite complex. Over the years, metallurgists have carefully tailored Ni-base superalloy compositions to provide an optimum range of properties for the intended application. Table 4.1 lists the nominal compositions of some commercial Ni-base superalloys.

'Minor' elemental additions also play a major role in superalloy metallurgy. During melting of Ni-based superalloys, unavoidable impurities present in the elemental additions or revert material in trace amounts are often found in the final product. The influence of alloy cleanliness on structural properties has been the subject of many detailed investigations.[4,5] Advances in vacuum melting technology have minimized the levels of undesirable low melting point elements[6] such as Pb, Bi, Se, Ag, As, Sb, Cu, Te and S. Sulfur levels, which are often the most difficult to control, are also the most deleterious impurity in terms of influencing high-temperature structural properties. Conventional methods of minimizing the content of S in the alloy consist of melting in MgO-Al_2O_3 lined crucibles to form MgS that can then be removed from the melt.

In addition to the unintentional presence of trace impurities, intentional additions of C, B, Hf, Zr, Y and La are often made to Ni-base superalloys. Some of the early superalloys, particularly the Co-based alloys, relied on carbide precipitation as the dominant strengthening mechanism. In Ni-based alloys, B, C, Zr and Hf strengthen grain boundaries and help control the presence of deleterious tramp elements during processing. When present within the alloy during vacuum induction melting, carbon reacts readily with oxygen to form carbon monoxide gas resulting in a 'carbon boil' reaction that deoxidizes the metal[7] and prepares the melt for alloying with Al, Ti and Hf (Fig. 4.2). These intentional additions of C, B, Hf and Zr result in the formation of carbides, and occasionally borides, at the grain boundaries of

Table 4.1 Compositions of commercial Ni-base superalloys (wt. %)

Alloy	Cr	Co	Mo	W	Ta	Re	Nb	Al	Ti	Hf	C	B	Y	Zr	Other
Conventionally cast alloys															
Mar-M246	8.3	10.0	0.7	10.0	3.0	–	–	5.5	1.0	1.50	0.14	0.02	–	0.05	–
IN-100	10.0	15.0	3.0	–	–	–	–	5.5	4.7	–	0.18	0.01	–	0.06	1.0 V
René 80	14.0	9.5	4.0	4.0	–	–	–	3.0	5.0	–	0.17	0.02	–	0.03	–
IN-713LC	12.0	–	4.5	–	–	–	2.0	5.9	0.6	–	0.05	0.01	–	0.10	–
C1023	15.5	10.0	8.5	–	–	–	–	4.2	3.6	–	0.16	0.01	–	–	–
Directionally solidified alloys															
IN792	12.6	9.0	1.9	4.3	4.3	–	–	3.4	4.0	1.00	0.09	0.02	–	0.06	–
GTD111	14.0	9.5	1.5	3.8	2.8	–	–	3.0	4.9	–	0.10	0.01	–	–	–
1st-generation single crystal alloys															
PWA 1480	10.0	5.0	–	4.0	12.0	–	–	5.0	1.5	–	–	–	–	–	–
René N4	9.8	7.5	1.5	6.0	4.8	–	0.5	4.2	3.5	0.15	0.05	0.00	–	–	–
CMSX-3	8.0	5.0	0.6	8.0	6.0	–	–	5.6	1.0	0.10	–	–	–	–	–
2nd-generation single crystal alloys															
PWA 1484	5.0	10.0	2.0	6.0	9.0	3.0	–	5.6	–	0.10	–	–	–	–	–
René N5	7.0	7.5	1.5	5.0	6.5	3.0	–	6.2	–	0.15	0.05	0.00	0.01	–	–
CMSX-4	6.5	9.0	0.6	6.0	6.5	3.0	–	5.6	1.0	0.10	–	–	–	–	–
3rd-generation single crystal alloys															
René N6	4.2	12.5	1.4	6.0	7.2	5.4	–	5.8	–	0.15	0.05	0.00	0.01	–	–
CMSX-10	2.0	3.0	0.4	5.0	8.0	6.0	0.1	5.7	0.2	0.03	–	–	–	–	–
Wrought superalloys															
IN 718	19.0	–	3.0	–	–	–	5.1	0.5	0.9	–	–	0.02	–	–	18.5Fe
René 41	19.0	11.0	10.0	–	–	–	–	1.5	3.1	–	0.09	0.005	–	–	–
Nimonic 80A	19.5	–	–	–	–	–	–	1.4	2.4	–	0.06	0.003	–	0.06	–
Waspaloy	19.5	13.5	4.3	–	–	–	–	1.3	3.0	–	0.08	0.006	–	–	–
Udimet 720	17.9	14.7	3.0	1.3	–	–	–	2.5	5.0	–	0.03	0.03	–	0.03	–
Powder processed superalloys															
René 95	13.0	8.0	3.5	3.5	–	–	3.5	3.5	2.5	–	0.065	0.013	–	0.05	–
René 88 DT	16.0	13.0	4.0	4.0	–	–	0.7	2.1	3.7	–	0.03	0.015	–	–	–
N18	11.2	15.6	6.5	–	–	–	–	4.4	4.4	0.5	0.02	0.015	–	0.03	–

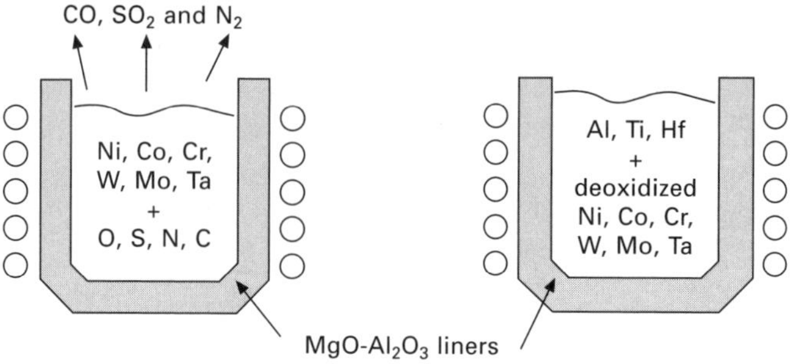

4.2 Schematic illustration of the VIM melting process of Ni-based superalloys. Reactive elements are added to the alloy melt after the levels of volatile trace elements are minimized and deoxidation is complete.

polycrystalline alloys. Carbon atoms exhibit a high affinity for elements such as Hf, Zr, Ta, Ti, Nb, W, Mo, V and Cr, and tend to form primary MC (where M = metal atom) carbides directly from the liquid during solidification of Ni-based superalloys. Some general characteristics of selected MC carbide phases are listed in Table 4.2. Although carbides are thought to affect the fatigue properties of the alloy, the presence of discrete carbides at the grain boundaries inhibits sliding and damage accumulation during high temperature creep.

4.2.2 Microstructural features and strengthening mechanisms

Despite the large number of elemental alloying additions contained within most commercial Ni-based superalloy compositions, the microstructure of these alloys is typically comprised of only two major phases, ordered intermetallic γ' precipitates contained within a austenitic FCC (A1) matrix.

Table 4.2 Characteristic properties of selected refractory carbides

Carbide	Crystal structure	ΔH at 298 K (eV/atom)	Density (g/cm^3)	Lattice parameter (Angstroms)
TiC	B1 (NaCl)	1.91	4.91	4.328
ZrC	B1 (NaCl)	2.04	6.59	4.698
HfC	B1 (NaCl)	2.17	12.67	4.640
VC	B1 (NaCl)	1.06	5.05	4.166
NbC	B1 (NaCl)	1.46	7.79	4.470
TaC	B1 (NaCl)	1.48	14.50	4.456
CrC	B1 (NaCl)	–0.01		
MoC	Hexagonal	0.13	9.06	a: 2.932 c: 10.97
WC	Hexagonal	0.42	15.80	a: 2.906 c:2.837

In multicomponent superalloys, solid-solution strengthening occurs within the γ' phase. When present in the intermetallic phase, refractory elements, W, Re, Mo, Nb, Ta and Ti all tend preferentially to displace the Al atoms that occupy the corner sites in the $L1_2$ structure, while additions of Cr, Co and Fe substitute directly for Ni in the ordered structure.[8,9] Although the γ' precipitates and austenitic matrix are structurally and compositionally different, enough similarity exists between the $L1_2$ and Al crystal structures to enable the precipitates to remain coherent within the matrix. The minor differences in lattice parameters result in misfit strains that are often quantified as:

$$\delta = \frac{a_{\gamma'} - a_{\gamma}}{\frac{1}{2}(a_{\gamma'} + a_{\gamma})}$$

where $a_{\gamma'}$ and a_{γ} are the lattice parameters of the γ' and γ phases respectively. Due to the misfit stresses and elastic interactions with neighbouring precipitates, alloys containing high volume fractions of γ' often exhibit cuboidal precipitate morphologies (Fig. 4.3).

Although some of the high-temperature strength of Ni-base superalloys can be attributed to solid-solution strengthening, the dominant strengthening mechanism occurs via precipitation hardening. Unlike engineered composite systems which often adhere to a conventional rule of mixtures, the structural properties of $\gamma - \gamma'$ microstructures are superior to either of the constituent γ or γ' phases considered individually[10] (Fig. 4.4). The mechanical properties of Ni-based superalloys are therefore strongly dependent on the various parameters influencing γ' volume fraction, size and morphology.

The volume fraction of the γ' phase contained within the γ' matrix is

4.3 Microstructure of a Ni-based superalloy single crystal CMSX-4 revealing the cuboidal morphology of the γ' precipitates (Courtesy of M. Hook).

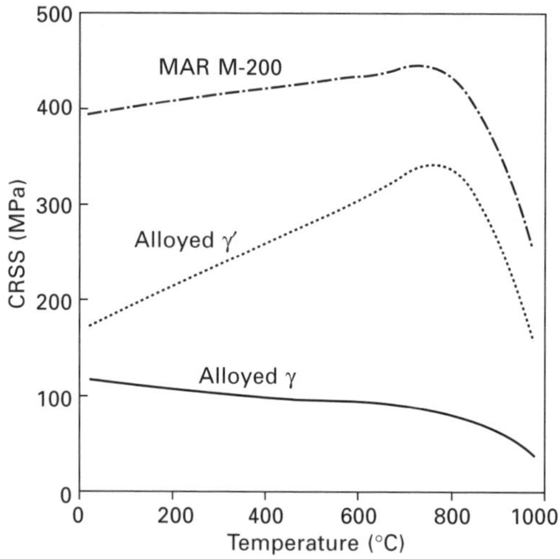

4.4 Comparison of the critical resolved shear stress (CRSS) corresponding to the Ni-based superalloy MAR M-200 and the individual constituent phases (Adapted from Ref. 10).

primarily dependent upon the content of Al and Ti within the alloy. Alloys with low overall levels of Al and Ti contain low volume fractions (~20%) of γ′. These superalloys are generally polycrystalline and used in applications where ductility and weldability are desirable attributes. The γ′ precipitates present in these alloys are not stable at elevated temperatures and typically exhibit γ′ solvus temperatures well below the solidus temperature of the alloy. When cooled at sufficiently high rates upon solidification, precipitation of γ′ can be suppressed resulting in a supersaturated single-phase microstructure. Post solidification thermal ageing treatments are required for precipitation of a uniform dispersion of spherical intragranular γ′, or precipitates contained within the grain (Fig. 4.5). In these low volume fraction γ′ alloys, the size or radius of the precipitates is often dependent upon the ageing temperatures and times.

To maximize high-temperature strength, the volume fraction of the intermetallic precipitates often exceeds the volume fraction of the matrix. At room temperature, the volume fraction of γ′ contained within the microstructure of advanced polycrystalline and single crystal Ni-base superalloys are typically of the order of 60% and 75%, respectively. γ′ precipitates in superalloys containing large volume fractions of the intermetallic phase are typically stable at temperatures approaching the solidus temperature of the alloy. Consequently, γ′ precipitation in cast alloys often occurs upon cooling over a range of temperatures during solidification.[11] As these alloys solidify dendritically, chemical segregation occurs as the solid-liquid interfaces and elements that partition preferentially into the γ′ phase are accumulated in the

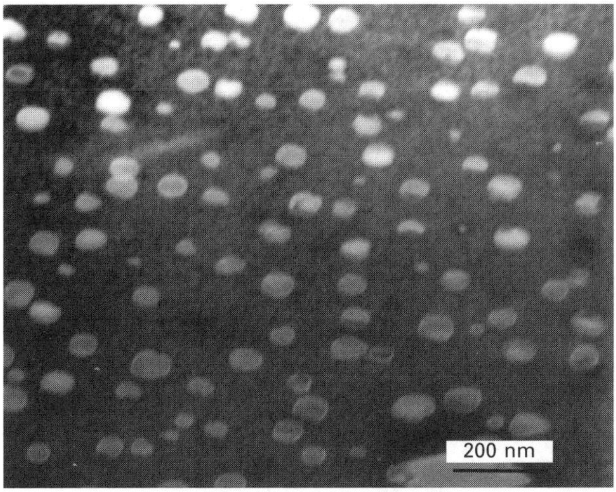

(a)

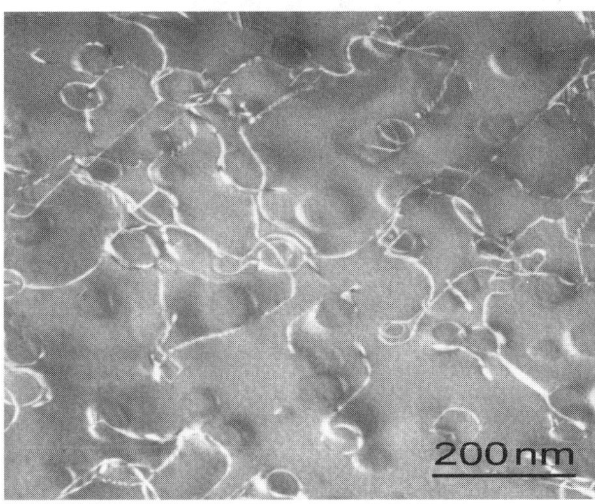

(b)

4.5 Spherical morphologies of γ′ precipitates in Ni-based superalloy C263 (a) in the aged condition and (b) after creep deformation.

solute. If levels of Al, Ti, Ta and Hf are sufficiently high, enrichment of the solute with these elements enables the formation of γ′ eutectic pools within the interdendritic regions of the microstructure (Fig. 4.6). Within the dendritic core, the γ′ sizes and morphologies in as-solidified components are dependent upon the cooling rate from the solutioning temperature, with high cooling rates resulting in finer precipitates and low cooling rates yielding coarse precipitates. To optimize the structural response of these materials, solution heat-treatments are applied to refine the γ′ precipitates and yield a uniform distribution.[12]

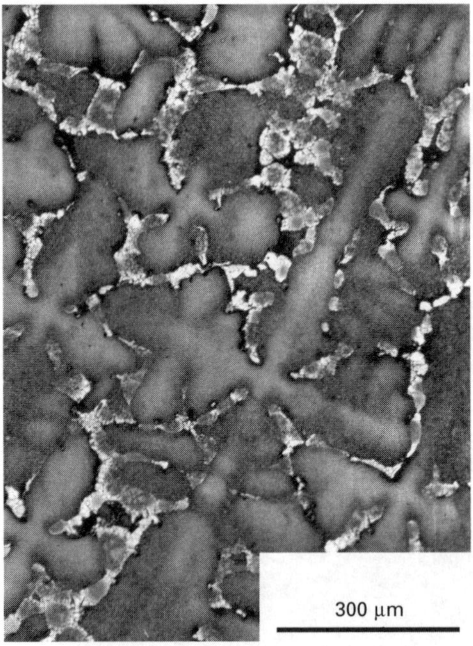

4.6 Chemical heterogeneities due to microsegregation exist within the dendritic structure of Ni-based superalloys. Eutectic pools of γ – γ' are observed within the interdendritic regions of the microstructure.

Carbides

As mentioned in the previous section, the presence of carbon in Ni-based superalloys often results in the formation of primary MC carbides. These MC carbides form directly from the liquid during solidification and are commonly located along the grain boundaries or within the interdendritic regions of the microstructure. Depending on the composition of the primary MC carbide and the constituent elements present in the alloy, a solid-state transformation may decompose the MC carbide into a variety of $M_{23}C_6$, M_6C and M_7C_3 carbides.[13] For example, Cr from the matrix can potentially react with the less stable TiC and NbC carbides to form a series of $M_{23}C_6$, M_6C and M_7C_3 carbides. As the primary MC carbides are consumed, the depletion of Cr from the matrix surrounding the carbides results in the formation of a layer of γ' at the carbide interface. Since these carbides occupy a significantly larger volume along the grain boundaries and are often interconnected, potential degradation of the mechanical properties may occur when the MC carbides undergo this phase transformation (Fig. 4.7). Primary HfC and TaC carbides are typically stable at elevated temperatures and are resistant to the phase transformations.

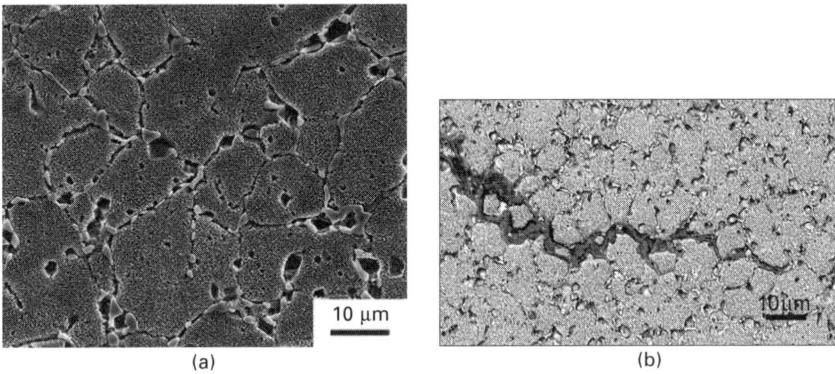

4.7 (a) Primary γ' and $M_{23}C_6$ carbides along the grain boundaries of a polycrystalline Ni-based superalloy. (b) Fatigue crack growth along the film of $M_{23}C_6$ carbides at the grain boundaries in a powder processed Ni-based superalloy.

The morphologies of the primary MC carbides are governed by both the composition and solidification rate of the bulk alloy.[14] As seen in Figs 4.8 to 4.10, three distinct carbide morphologies have been observed in Ni-based superalloys: blocky, script and nodular.[15] Blocky and script morphologies are common to all Ni-based superalloys, while the nodular carbides are unique to Ta-bearing superalloys. When present in the as-cast structure, script and nodular carbides are confined to the interdendritic regions and exhibit a strong orientation relationship with the $\gamma - \gamma'$ matrix, while blocky carbides are randomly orientated along the grain boundaries or within the interdendritic regions. Large thermal gradients during solidification tend to refine the size of the blocky carbides and suppress the formation of the intricate script and nodular carbide morphologies.

4.8 Primary MC carbides found within Ni-based superalloys exhibiting a faceted blocky morphology.

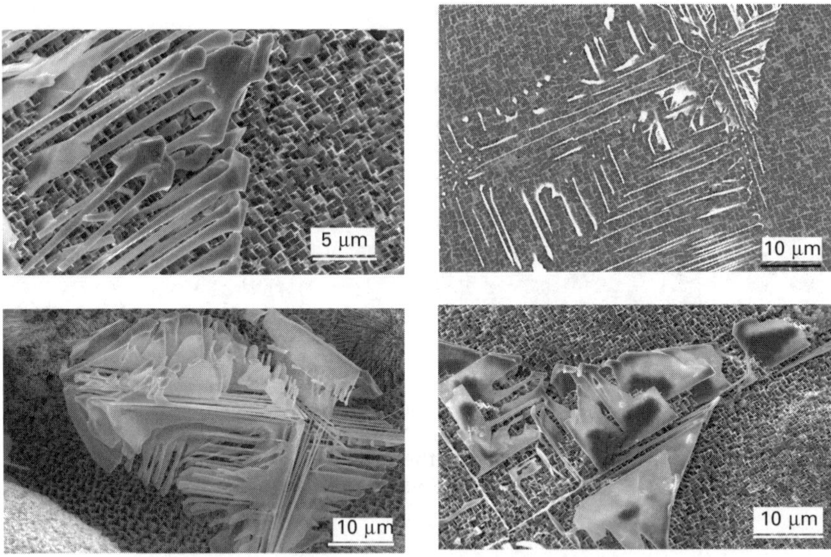

4.9 SEM micrographs of lacy script carbides contained within the interdendritic regions of the carbon-containing superalloys.

4.10 SEM micrograph of carbides with uniformly spaced submicron nodules on the surfaces.

Topologically close packed phases

In addition to the phase transformations encountered in certain carbides after long-term exposures to high temperature and high stress environments, equilibrium intermetallic phases may also form within the $\gamma - \gamma'$ microstructure of these complex multicomponent Ni-base alloys. It was first observed in the 1960s that precipitation of intermetallic sigma platelets in alloy IN-100 resulted in significant reductions in mechanical properties and led to premature failure of the component.[16–18] These observations triggered a series of investigations that provided a greater understanding of phase stability in these alloys. Suppressed during solidification, the high-temperature equilibrium phases are generally intermetallics exhibiting a topologically close-packed (TCP) crystal structure. The TCP phases (P, μ, R and σ) are typically characterized by close-packed layers of atoms (atomic coordination number >12) forming 'basket weave' sheets that are often aligned with the octahedral planes in the FCC nickel matrix (Fig. 4.11). Since these precipitates are stabilized by the refractory alloying elements (W, Re, Mo), the precipitation of these brittle intermetallic phases results in the depletion of potent solid solution strengtheners from the surrounding matrix. Similarities in the composition and crystallography of the various TCP phases allows these precipitates to develop as mixed structures consisting of a number of different phases.[19] Although the chemistries of the TCP phases are dependent upon the composition of the parent alloy, these phases can be characterized based upon their crystallographic structures. The μ and R phase precipitates have a rhombohedral crystal structure while the σ phase is tetragonal and the P phase is orthorhombic. Serving as effective crack initiators, the presence of these brittle intermetallic TCP phases in the $\gamma - \gamma'$ microstructure is highly undesirable.

A unique sub-class of superalloys are the nickel-iron superalloys which typically rely on additions of Nb for high temperature strength. Alloys such as Inconel 718 and 706 contain significant levels of iron that are intentionally added to reduce the overall levels of nickel and cobalt. Since this results in a significant reduction in the cost of the alloy, many of the Ni-Fe superalloys are used in applications where the need for extreme high-temperature strength is not required. In terms of volume, a large majority of the commercial superalloys market is comprised of these iron-containing Ni-base superalloys.

The microstructures of these Ni-Fe-Nb alloys are highly complex and multiple intermetallic phases may exist within the microstructure of the

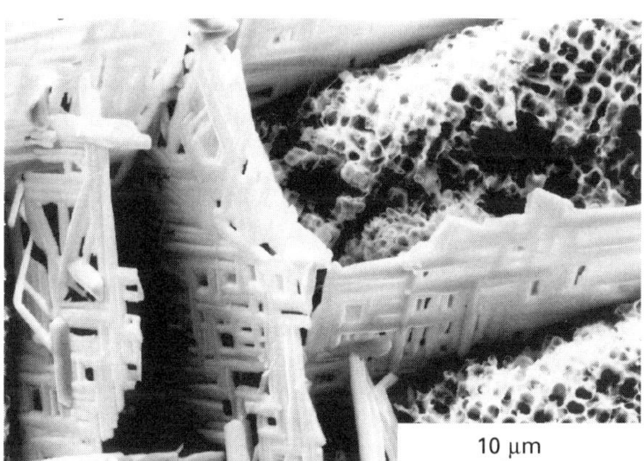

4.11 Ni-base superalloys containing elevated levels of refractory elements are prone to the precipitation of various TCP phases when exposed to elevated temperatures.

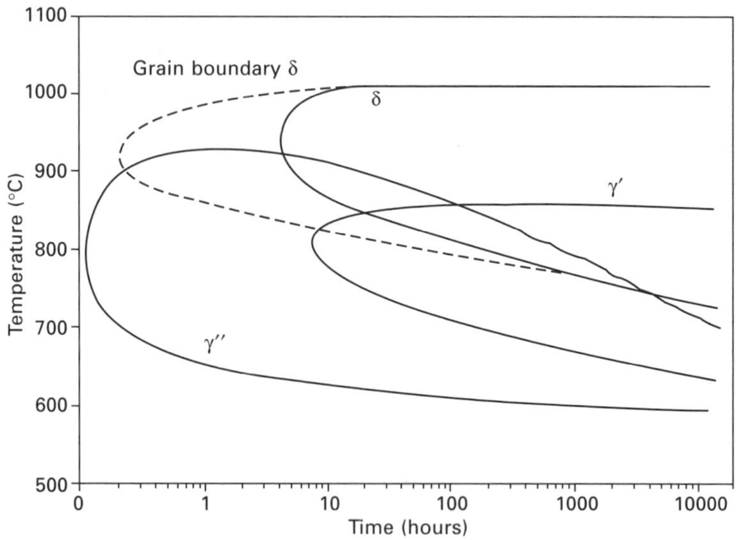

4.12 Time-temperature-transformation diagram showing the precipitation kinetics of the major phases in IN-718.

alloy (Fig. 4.12). The combination of Fe and Nb present within the austenitic Ni-matrix leads to the precipitation of both γ′ and Ni_3Nb. Interestingly, these Ni_3Nb precipitates may exist as two distinct phases. At relatively high temperatures, Ni_3Nb precipitation occurs at the grain boundaries with the δ precipitates exhibiting an orthorhombic crystal structure. Lower temperatures promote the formation of coherent disc-shaped precipitates, γ″, with a body-centred-tetragonal (BCT) $D0_{22}$ crystal structure within the γ matrix. Despite the low overall volume fraction, the low γ″ coarsening rates at service temperatures combined with the misfit strains provide an unusually high degree of strengthening in Ni-Fe superalloys.[20] Due to the phase instabilities associated with the high levels of Fe present within these alloys, useful service temperatures are limited to below 650 °C. Above this temperature, the structural properties rapidly degrade as TCP phases form and microstructural changes occur.[18]

4.3 Polycrystalline superalloys

As high-temperature structural materials, Ni-based superalloys exhibit a balanced set of inherently desirable properties. Although selected ceramic matrix composites and advanced intermetallics may offer improved resistance to environmental degradation and high-temperature strength, the implementation of these materials as substitutes for Ni-based superalloys in high-temperature structural applications becomes difficult to justify when other important design criteria, such as ductility, thermal conductivity and cost, are taken into consideration. Consequently, Ni-base superalloys are extensively used in applications where resistance to high-temperature deformation and/or environmental interactions are required.

4.3.1 Applications of polycrystalline superalloys

The use of polycrystalline Ni-base superalloys is widespread over a range of applications and many alloys have been exclusively engineered to yield specific properties. Today, there are hundreds of distinct wrought and cast Ni-base superalloy compositions that are commercially available (Table 4.1). Having been developed exclusively for gas turbine engines, the largest application of Ni-base superalloys is still for components used for aircraft and industrial gas turbines. However, these materials are also used in space vehicles, nuclear reactors, steam power plants, petrochemical processing facilities and any other applications where the structural and/or environmental properties of stainless steel are not sufficient.

4.3.2 Processing

Processing of polycrystalline Ni-based superalloy components begins with vacuum induction melting (VIM) in a refractory crucible to consolidate elemental or revert materials to form a base alloy. Although selected alloys can potentially be melted in air/slag environments using electric arc furnaces, VIM melting of superalloys is much more effective in the removal of low melting point trace contaminants. Following the vaporization of the contaminants, the carbon boil reaction is utilized to deoxidize the melt prior to the addition of the reactive γ' forming elements, Ti, Al and Hf. Once the desired alloy composition of the VIM ingot is attained, the solidified ingot is then subsequently subjected to additional melting or consolidation processes that are dependent upon the final application of the material.

Cast and wrought alloys

Conventionally cast

A variety of polycrystalline components are manufactured using conventional casting techniques. Conventional casting enables the production of complex near net-shape part geometries for structural applications, such as low-pressure turbine blades and nozzle guide vanes. Production of these components involves remelting of the alloyed VIM ingot and casting into a ceramic investment mould under vacuum. Due to the reactive nature of molten nickel and many of the constituent alloying elements, a vacuum or inert gas environment is required when casting Ni-base superalloys. This prevents the formation of oxides and/or nitrides, and assists in maintaining the compositional specifications of the alloy. Upon solidification, the radial heat extraction from the surfaces of the mould result in an equiaxed polycrystalline structure. Although cast polycrystalline components exhibit a substantially reduced temperature capability when compared to single crystal components, these components are much more cost-efficient and are suitable for a wide range of applications throughout the gas turbine engine.

Vacuum arc remelting

The high-temperature structural properties of Ni-base superalloys are highly sensitive to microstructural variations, chemical inhomogeneities and inclusions. As ingot sizes increase, VIM melting often results in macrosegregation or the formation of large shrinkage cavities during solidification. The formation of these solidification defects is caused by large-scale solute segregation associated with dendritic solidification under low thermal gradients. Since heat transfer during solidification of VIM ingots is limited by the low intrinsic thermal conductivity of the solidifying mass, large ingots are highly prone to the formation of these features. The charge weights of VIM ingots range from ~2,500 kg to in excess of 27,500 kg.[1] Moreover, vacuum levels associated with VIM melting are often sufficiently low that the trace element contaminants can further be reduced. For the production of critical rotating components, such as turbine discs, vacuum arc remelting (VAR) is used to refine the ingot and eliminate macrosegregation. Consumable electrodes (30 cm to 50 cm in diameter) cast from the VIM charge are remelted into a water-cooled copper crucible. Opposed to the VIM process in which the entire charge of the alloy is molten and allowed to solidify, VAR involves only localized melting of the electrode tip (Fig. 4.13). Melt rates of VAR are in the order of ~0.5 to 1 kg/s. Defect features, such as macrosegregation and shrinkage, are effectively minimized as high thermal gradients are maintained during solidification of the comparatively smaller melt pool. Processing parameters are selected such that the melt pool exhibits a steady state in size and shape.

Although VAR can effectively eliminate certain undesirable features in the VIM ingot, the remelt process may introduce potential inclusions into the finished ingot. Inclusions in the VAR process may be classified into two groups, extrinsic and intrinsic. Extrinsic inclusions can come from a variety of sources. Incomplete removal of refractory ceramic particles and agglomerates of oxides and nitrides in the revert material used during VIM melting may enable these inclusions to be present in the remelted ingot. As the surface of the VIM ingot is machined to form a consumable electrode, fragments of tungsten carbide cutting tools can be embedded within the ingot. Steel shot

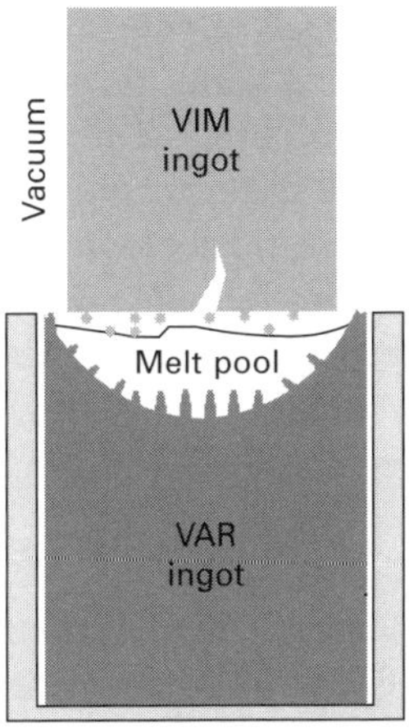

4.13 Schematic of the VAR process.

used to clean the copper crucible and splash from the previous melt in the VIM crucible may also potentially serve as extrinsic inclusion. With clean melting practices and stringent quality control measures, many of these extrinsic inclusions can be minimized. Intrinsic inclusions, however, are much more difficult to control during processing and are often dependent upon the chemistry of the alloy.

Thermal and compositional perturbations in the mushy zone during solidification lead to the formation of microstructural defects, such as freckles and white spots (Fig. 4.14). In VAR ingots, freckle defects consist of chains of equiaxed grains aligned parallel to the melt pool profile or solidification direction. Highly enriched with solute, freckle chains are compositionally different from the bulk alloy and form as a result of thermosolutal convection.[21] As with many multi-component superalloys, as solute accumulates within the mushy zone during dendritic solidification, the subsequent density imbalance between the solute and bulk liquid serves as a driving force for

4.14 Freckles and white spot defects in IN718.

the onset of convective fluid flow. Upon cooling, the solute-enriched convective instabilities solidify as isolated regions of equiaxed grains. White spots are discrete features in the superalloy billet that are observed after chemical etching.[22] Although compositionally similar, these features are typically less heavily alloyed than the superalloy matrix. Compositions of white spots and freckles in IN718 are listed in Table 4.3. White spot formation is commonly attributed to the entrapment of fragments from the melting electrode or crown of the solidifying shelf.

Electro-slag remelting

For the production of high-grade superalloy billets with minimal sulfur levels and inclusion content, the VIM ingot can be further refined using an electro-slag remelting (ESR) process.[23,24] Similar to the VAR process, consumable electrodes, measuring 60 cm to 80 cm in diameter, are cast from the VIM melt. Instead of electrical arcing, however, resistance heating from the molten slag is used to melt the electrode. Macrosegregation and chemical heterogeneities are minimized when solidification is restricted to a comparatively small volume of molten metal. During ESR, droplets of molten metal are passed through a layer of CaO-MgO-CaF_2-Al_2O_3 slag prior to accumulating in the melt pool (Fig. 4.15). The slag resides on the surface of the melt pool and effectively removes the residual sulfur and traps ceramic inclusions that are drawn out to the surface. Unlike VAR processes that are restricted to round ingot geometries, ESR processes can be adapted to yield shaped ingots, such as rectangular slabs for sheet production. ESR can also be used to refine alloy ingots melted in air.

In addition to the potential for the entrapment of slag within the ingot, a number of limitations are associated with ESR processing. The protective layer of slag resident on the surface of the melt pool enables the melting process to occur in atmosphere and a vacuum environment is unnecessary. Although protective, the layer of hot molten slag is also highly insulating. Consequently when compared to VAR, solidification within the melt pool occurs under slightly lower thermal gradients as the pool volume increases. As the size of the melt pool increases, the ESR ingot becomes increasingly susceptible to the onset of thermosolutal convection and the formation of freckle defects. In practice, ESR is limited to the production of moderately

Table 4.3 Compositions of freckles and white spot defects in IN718

	Al	Ti	Cr	Fe	Ni	Nb	Mo	Si
Nominal IN718	0.5	0.9	19.0	18.5	Bal	5.1	3.0	0.2
Freckle	0.43	1.33	17.4	15.2	Bal	9.43	3.51	0.16
Whites spot	0.41	0.62	17.7	19.2	Bal	2.96	3.2	0.19

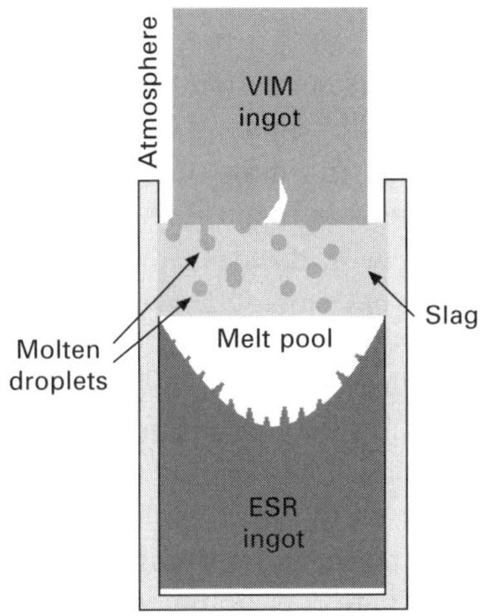

4.15 Schematic of the ESR process.

sized ingots or slabs for the manufacture of bar or sheet products. The associated changes in the melt pool shape also induce the formation of textured columnar grains within the resultant microstructure of the ESR ingot. The anisotropy associated with the aligned microstructure causes ESR ingots to be less compliant during forging and other hot working processes. Moreover, compositional changes also occur during ESR melting. Volatile elements, such as Ti, Al and Hf, can react with the atmosphere or constituents in the slag and result in appreciable losses.

Electron beam cold hearth refining

Due to the low operational costs associated with VIM, VAR and ESR, these processes are widely used for most commercial applications. In recent years, advances in melting technology have led to the development of electron beam cold hearth refining (EBCHR).[25] In this complex melting process, alloy ingots are placed into a high vacuum chamber and a series of electron beams are used to melt the ingot. Before the molten metal is allowed to fill the casting mould, the melt flows down a water-cooled copper hearth. As the depth of the flowing liquid pool is shallow, inclusions in the melt rapidly rise to the surface and can be skimmed off prior to filling of the mould. This process is extremely effective in minimizing inclusion sizes as ingots produced using EBCHR have been analyzed to contain oxide particles with a maximum size of ~25 μm. For comparison, inclusions in VIM/VAR ingots are often an order of magnitude larger at ~230 μm.

Homogenization of cast components

Solidification under conventional processing conditions results in segregation within the dendritic microstructure of all multicomponent Ni-based superalloys. Microsegregation, or the preferential partitioning of refractory alloying additions and γ' forming elements to microstructural features such as the dendritic or interdendritic regions, results in localized chemical heterogeneities that impact the resultant mechanical properties. Often used in the as-cast condition, conventionally cast near net-shape components are occasionally solution heat-treated prior to use in service. A solution-heat treatment above the γ' solvus temperature tends to minimize the compositional differences in the microstructure. Eutectic features, such as pools of γ' eutectic and laves phases are reincorporated into the bulk alloy. Upon cooling or subsequent ageing treatments, precipitation of the intermetallic strengthening phases is controlled such that the size and distribution is optimized throughout the microstructure of the conventionally cast component.

Homogenization is also crucial for the conversion of unfinished cast ingots into billet, sheet or rod via hot working. Variations in composition and constituent phases within the microstructure lead to anisotropic behaviour during deformation. As the main aim of hot working processes is to attain a uniform microstructure, the chemical heterogeneity of the as-cast ingot is not conducive to this transformation. Thus, despite the large degree of grain growth experienced in the ingot during homogenization, minimization of the microsegregation features results in the ingot being more amenable for hot working.

Powder metallurgy alloys

In order to increase the strength of polycrystalline Ni-base superalloys, levels of refractory alloying additions and γ' forming elements have gradually increased to levels that make conventional processing routes deficient.[26] Elements such as W, Mo, Ti, Ta, Nb effectively strengthen the alloy, but also result in severe segregation within the ingot upon solidification. Additionally, the limited ductility of the high strength alloys renders the ingot susceptible to cracking as thermally induced stresses evolve during cooling. Powder processing routes have been developed to overcome the difficulties associated with melt-related defects and are viable for the production of advanced high strength polycrystalline superalloy components. Also listed in Table 4.1 are the compositions of some commercially available powder processed Ni-based superalloys.

Powder processing begins with gas or vacuum atomization of a highly alloyed VIM ingot. Rapid solidification of the fine powders effectively suppresses macrosegregation within the alloy. Since the low ductility associated with the correspondingly high strength causes many of these advanced superalloys to be highly sensitive to initial flaw sizes, the atomized powders are separated based on particle size. Standard 150 or 270 meshes are used to separate the powders into sizes >100 μm and >50 μm respectively. Powder sizes directly influence the initial potential crack size present in the finished component. Although finer powder sizes are desired to minimize initial defect sizes, costs increase substantially as yields are substantially reduced.

Once powders are collected into steel cans, the cans are evacuated under vacuum and sealed. The cans are then hot isostatically pressed (HIP) or extruded to consolidate the powder. The HIP process consists of heating the alloy to just below the γ' solvus temperature under a hydrostatic pressure of up to 310 MPa. After four to five hours, diffusion bonding and sintering of the powders under pressure yields a fully dense superalloy billet. Billet sizes are limited by the capacity of the HIP furnace, however, systems capable of forming billets up to 150 cm diameter and 300 cm height are available. Consolidation under hot extrusion is often preferred over HIP due to the ability to produce fine grained structures (ASTM 12) and reduce effects associated with prior particle boundaries. The evacuated can containing the superalloy powder is hot extruded through a set of dies that greatly reduces the diameter. During this thermo-mechanical process, the individual powder particles are subjected to deformation during sintering and any oxide films initially present on the surfaces of the powder are broken up. Since substantial plastic deformation and adiabatic heating occurs during this process, hot extrusion temperatures are selected such that temperatures are maintained below the γ' solvus temperature.

4.3.3 Deformation

Hot working

Forging and cogging are common hot working processes by which superalloy ingots are converted into useful structural components. Due to the high intrinsic strength of Ni-based superalloys, forming of these materials generally occurs at high temperatures (~1000 °C). Hot working processes are primarily designed to refine the microstructure to yield isotropic properties and attain a near net shape component. Microstructures in the homogenized ingot are typically extremely large (grain sizes >10 mm) and often have a residual columnar-grained structure. Ideally, depending on the application, uniform equiaxed grain sizes of the order of ASTM 12 to 6 (5–50 μm diameter) are desired in the forged components. Conversion of the original microstructure into the fine-grained structure is achieved via dynamic and meta-dynamic recrystallization during and post-hot working, respectively.[27,28] Process variables, such as strain, strain rate, die and workpiece temperature are carefully controlled such that complete recrystallization occurs throughout the material and a uniform microstructure is attained (Fig. 4.16). Superalloy sheets or small diameter billets (up to 13 cm) can be rolled or forged directly from cast slabs or bars. Prior to the forging of large net shape superalloy discs for turbine engine applications, large homogenized ingots (36–60 cm diameter) are cogged, or hot worked through a set of open dies, to form billets measuring 15–40 cm. The cogging process assists in breaking down the initial grain structure such that complete recrystallization occurs in the final forging.

One of the major advantages of powder processed superalloy billets is the initial starting microstructure. Due to the rapid solidification of the powder particles, homogenization heat treatments are not required due to the negligible amount of elemental segregation within the microstructure. Powder consolidation processes also tend to yield a fine equiaxed microstructure in the billet ideal for the direct production of net shape components (Fig. 4.17). Conventional forging practices, however, are not ideal for powder processed Ni-based superalloys. Due to the limited ductility associated with the high levels of refractory alloying elements, isothermal forging of these alloys is generally required. Opposed to conventional forging processes where dies are often cold or warm and strain rates are high, dies used for isothermal forging are at the same temperature as the billet. In most instances, isothermal

4.16 Images showing the volume fraction of dynamically recrystallized microstructure for IN718 as a function of strain (a) $\varepsilon = 0.1$ (b) $\varepsilon = 0.3$ (c) $\varepsilon = 0.5$ (d) $\varepsilon = 0.8$ at 1253 K and $\dot{\varepsilon} = 0.1\text{s}^{-1}$.

(a)

(b)

4.17 (a) Sub-solvus heat-treated and (b) super-solvus heat-treated microstructures of powder processed Ni-based superalloys containing elevated levels of refractory alloying additions.

forging temperatures are maintained just below the γ' solvus temperature. This enables the fine initial grain sizes to be retained throughout the forging process. The high temperatures combined with the fine grain size and characteristically low strain rates enable the high strength superalloy to be formed via superplastic flow during isothermal forging.

4.3.4 Properties

The high-temperature mechanical properties of polycrystalline Ni-based superalloys are greatly dependent upon the microstructure of the component. In addition to the volume fraction, the shape and morphology of the γ' or γ'' precipitates play a major role in determining the strength of the alloy. While forging and hot deformation parameters control the grain size of the component, the shape and distribution of the strengthening precipitates within the individual grains are carefully controlled via a series of thermal treatments. Precipitation of intermetallic phases typically occurs preferentially at the grain boundaries in superalloy forgings. The grain boundary precipitates play an important role during processing since they effectively pin the boundaries and prevent undesirable growth of the recrystallized grain structure at elevated temperatures. Hence, final heat treatment temperatures are often specified to occur just below the solvus temperature of the grain boundary precipitates followed by a rapid quench. This process optimizes the strength of the alloy since it enables precipitation of γ' or γ'' within the grains and minimizes the extent of grain growth. Due to typical sub-surface cooling rates of 1–2 °C/s in large structural forgings, a bimodal distribution of intragranular γ' is often observed in polycrystalline superalloys with high levels of γ' formers.[29] When high-temperature creep resistance is desired, the forged components are solution treated above the solvus temperature to produce a coarse grained structure with a larger fraction of intragranular γ'. Following the quench, all components are subject to an ageing or annealing treatment to refine further the shape of the intermetallic precipitates and relieve any residual stresses induced by the thermal contractions associated with the high cooling rates.

Properties of polycrystalline Ni-base superalloys are often tailored specifically for the intended application. Many of the Ni-Fe superalloys are limited to temperatures below ~650 °C, while some of the advanced powder processed alloys are capable of operating at peak temperatures of ~750 °C. At higher temperatures, resistance to creep deformation in these alloys becomes limited due to precipitation of TCP phases or damage accumulation along the grain boundaries. Thus, polycrystalline alloys are commonly used in turbine engine components where temperatures remain relatively low, but high strength and resistance to hot corrosion resistance is required. After hot working, components also exhibit isotropic mechanical properties that are highly important when used as rotating components, such as turbine discs.

Over the range of temperatures and stresses incurred in service applications, the strength of Ni-based superalloys is largely dependent upon the interaction of dislocations with microstructural features. As dislocations along the active <110>{111} slip systems in the FCC matrix encounter the ordered coherent particles, differences in the magnitude of the Burgers vector between the precipitate and matrix prevent them from readily penetrating and shearing the intermetallic precipitates. Sufficiently high stresses are required for dislocations to penetrate the particles since the process of dislocation shearing disrupts the stacking sequence of the ordered structure and results in the formation of an anti-phase boundary (APB). Passage of another dislocation along the same slip plane will restore the ordering sequence in the intermetallic structure and lower the overall energy state of the system. Thus, shearing of the precipitates often occurs via 'strongly or weakly coupled' dislocation pairs.[30]

During deformation dislocations pile up at the matrix-particle interface if the resolved shear stresses are not sufficient to cause shearing of the γ' precipitates. As dislocations accumulate, an internal stress associated with the pile-up eventually enables shearing of the particle and the formation of an APB. With the magnitude of the APB energy being dependent on the size of the γ' precipitates, small precipitates (~10 nm) are more easily sheared than large precipitates (~300 nm). As particle sizes increase for a constant volume fraction, the mechanism of deformation also changes. Since the spacing between the particles also increases, it becomes more energetically favourable for dislocations to bow between the precipitates. Misfit or coherency strains between the matrix and precipitate also contribute to the strength of Ni-based superalloys.[31] When large negative coherency strains are present at the precipitate-matrix interface, this tends to repel dislocations and slow down the rate of recovery at low temperatures. Compared to other superalloys containing similarly low levels of γ', the unusually high strength and low precipitate coarsening kinetics are often attributed to the large coherency strains at $\gamma - \gamma''$ interfaces in IN718. As temperatures increase, however, the dominant deformation mechanisms change and microstructural stability becomes critical.

At high temperatures ($T > \sim 0.6\ T_M$), dislocations are sufficiently mobile that by-pass mechanisms, such as dislocation climb and bowing, may potentially be more energetically favourable than cutting of the precipitate. With the lower applied stresses typically associated with high-temperature deformation generally insufficient to result in shearing of the γ' and Orowan stresses providing resistance to bowing of the dislocation between the precipitates, dislocations are forced to climb. As climb is a thermally controlled process and a rate-limiting step, strain rates are correspondingly low when dislocations are not able to cut the γ' precipitates. Within the γ' precipitate, thermal activation of cross slip from the {111} octahedral to {001} cube planes may occur and contribute to the high-temperature strength of the alloy.

With the deformation characteristics of these polycrystalline Ni-base superalloys highly sensitive to the APB energies of the precipitate and temperature, substantial increases in strength and creep resistance can be incurred by increasing the volume fraction of intermetallic precipitates.[32] As the volume fraction of γ' increases to over 50%, optimum structural properties are attained when cuboidal γ' precipitate morphologies measuring approximately 0.5 μm in diameter are dispersed within the γ matrix. Under these conditions, the microstructure is highly resistant to deformation since the stresses required for dislocation bowing are approximately the same as the stresses required for dislocation cutting of the γ′ precipitates.

4.4 Directionally solidified superalloys

For many structural applications, equiaxed polycrystalline microstructures are highly desirable because they exhibit isotropic properties, high proof stresses and are relatively simple to process into complex shapes. At elevated temperatures, however, accumulation of creep damage at grain boundaries transverse to the direction of the applied load results in premature failure of the material. To negate the mechanism by which these polycrystalline components fail, manufacturing processes were developed to minimize effectively or eliminate the formation of these transverse grain boundaries during solidification.

4.4.1 Applications of directionally solidified superalloys

Although Ni-base superalloys have inherently good high-temperature properties to begin with, these properties can be improved by processing. The creep rupture resistance of Ni-based superalloys can be enhanced by orientating the grain boundaries parallel to the applied stress direction or by removing the grain boundaries entirely. Due to the anisotropic FCC structure of the Ni-base γ matrix and the $L1_2$ crystal structure of the γ′ precipitates, the <111> orientations exhibit a high elastic modulus while the 'soft' <001> orientations exhibit a low elastic modulus. During thermal cycling, the resulting strains associated with the thermal expansion lead to the formation of stresses proportional to the elastic modulus. For many turbine blade applications, components exhibiting a <001> low-modulus texture tend to have improved resistance to thermal fatigue.

4.4.2 Processing

Directionally solidified alloys

Directionally solidified (DS) components, are produced using a Bridgman furnace. Operating in a vacuum or inert gas atmosphere, alloyed VIM ingots are remelted and poured into an investment mould heated to a temperature approximately 150 °C above the liquidus temperature of the alloy. By withdrawing the mould from the furnace while maintaining a unidirectional thermal gradient, preferential, orientated grain growth occurs. Since Ni-base superalloys exhibit cubic symmetry, these processing conditions encourage solidification along the orthogonal <001> crystallographic orientations. The [001] dendrites aligned most favourably with the thermal gradient tend to grow rapidly while the superheated bulk liquid prevents the formation of equiaxed grains ahead of the solidifying interface. Consequently, directionally solidified components typically consist of a number of columnar <001> grains aligned parallel to the solidification direction (Fig. 4.18).

4.18 Grain structures of single crystal, directionally solidified and conventionally cast turbine blades.

Single crystal alloys

Solidification of single crystal components occurs in a manner similar to the processing of DS components. In the case of single crystal solidification, however, a seed crystal or starter block is employed to allow only one grain orientated most favourably with respect to the thermal gradient to grow and form the component. VerSnyder and coworkers[33] were the first to demonstrate that directionally solidified and single crystal superalloy components possessed superior creep rupture properties compared to conventionally cast components. At elevated temperatures, sliding and damage accumulation occurs along the high angle boundaries. As shown in Fig. 4.19, conventionally cast components that contain grain boundaries transverse to the principal stress direction exhibit the lowest creep rupture lives. Removal of these transverse boundaries through directional solidification delays failure initiation and enhances the creep rupture life. Creep properties are further enhanced with the elimination of all high angle grain boundaries in single crystal components.

With the commercial introduction of single crystal turbine blades to aircraft engine applications in the early 1980s, the demand for higher operating temperatures led to the development of improved Ni-base superalloys.[34] The initial, or '1st generation', single crystal Ni-base superalloys[35] were essentially

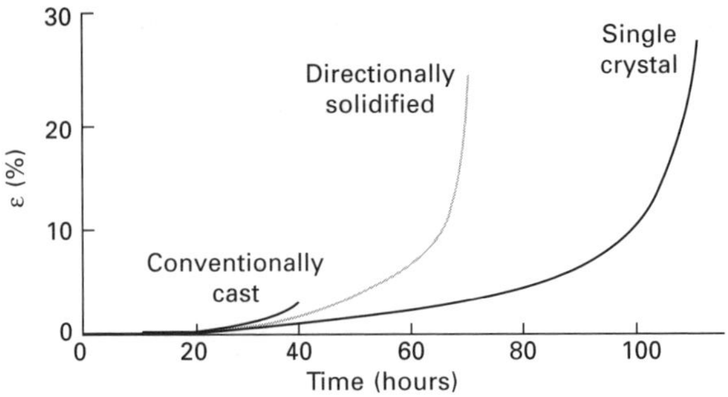

4.19 Comparison of the creep-rupture properties of MAR-M200 processed as conventionally cast, directionally solidified and single crystal components (Adapted from ref. 33).

derived from the compositions of conventional DS alloys (Table 4.1). Since there were no grain boundaries to strengthen in these single crystal components, the levels of the grain boundary strengthening elements, C, B, Hf and Zr, were minimized to increase the incipient melting temperatures and allow for complete solutioning of the γ′ precipitates during heat treatment. Accompanying the development of more efficient and more powerful turbine engines, the temperature requirements for turbine blade alloys increased and subsequently led to the development of a '2nd generation' of Ni-base superalloys. These 2nd generation Ni-base superalloys are typically characterized by an addition of ~3.0 wt.% rhenium to improve the high-temperature creep properties of the alloy.[36,37] To further extend the high-temperature capabilities of these materials, a '3rd generation' of Ni-base single crystal superalloys.[38–40] was developed in conjunction with advanced high-output turbine engines. Relative to the 2nd generation alloys, these 3rd generation single crystal alloys contain less chromium and even higher levels of refractory alloying constituents (Re>~3.0 wt.%) to further enhance solid-solution strengthening.

Grain defects

Manufacturing of DS and single crystal components requires control of solidification parameters to prevent the development of grain defects during solidification. The two most prevalent problems associated with the directional solidification process are the development of misorientated grains and freckle defects (Fig. 4.20). In single crystal components, misorientated grains are defined as high angle grains that exhibit an orientation significantly different (>10°) from the bulk [001] texture. Similar to the freckles in VAR and ESR ingots, freckles in DS and single crystals consist of channels comprised of small equiaxed grains or dendrite fragments that are aligned parallel to the solidification direction. Each of the individual grains in a freckle chain is

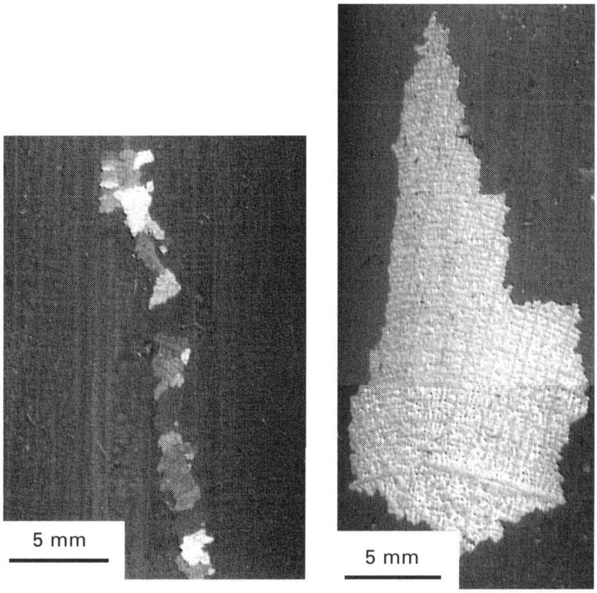

4.20 Macroscopic grain defects present on the surface of single crystal Ni-based superalloy castings.

misorientated with respect to the <001> direction. Additionally, the composition of the equiaxed grains present in the freckle chains is significantly different from the bulk crystal(s) (Table 4.4). Because of the adverse effects of high angle grain boundaries on mechanical properties, the presence of either of these unacceptable defects in DS or single crystal components causes a significant decrease in the mechanical properties of the component.

Table 4.4 Composition of freckle chains in high refractory single crystal superalloys

	Al	Cr	Co	Hf	Re	Ta	W	Ni
Nominal	6.0	4.5	12.5	0.16	6.3	7.0	5.8	Bal.
Freckle	8.2	3.6	11.0	0.18	2.3	10.0	2.9	Bal.
Interdendritic	7.6	4.4	12.2	0.13	3.4	8.2	3.9	Bal.

Misoriented grains

The columnar to equiaxed transition in DS and single crystal superalloys involves the nucleation of misorientated grains due a change in the magnitude of thermal gradient resulting from the latent heat expelled at the solid/liquid interface. These types of defect can typically be prevented by maintaining solidification parameters such that the magnitude of the latent heat remains proportional to some critical ratio of the thermal gradient (G) and the withdrawal rate (R):

$$\frac{G}{R} \propto \frac{K_T}{\Delta H}$$

where K_T = the thermal conductivity of the solid and ΔH = the latent heat of solidification. This relationship helps sustain the unidirectional thermal gradient and prevents the breakdown of directional or single crystal solidification.

Preventing the formation of grain defects during solidification becomes more complicated when dealing with castings where heat extraction is asymmetric. Rapid transitions in geometry, such as those seen in blade-shaped components, can often result in the curvature of the liquidus isotherm that leads to thermal undercooling and/or localized changes in the solidification parameters.[41] In single crystal alloys that exhibit a low undercooling capacity, the undercooling may nucleate misorientated grains. Although macroscopic criteria for predicting grain defect formation assume uniform thermal gradients, transitions in geometry may result in localized changes of the solidification parameters, which can lead to the formation of grain defects. Misorientated grains can also develop as a consequence of convective instabilities transporting dendrite fragments out of the mushy zone and depositing them ahead of the solidifying interface.[42] In certain instances, these dendrite fragments may serve as nuclei for misorientated grains. Currently, through solidification modelling, mould withdrawal profiles can be optimized to a certain degree to prevent the formation of defects due to component geometry.

Freckles

As the levels of refractory alloying additions to these superalloy single crystals have increased to improve high-temperature mechanical properties, freckle formation during directional solidification has become an increasingly important problem. Typically, freckle formation in these multicomponent alloys is caused by the onset of thermosolutal convective instabilities due to dendritic segregation. During unidirectional solidification, solute is gradually accumulated in the mushy zone. Considering the gravitational forces that are generally aligned anti-parallel to the solidification direction, the density variation due to the thermal and compositional differences between the bulk liquid and solute can potentially result in convective fluid flow. Consequently, certain multicomponent systems, particularly the ones that exhibit strong segregation characteristics and form large liquid density differences between the solute and bulk liquid, are more prone to the development of these convective instabilities.

Freckle formation has been a persistent problem in the solidification of single crystal Ni-base superalloys.[43,44] The presence of these undesirable grain defects, which lower the creep and fatigue properties, could potentially result in the premature failure of critical components. Early efforts taken to prevent these grain defects included the modification of gating designs and the application of high thermal gradients at the solidifying interface. This results in a refined dendritic microstructure incapable of supporting the formation of freckles due to the narrow 'channels' and low permeabilities. Maintaining a large thermal gradient during conventional Bridgman solidification, however, becomes increasingly difficult as the size of the casting increases. Once a freckle chain has formed in a fully solidified crystal, the inability to remove such defects through thermal treatments or mechanical means poses a number or serious implications.

4.5 Creep deformation

Single crystal and directionally solidified components are used exclusively in high-temperature applications where resistance to creep deformation is required. For single crystal components, structural integrity is required at metal temperatures ranging from ~40–90% of the melting point. Depending on the application, the stress states associated with components such as shroud segments, nozzle guide vanes and turbine blades are also highly complex. Turbine engine components are designed such that the load-bearing capacity and temperature capability is maximized. Stresses in regions exposed to extremely high temperatures are maintained at comparatively low levels, while high stresses are partitioned to areas where temperatures are low. Hence, the deformation characteristics of these alloys vary greatly under these conditions.

4.5.1 Low temperature creep

Single crystal and directionally solidified superalloys are highly resistant to deformation at low temperatures. Although dislocations within the alloy are inherently sessile at temperatures $<0.6\ T_M$, high uniaxial stresses and crystal anisotropy may result in the activation of $<112>\{111\}$ slip systems. As dislocations accumulate at the $\gamma - \gamma'$ interface in [001] orientated crystals, $1/2<110>$ dislocations dissociate and form $<112>$ dislocations which are able to penetrate the γ' precipitate. Since an energetically unfavourable superlattice intrinsic stacking fault is left with the initial passing of the 1/3[112] dislocation through the γ', subsequent dislocations forced to the interface tend to dissociate in a similar manner and form a superlattice extrinsic stacking fault which partially restores order to the precipitate. Unlike conventional $<110>\{111\}$ slip systems where cross slip may occur as multiple slip planes become active, slip along $<112>$ is constrained onto one $\{111\}$ plane. As shearing of the γ' precipitates occurs (Fig. 4.21), strain is accumulated rapidly

4.21 Low temperature creep results in deformation of both the γ and γ′ phases. Stacking faults are present within the γ′ precipitates (Courtesy of T.M. Pollock).

as a result of the planar nature of the slip system. The characteristic primary creep transient, or extent of the rapid strain accumulation, is highly dependent on crystal orientation and stress level. [110] and [111] orientated crystals are less prone to this type of deformation, while in [001] crystals, planar slip along <112> continues until crystal rotations enable the resolved shear stresses to activate slip on other planes. This effectively suppresses planar <112> slip and activates the <110> slip systems. Other factors including alloy composition, and γ′ size and volume fraction also influence the anisotropy of the components.

4.5.2 Intermediate temperature creep

At intermediate temperatures ($0.6 < T_M < 0.75$), stress levels are typically insufficient to result in shearing of the γ′ precipitates. Thus, deformation within the microstructure is generally confined to the γ matrix and results in three distinct stages of creep deformation: incubation, primary and secondary creep.[45] No macroscopic straining occurs during the incubation period that involves the distribution of grown-in dislocation networks. These initial dislocations serve as sources from which dislocations in the γ matrix are able to multiply (Fig. 4.22). When an external uniaxial stress is applied, the misfit stresses between the γ matrix and γ′ precipitate are unbalanced and the effective stresses enable preferential flow of dislocations within the horizontal channels. As the Orowan stresses are sufficiently high that dislocations are unable to bow between the vertical channels formed aligned precipitates,

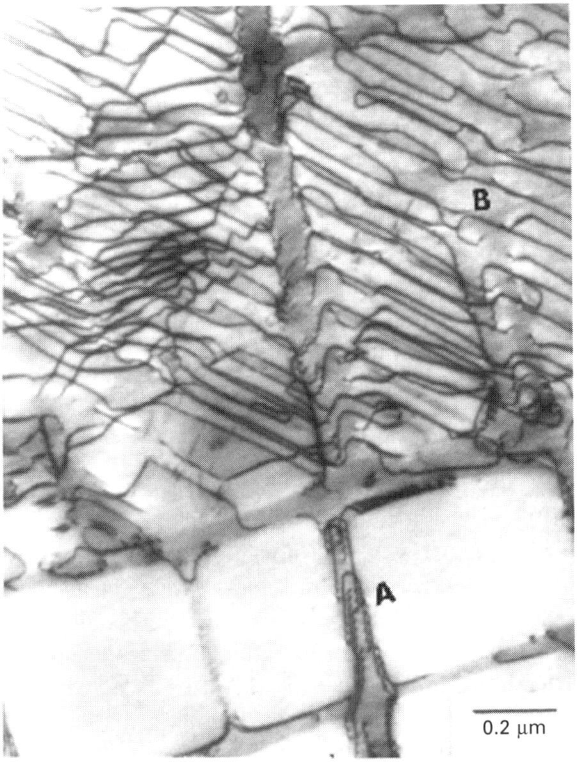

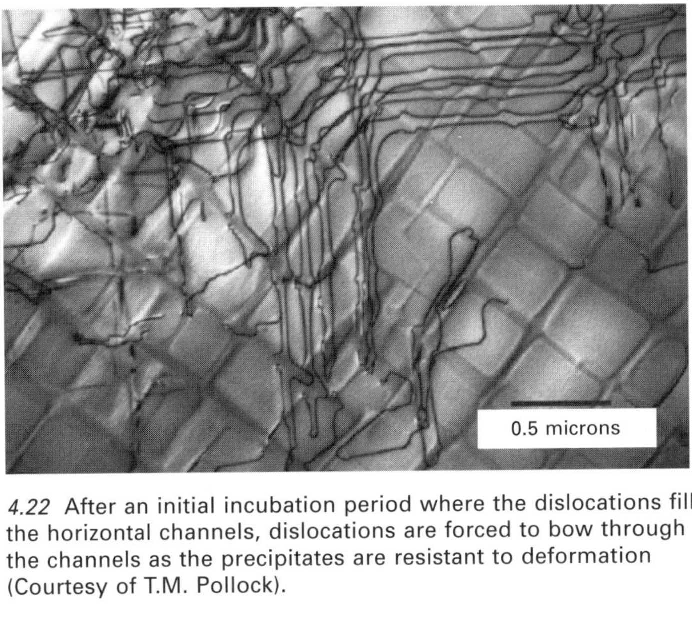

4.22 After an initial incubation period where the dislocations fill the horizontal channels, dislocations are forced to bow through the channels as the precipitates are resistant to deformation (Courtesy of T.M. Pollock).

dislocation glide in the horizontal channels continues until the percolation process is complete.

The deformation mechanism associated with the primary creep transient during intermediate temperature creep is distinctly different from that of low temperature creep. Unlike dislocation shearing of the γ' precipitates along the <112> direction at high stresses and low temperatures, the primary creep transient at intermediate temperatures can be attributed to the relief of coherency stresses at the $\gamma - \gamma'$ interface as dislocations are accommodated. At the end of primary creep, a three-dimensional network of dislocations forms around the precipitates. Despite the lack of a steady-state strain rate, these dislocation networks surrounding the precipitates are extremely stable and contribute to the gradual progression into tertiary creep.

4.5.3 High temperature creep

The enhanced diffusivity associated with deformation at extremely high temperatures ($0.75 < T_M$) results in morphological changes within the microstructure. With the application of an external stress, the discrete cuboidal γ' precipitates coalescence into rafts or rods aligned perpendicular or parallel to the applied stress direction. The kinetics of directional coarsening is strongly influenced by the temperature as well as the stresses associated with the coherency strains at the $\gamma - \gamma'$ interface. Alignment of the rafts or rods, however, is dependent upon whether the external and misfit stresses are compressive or tensile. For example, uniaxial tensile stresses cause alloys with negative misfit to form rafts perpendicular to the applied stress direction, while compressive stresses applied to the same alloy will result in the formation of rods aligned parallel to the direction of applied stress. Since most commercial directionally solidified and single crystal alloys exhibit a negative misfit and are used to sustain tensile loads, rafts are generally formed perpendicular to the applied stress direction (Fig. 4.23).

With single crystal and directionally solidified Ni-based superalloys containing in excess of 60% γ' by volume, directional coarsening causes an inversion of the microstructure. Once rafting is complete, rafts of γ are contained within an intermetallic γ' matrix. As the rafts of γ are discrete, a continuous path for dislocation motion along the matrix channels no longer exists. In order for deformation to continue, dislocations must shear the γ' phase. Provided that the rafted structure remains phase stable, rafted structures are highly resistant to deformation at low stresses where dislocations are unable to penetrate into the γ' matrix. When stresses are sufficient to cause shearing of the γ' phase, microstructural damage is able to accumulate rapidly and tertiary creep rates are accelerated.

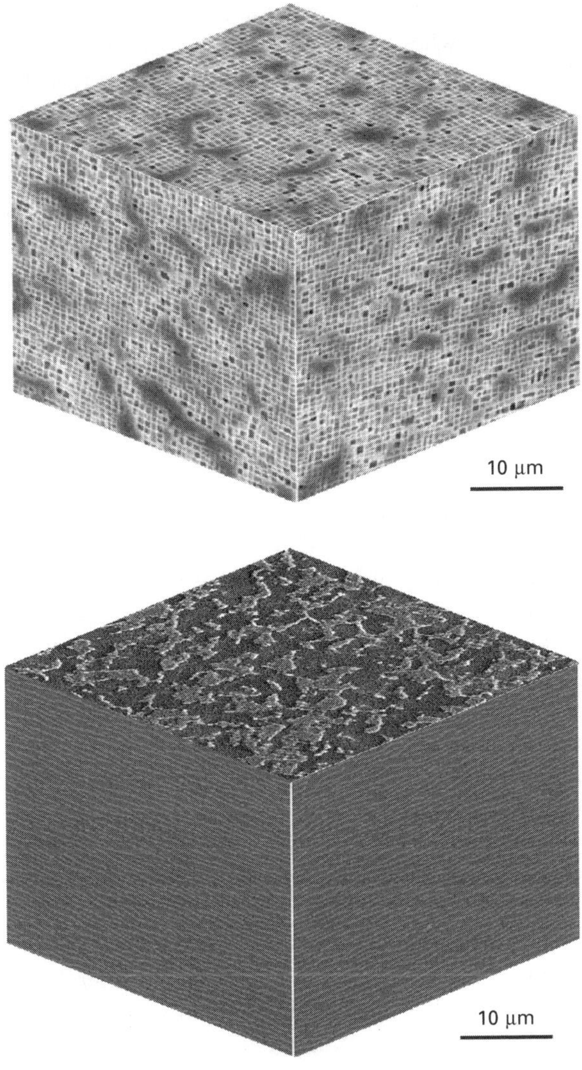

4.23 Directional coarsening at elevated temperatures results in the formation of rafts of γ aligned perpendicular to the direction of applied stress. The matrix inverts as the γ rafts are contained with a γ′ matrix.

4.6 Properties

Continued modification of single crystal superalloy compositions has significantly enhanced the strength and temperature capabilities when compared to columnar-grained or polycrystalline components. With the removal of grain boundaries, dislocation slip is able to occur over greater distances in single crystals. Consequently, single crystals are able to accommodate significantly more strain when compared to polycrystalline components. As knowledge of the fundamental deformation mechanisms is revealed, intrinsic parameters, such as alloy composition, γ' size, morphology and volume fraction, have been carefully modified to engineer creep resistant materials. The advances in strengthening through refractory alloying additions, however, has created new challenges with regard to processing of single crystal components. Increases in the refractory alloying content of these alloys have enhanced the propensity for the formation of grain defects during solidification. Since the presence of these macroscopic grain defects results in the subsequent rejection of the component, these manufacturing issues present a serious limitation to the continued development of single crystal Ni-base superalloys.

4.7 Conclusion

As high-temperature structural materials, Ni-based superalloys remain crucial to the continued development of high-performance turbine engines for aerospace, marine and power-generation applications. For high-performance applications, advanced single crystal and powder-processed alloys are required for load-bearing applications where the temperatures of combustion gases are in excess of 1700 K. By enabling the engine to operate under higher temperatures, significant gains in efficiency and power output are achieved. Physical metallurgy can be applied to improve the properties of these alloys, but the constraints associated with commercial processing procedures must also be addressed when investigating new alloys.

Processing and properties of Ni-base superalloys are important in the manufacture of physically large industrial gas turbine engines for power-generation applications. Due to the sheer size of the components, processing constraints pose a significant barrier to the development of these advanced industrial gas turbines. The structural response of superalloys to high-temperature deformation is dependent upon the microstructural features within the alloy. Since physically large components tend to exhibit a variable microstructure, the resulting mechanical properties may be significantly different when compared to a smaller component with an identical alloy composition. Moreover, the high processing costs and low yields associated with the manufacture of these components results in a large economic incentive to meet the demand for these advanced gas turbines. Increases in power consumption combined with the decommissioning of nuclear power plants in North America and Europe has necessitated a great need for highly efficient power-generation turbines capable of generating in excess of 450 MW.[46]

The increase in performance that has been realized in Ni-base superalloys in the last 50 years has been remarkable, but the continued optimization of turbine engines has pushed the operating temperature of advanced gas turbine engines above the melting point of pure nickel (1455 °C). While the use of thermal barrier coatings (TBCs) and internal cooling channels allows these heavily alloyed Ni-base components to retain their structural integrity at these temperatures, future demands for engines with greater power output/ efficiency will require materials with even higher temperature capabilities.

4.8 References

1. Tien J.K. and Caulfield T., *Superalloys, Supercomposites and Superceramics*. 1988, New York: Academic Press, Inc.
2. Decker R.F., *Strengthening Mechanisms in Nickel-Base Superalloys*. 1969, Climax Molybdenum Company.
3. Sims C.T., Stoloff N.S. and Hagel W.C., *Superalloys II*. 1986, New York: John Wiley and Sons.
4. Meetham G.W., 'Trace Elements in Superalloys – An Overview.' *Metals Technology*, 1984. **11**. p. 414–418.
5. McLean M. and Strang A., 'Effects of Trace Elements on the Mechanical Properties of Superalloys'. *Metals Technology*, 1984. **11**. p. 454–464.
6. Machlin E.S., *Trans. AIME*, 1961. **218**. p. 314–326.
7. Simkovich A., *Journal of Metals*, 1966. **18**(4). p. 504–512.
8. Enomoto M. and Harada H., 'Analysis of γ'/γ Equilibrium in Ni-Al-X Alloys by the Cluster Variation Method with the Lennard-Jones Potential'. *Metall. Trans.*, 1989. **20A**. p. 649–664.
9. Blavette D. and Bostel A., 'Phase Composition and Long Range Order in γ' Phase of a Single Crystal Superalloy CMSX2: An Atom Probe Study'. *Acta. Metall.*, 1984. **32**(5). p. 811–816.
10. Copley S.M. and Kear B.H., 'A Dynamic Theory of Coherent Precipitation Hardening with Application to Nickel-Based Superalloys'. *Trans. AIME*, 1967. **239**. p. 984–992.
11. Grosdider T., Hazotte A. and Simon A., 'Precipitation and Dissolution Processes in γ/γ' Single Crystal Nickel-Based Superalloys'. *Mater. Sci. Eng*. A, 1998. **A256**. p. 183–196.
12. Khan T. and Caron P., 'Effect of Processing Conditions and Heat Treatments on Mechanical Properties of Single-Crystal Superalloy CMSX-2'. *Mater. Sci. and Tech.*, 1986. **2**. p. 486–492.
13. Collins H.E., 'Relative Stability of Carbide and Intermetallic Phases in Ni-Base Superalls'. In *Superalloys*'. 1968. Metals Park, OH ASM. p. 171–198.
14. Fernandez R., Lecomte J.C. and Kattamis T.Z., 'Effect of Solidification Parameters on the Growth Geometry of MC Carbide in IN-100 Dendritic Monocrystals'. *Metall. Trans.*, 1978. **9A**. p. 1381–1386.
15. Tin S., Pollock T.M. and Murphy W., 'Stabilization of Thermosolutal Convective Instabilities in Single Crystal Ni-Base Superalloys: Carbon and Freckles'. *Metall. Mater. Trans.*, 2001. **32A**. p. 1743–1753.
16. Wlodek S.T., 'The Structure of IN-100'. *Trans. ASM*, 1964. **57**. p. 110–119.
17. Ross E.W., 'Rene 100: A Sigma Free Turbine Blade Alloy'. *J. Met.*, 1967. p. 12–14.
18. Wlodek S.T., 'The Stability of Superalloys'. In *Long Term Stability of High Temperature Materials*. 1999. Warrendale, PA TMS. p. 1–38.

19. Darolia R. Lahrman D.F. and Field. R.D., 'Formation of Topologically Closed Packed Phases in Nickel Base Single Crystal Superalloys'. In *Superalloys 1988*. Warrendale, PA TMS. p. 255–264.
20. Decker R. and Mihalisin J.R., *Trans. ASM*, 1969. **62**. p. 481–490.
21. Auburtin P., Cockcroft S.L. and Mitchell A., 'Liquid Density Inversions during the Solidification of Superalloys and Their Relationship to Freckle Formation in Castings'. In *Superalloys 1996*. Warrendale, PA TMS. p. 443–450.
22. Jackman L.A., Maurer G.E. and Widge S., 'New Knowledge About White Spots in Superalloys'. *Advanced Materials and Processes*, 1993. **5**. p. 18–25.
23. Durber G.L.R., Jones C.L. and Dykes A.J., 'VIM+ESR Alloy 718 – An Assessment of Chemistry Control, Alloy Cleanliness and Mechanical Properties'. In *Superalloys 1984*. Warrendale, PA TMS/AIME. p. 433–442.
24. Shamblen C.E., Chang D.R. and Corrado J.A., 'Superalloy Melting and Cleanliness Evaluation'. In *Superalloys 1984*. Warrendale, PA TMS/AIME. p. 509–520.
25. Quested P.N., McLean M. and Winstone M.R., 'Evaluation of Electron-Beam Cold Hearth Refining of Virgin and Revert IN738LC'. In *Superalloys 1988*. Warrendale, PA TMS/AIME. p. 387–396.
26. Maurer G.E., *et al.* 'Development of HIP consolidated P/M Superalloys for Conventional Forging to Gas Turbine Engine Components'. In *Superalloys 1996*. Warrendale, PA TMS. p. 645–652.
27. Shen S., Semiatin S.L. and Shivpuri R., 'Modeling Microstructural Development during Forging of Waspaloy'. *Metall. and Mater. Trans.*, 1995. **26A**. p. 1795.
28. Luton M.J. and Sellars C.M., 'Dynamic Recrystallization in Nickel and Nickel-Iron Alloys During High Temperature Deformation'. *Acta. Metall.*, 1969. **17**. p. 1033–1043.
29. Wlodek S.T., Kelly M. and Alden D.A., 'Structure of Rene 88DT'. In *Superalloys 1996*. Warrendale, PA TMS. p. 129–136.
30. Nembach E. and Neite G., 'Precipitation Hardening of Superalloys by Ordered γ'-Particles'. *Progress in Materials Science*, 1985. **29**. p. 177–319.
31. Law C.C. and Blackburn M.J., 'Creep Rupture in Powder Metallurgical Nickel-Base Superalloys at Intermediate Temperatures'. *Metall. Trans.*, 1980. **11A**. p. 495–507.
32. Jackson J.J. *et al.*, 'Effect of Volume Fraction of Fine Gamma Prime on Creep in DS Mar-M200 Plus Hf'. *Metall. Trans.*, 1977. **8A**. p. 1615–1620.
33. VerSnyder F.L. and Shank M.E., 'The Development of Columnar Grain and Single Crystal High Temperature Materials Through Directional Solidification'. *Mater. Sci. Eng.*, 1970. **6**. p. 213–247.
34. Gell M., Duhl D.N. and Giamei A.F., 'The Development of Single Crystal Superalloy Turbine Blades'. In *Superalloys*. 1980. Metals Park, OH American Society for Metals. p. 205–214.
35. Ross E.W. and O'Hara K.S., 'Rene N4: A First Generation Single Crystal Turbine Airfoil Alloy with Improved Resistance, Low Angle Boundary Strength and Superior Long Time Rupture Strength'. In *Superalloys 1996*. Warrendale, PA TMS. p. 19–25.
36. Harris K. *et al.* 'Development of the Rhenium Containing Superalloys CMSX-4 and CM186LC for Single Crystal Blade and Directionally Solidified Vane Applications in Advanced Turbine Engines'. In *Superalloys 1992*. Warrendale, PA TMS. p. 297–306.

37. Giamei A.F. and Anton D.L., 'Rhenium Additions to a Ni-base Superalloy: Effects on Microstructure'. *Metall. Trans*, 1985. **16A**. p. 1997–2005.
38. Walston W.S. *et al.,* 'Rene N6: Third Generation Single Crystal Superalloy'. In *Superalloys 1996*. Warrendale, PA TMS. p. 27–34.
39. Erickson G.L., 'The Development and Application of CMSX-10'. In *Superalloys 1996*. Warrendale, PA TMS. p. 35–44.
40. Koizumi Y. *et al.* 'Third Generation Single Crystal Superalloys with Excellent Prossability and Phase Stability'. In *Materials for Advanced Power Engineering*. 1998. Universite De Liege Forschungszentrum Julich. p. 1089–1098.
41. Meyer ter Vehn M. *et al.* 'Undercooling Related Casting Defects in Single Crystal Turbine Blades'. In *Superalloys 1996*. Warrendale, PA TMS. p. 471–479.
42. Gu J.P., Beckermann C. and Giamei A.F., 'Motion and Remelting of Dendrite Fragments during Directional Solidification of a Nickel-Base Superalloy'. *Metall. and Mat. Trans.*, 1997. **28A**. p. 1533–1542.
43. Copley S.M. *et al.*, 'The Origin of Freckles in Unidirectionally Solidified Castings'. *Metall. Trans.*, 1970. **1**. p. 2193–2204.
44. Giamei A.F. and Kear B.H., 'On the Nature of Freckles in Nickel-Base Superalloys'. *Metall. Trans.*, 1970. **1**. p. 2185–2192.
45. Pollock T.M. and Argon A.S., 'Creep Resistance of CMSX-3 Nickel Base Superalloy Single Crystals'. *Acta. Metall. Mater.*, 1992. **40**(1). p. 1–30.
46. Seth B.B. 'Superalloys – The Utility Gas Turbine Perspective'. In *Superalloys 2000*. Warrendale, PA TMS. p. 3–16.

5 Plastic deformation of ordered intermetallic alloys

K XIA, University of Melbourne, Australia

5.1 Introduction

In a simple binary system consisting of two metallic components, the two elements may form a complete solid solution, as in the case of Cu and Ni, or there may exist two terminal solid solutions with limited solubility of the minor component atoms (e.g. Pb-Sn). In more complicated systems, however, intermediate phases may form between the two components as in the example of Ti-Al. These intermediate phases are referred to as intermetallic compounds. Many of these compounds possess an ordered structure in which the component atoms occupy specified sites rather than being randomly distributed. The ordered structure may be observed only below a certain transition temperature or it may exist up to the melting point of the compound. A similar description can be used for systems of more than two components although the structures may become more complicated. Such ordered intermetallic compounds (alloys) possess unique properties and are the subject of this chapter.

Scientific studies of ordered intermetallic compounds started over one hundred years ago (Westbrook, 1995). Ordered intermetallic phases in the form of precipitates have been used very effectively in strengthening. For instance, $CuAl_2$ and $MgZn_2$ (and their non-equilibrium preforms) are the major strengthening phases in high strength Al-Cu and Al-Zn-Mg alloys. The most prominent example is perhaps Ni_3Al in Ni based superalloys. More recently, their uses as key engineering materials by themselves have been attracting increasing attention, as evidenced by a series of major international conferences (e.g. the *High-Temperature Ordered Intermetallic Alloys* series published by the Materials Research Society, the *Structural Intermetallics* series and the *Gamma Titanium Aluminides* series by The Minerals, Metals and Materials Society) and the publication of a dedicated journal *Intermetallics*. The development has mainly been focused on high-temperature structural applications, in particular, components inside a jet engine or a turbocharger (Pope and Darolia, 1996, Yamaguchi *et al.*, 2000) although other applications are also being pursued (Westbrook, 1996, Masumoto and McGahan, 1996, Schetky, 1996). As structural materials, intermetallic alloys behave in a manner between metallic materials and ceramics. They show much higher heat resistance (e.g. strength and oxidation resistance at elevated temperatures) than many conventional metals. On the other hand, unlike ceramics, many intermetallics can deform readily at high temperatures and show much improved ductility at room temperature. Intermetallic alloys, therefore, represent a new class of materials with a unique combination of many desirable mechanical properties.

In this chapter, unique features of bonding, crystal structure and defects in ordered intermetallic compounds will be discussed first since they are related to the unique mechanical behaviours observed. Typical plastic deformation behaviour at room and elevated temperatures in selected intermetallic alloys will be described and mechanisms analysed. These will lead to deformation processing and applications. The alloys selected for discussion are mostly those based on nickel, titanium and iron aluminides as they represent some typical behaviours and microstructures and are closest to practical uses. Although important, fracture behaviour will not be addressed other than in the context of ductility. It is also not the intention to review the topics thoroughly. Instead, selected references are cited to illustrate the main points.

5.2 Bonding, crystal structures and defects

5.2.1 Bonding

It is safe to say that bonding between atoms determines all the properties in a particular solid. A good physical understanding of simple chemical bonding such as covalent, ionic and metallic bonds has emerged thanks to the application of advanced quantum mechanics models (see, for example, (Harrison, 1989)). Since intermetallics are narrowly defined in this chapter as compounds between element metals, it is tempting to classify bonding in intermetallic compounds as simply of metallic type, i.e., with valence electrons freely moving among the lattice of positively charged ions (Rohrer, 2001). However, intermetallic alloys often show behaviours that are not metal-like, e.g., high melting points and low ductility at room temperature. In ordered intermetallics, bonding between unlike atoms is preferred to that between like atoms, and the resulting structure resembles more the structure of ceramic compounds with ionic and covalent bonds than that of typical metallic alloys. Although the exact details are not entirely clear, bonding in ordered intermetallic compounds is believed to be complex and involves a hybrid of metallic, covalent and ionic bonds. It is difficult to classify the bonds using some simple rules, as concluded by Sauthoff (1995). However, it is possible to gain useful insight by analysing bonding in several intermetallic compounds with a view to understanding their brittle or ductile behaviour.

In general, for a material to behave in a ductile manner the following two conditions are necessary. First of all, dislocation movement (emission) caused by the resolved shear stress ahead of a sharp crack tip should proceed before the resolved normal tensile stress at the crack tip reaches the cohesive strength to cause cleavage fracture (Rice and Thomson, 1974). Secondly, in the case of polycrystalline materials, at least five independent slip systems should be available to ensure continuity of material being deformed (von Mises, 1928). In addition, it is required that cross slip can be easily achieved to satisfy the so-called *slip flexibility* (Kelly and Macmillan, 1986). For monolithic metals, the bonding between atoms is of metallic character which gives rise to mostly close-packed or near close-packed structures (i.e. FCC, HCP and BCC) with moderate resistance to dislocation movement and the first requirement above is met for many. The ductility is thus largely determined by the crystal structure. Highly symmetric cubic structures with plenty of independent slip systems such as FCC and BCC normally lead to good ductility whereas materials with hexagonal and other less symmetric structures may behave in a brittle manner owing to a lack of sufficient slip systems that can operate with comparable ease.

The situation is more complicated in intermetallics as a result of complex electronic structures which determine bonding. For example, a component of covalent bonding was detected between Al and transition metals Fe, Co, Ni and Ru in FeAl, CoAl, NiAl and RuAl based on electron energy loss spectroscopy (EELS) as well as theoretical calculations (Botton *et al.*, 1996, Lin *et al.*, 1992, Botton and Humphreys, 1999) whereas simple metallic bonds exist in monolithic metals. The covalent character of the bonds results from the hybridisation between the *d* electrons in the transition metals and the *p* electrons in Al (Lin *et al.*, 1992). The covalent nature of the bonds in NiAl was also shown by analysis of the deformation electron density distribution map generated using high-energy electron diffraction (Fox and Tabbernor, 1991) although a later report by the same investigator appeared to suggest ionic bonding was more important in NiAl (Fox and Menon, 1997). Although all the compounds had the same structure (B2), different compounds showed different detailed electronic structures so that they behaved differently. The inherently ductile behaviour in FeAl was attributed to its higher cleavage strength from directional *d*-bonding which was not apparent in NiAl (Fu and Yoo, 1992) although external factors such as water vapour might cause embrittlement in FeAl (Liu *et al.*, 1989). Near the Fermi energy, the *d-d* bonding between Ru and Ru along <100> in RuAl was dominating while the *d-p* hybridisation between Ni and Al along <111> was strong in NiAl (Lin *et al.*, 1992). This may explain the relative brittleness in NiAl compared with RuAl which showed inherent ductility (Fleischer *et al.*, 1991) because the strong directional electron distribution in NiAl may be disturbed by slip, posing obstacles to dislocation movement.

In the case of TiAl which has a $L1_0$ structure, interaction between Ti-*d* and Ti-*d* electrons and Al-*p* and Ti-*d* hybridisation are observed from calculations of electronic structures (Morinaga *et al.*, 1990, Siegl *et al.*, 1997, Fu and Yoo, 1993, Sikora *et al.*, 1999, Lu *et al.*, 1994). These highly directional bonds are covalent in nature and may have resulted in the inherent brittleness in TiAl (Morinaga *et al.*, 1990, Sikora *et al.*, 1999), although bonding between Al atoms are of mostly metallic character (Sikora *et al.*, 1999, Lu *et al.*, 1994). Another important parameter in determining the behaviour of plastic deformation is shown elegantly by Eberhart *et al.* (1993) by making use of the topology of the charge density. The charge density reflects the density of electrons in space (a scalar field) and a plot of iso-density lines gives its topology. Although both CuAu and TiAl have the same $L1_0$ structure based on FCC (both having $c/a \neq 1$), the topology of the charge density in the (100) plane for CuAu resembles closely that of the FCC Cu whereas that for TiAl is more complicated. In other words, CuAu is not only structurally close to FCC but also topologically FCC while TiAl is topologically different. Through calculating the redistribution of electrons in response to a strain imposed, they showed that a redistribution resulting in decohesion along a (100) plane upon the application of a strain in [100] direction was reached at a much smaller strain in TiAl than in CuAu. This may explain the ductile behaviour in CuAu and brittle behaviour observed in TiAl.

It is clear from the above examples that the deformation behaviour in ordered intermetallic compounds depends strongly on the detailed electronic structures which determine the bonding characters. Materials with apparently identical structures may behave very differently, thanks to the presence of strong covalent and/or ionic components in bonding. If the covalent/ionic bonding causes a significant increase in shear strength and/or decrease in decohesion strength, the intermetallic compound would behave in a brittle manner. On the other hand, ductile behaviour would result from dominance of the metallic component in bonding. In general, bonding in intermetallics has a mixture of metallic, covalent and ionic characters and the relative importance of each type has considerable consequence on mechanical and other properties. Although there are still discrepancies about the details from both experiments and theoretical calculations, a good knowledge of the electronic structures appears to be important to our understanding.

5.2.2 Crystal structures

As a result of complex bonding, a large number of different crystal structures exist for intermetallic compounds (Villars and Calvert, 1985) (the Handbook actually contains ceramic phases that do not fit in our narrow definition of intermetallic phases between metals). In this section, several simple but technologically important ordered structures based on FCC, FCT (face-centred tetragonal), BCC and HCP structures are presented. They form the basis of analysis in several later sections. The data used have been obtained mostly from (Villars and Calvert, 1985).

FCC based ordered structure $L1_2$

An important structure based on FCC is $L1_2$ with a formula of A_3B. The B atoms occupy the corner sites and the A atoms the face centres, as shown in Fig. 5.1. The most studied $L1_2$ compound is perhaps Ni_3Al.

FCT based ordered structure $L1_0$

One structure that is face-centred but tetragonal rather than cubic is $L1_0$ with a formula of AB. The A atoms occupy the (001) planes whereas the B atoms occupy the (002) planes, as shown in Fig. 5.2. One of the most important intermetallic compounds, TiAl, possesses such a structure. The *c* axis may be greater than the *a* axis as in TiAl, or shorter as in CuAu.

BCC based ordered structure B2

The simplest structure based on BCC is B2 which is shown in Fig. 5.3. The formula of compounds possessing the B2 structure is AB, with A occupying the body centre and B the corner sites in a unit cell. B2 structured compounds of engineering importance include FeAl, NiAl and NiTi.

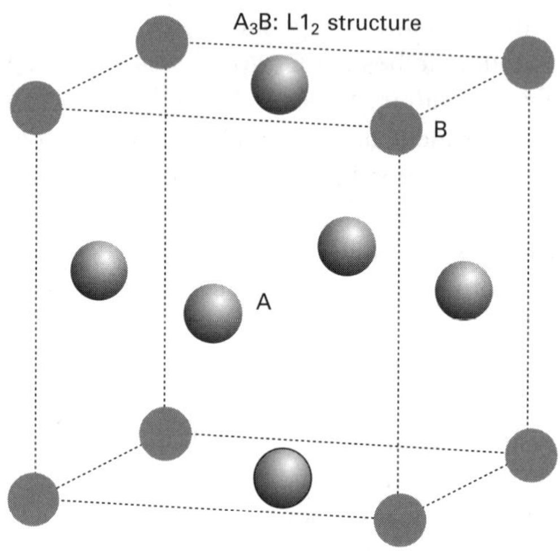

5.1 The unit cell of an $L1_2$ structured compound A_3B.

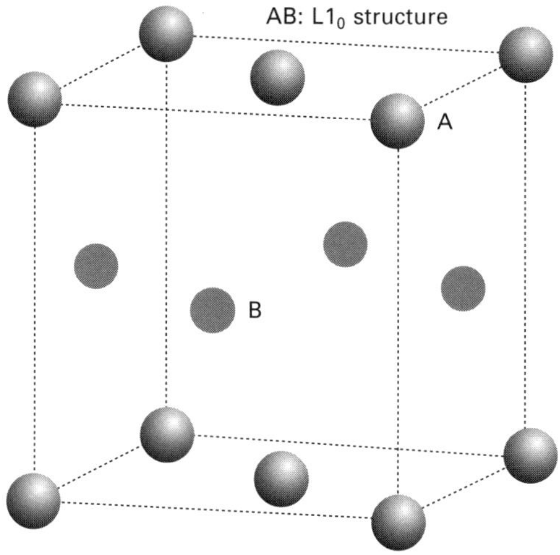

5.2 The unit cell of an L1$_0$ structured compound AB.

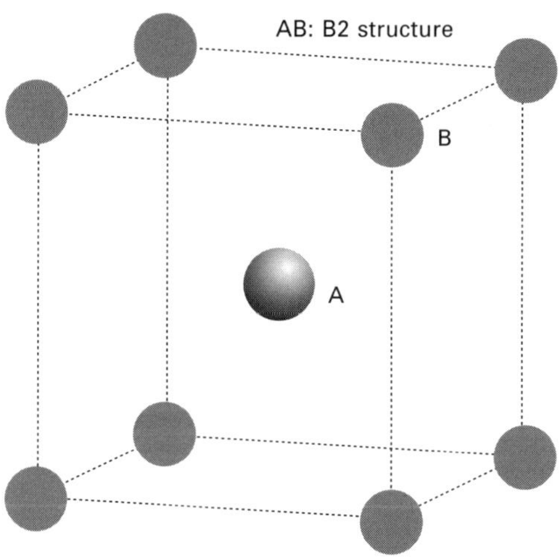

5.3 The unit cell of a B2 structured compound AB.

HCP based ordered structure D0$_{19}$

The ordered structures based on HCP lattice are visually more complicated. A unit cell of the $D0_{19}$ structure with a formula of A_3B is shown in Fig. 5.4a and a more familiar hexagonal unit cell is displayed in Fig. 5.4b. A representative compound with $D0_{19}$ is Ti_3Al.

5.2.3 Defects

The defect structures in ordered intermetallic compounds are more complicated. In addition to such point defects as vacancies and interstitial and substitutional

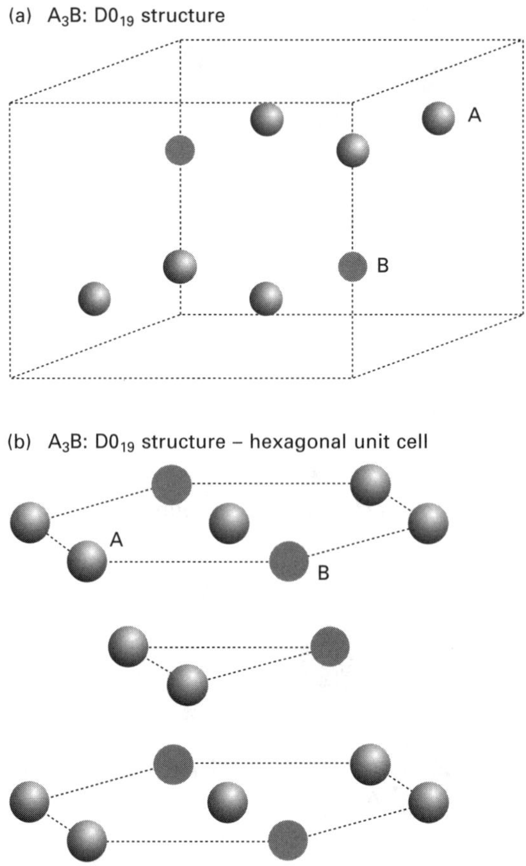

5.4 (a) The unit cell and (b) the hexagonal unit cell of a $D0_{19}$ structured compound A_3B.

atoms, an antisite is formed when an atom occupies a lattice site that belongs to a different species. A new planar defect, the antiphase boundary (APB), is created with the shifting of one part against the other part of a crystal that results in wrong neighbours (in the original ordered structure) across the plane on which the shifting has taken place. This, in turn, introduces many different types of dislocation configurations. In this section, the focus is on APB and associated stacking faults and dislocation structures since they have a significant influence on the deformation behaviour of ordered intermetallics.

Antiphase boundaries and stacking faults

The formation of an APB in a two-dimensional representation is shown in Fig. 5.5 for a binary intermetallic alloy AB. The top half of the ordered

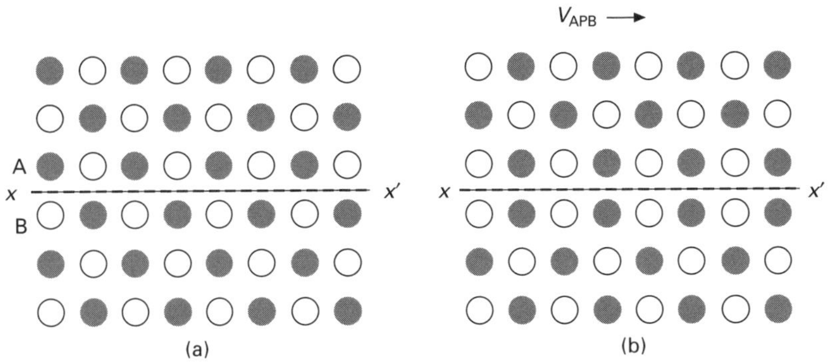

5.5 (a) A two-dimensional ordered structure of A and B atoms with A-B bonds across x–x', and (b) the structure after shifting the top half over the bottom half along x–x' by V_{APB}. An antiphase boundary (APB) is created along x–x' with A-A and B-B bonds.

structure in Fig. 5.5a is shifted along x–x' to create a new arrangement in Fig. 5.5b. The original A–B nearest neighbour bonds are replaced by A–A and B–B bonds along x–x' after shifting, i.e., x–x' has become an APB separating the two antiphase domains (APDs). The vector of shifting is often called the antiphase vector, $\boldsymbol{V_{APB}}$. When the APB is formed by dislocation dissociation on a slip plane, $\boldsymbol{V_{APB}}$ lies in the plane of shifting (i.e. the slip plane). This situation is of most interest in the present context.

The APB has an energy which is the difference in the atomic interaction energy between the structures before and after the shifting (Sun, 1995). Based on the classical model, the APB energy, γ_{APB}, can be conveniently calculated by considering only the nearest and second nearest neighbouring atoms (Yamaguchi and Umakoshi, 1990). As the nearest neighbours generally change with the orientation of an APB, the energy of the APB also varies with orientation, i.e., the APB energies on different crystallographic planes have different values. APB energies can also be calculated using a number of quantum mechanical methods or determined by experiments (Sun, 1995). Table 5.1 lists the ranges of the APB energies corresponding to different crystallographic planes as determined by various methods for selected intermetallic compounds with the structures described in 5.2.2. It is noted that the values vary considerably between different investigations and methods.

The shifting along $\boldsymbol{V_{APB}}$ does not lead to wrong stacking of layers if the structure is disordered. When stacking faults are formed in an ordered structure, it is necessary to distinguish between those which do not involve wrong bonds and those which involve wrong bonds. The former is called the superlattice intrinsic stacking fault (SISF) and the latter the complex stacking fault (CSF). They are formed by shifting along a SISF vector, $\boldsymbol{V_{SISF}}$, and a CSF vector, $\boldsymbol{V_{CSF}}$, respectively, as illustrated in Fig. 5.6 on the (111) plane of an $L1_2$ structured compound A_3B. Superlattice extrinsic stacking faults, SESF,

Table 5.1 Antiphase boundary energies (γ_{APB}) in selected intermetallic compounds as determined by various methods (data taken from values compiled in (Sun, 1995; Paidar and Vitek, 2002) except for those referenced otherwise)

γ_{APB} (mJ/m^2)	NiAl (B2)	FeAl (B2)	Ni_3Al ($L1_2$)	TiAl ($L1_0$)	Ti_3Al ($D0_{19}$)
{111} with $\boldsymbol{V_{APB}} = \frac{1}{2}$<110>			95–250	145–670	
{010}			30–230		
{100}				65–435	
{110} with $\boldsymbol{V_{APB}} = \frac{1}{2}$<111>	225–1000	180–650			
{112} with $\boldsymbol{V_{APB}} = \frac{1}{2}$<111>	260–1230	210–1040			
(0001) with $\boldsymbol{V_{APB}} = \frac{1}{3} < \bar{1}2\bar{1}0 >$					60–300 (Fu *et al.*, 1995; Legros *et al.*, 1996; Van de Walle and Asta, 2002)

may also be formed if two adjacent layers are involved (Pope and Ezz, 1984).

Dislocations

A perfect dislocation in an ordered intermetallic compound often has a much greater Burgers vector than that in a disordered metal and is called a superlattice dislocation or superdislocation. For example, a perfect dislocation in an FCC structured disordered alloy has a Burgers vector of $\frac{1}{2}$ <110> whereas a superdislocation in an $L1_2$ structured alloy has a Burgers vector of <110>. Similar to the dissociation of a perfect dislocation into partial dislocations with stacking faults between them in disordered alloys, a superdislocation is often dissociated into superpartials with APB, SISF and CSF between them. The configuration of these superpartials and faults plays a significant role in the deformation of ordered intermetallic alloys. Geometrically possible superdislocations and partial dislocations on likely slip planes in several ordered structures will be described and analysed in this section, although it is noted that the actual occurrence of a particular dissociation depends on the stability of the relevant fault (i.e. a local minimum fault energy state has to exist in the vicinity of the fault vector) (Vitek, 1992). Also, a dissociation would not be observed if the planar defect separating the partials has a high energy (Veyssière and Douin, 1995).

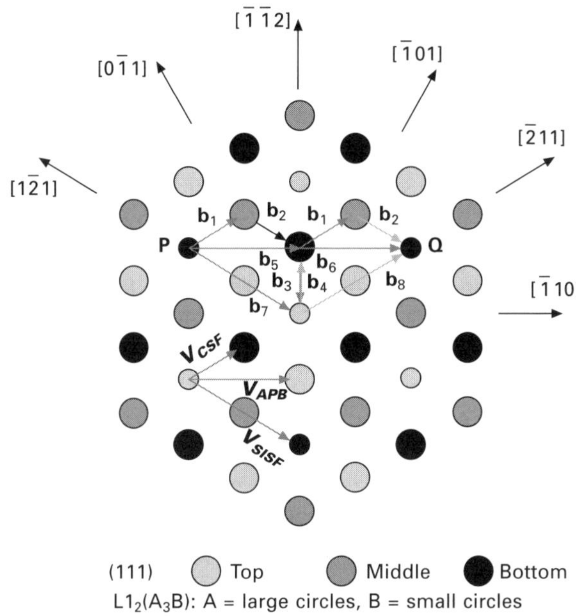

5.6 The atomic arrangement of three consecutive (111) planes in an $L1_2$ structured A_3B compound. The top layer is represented by light grey circles, the middle layer by dark grey ones and the bottom layer by black ones. The larger circles are A atoms and the smaller circles B atoms. The shifting vectors corresponding to the creation of an antiphase boundary (V_{APB}), a superlattice intrinsic stacking fault (V_{SISF}) and a complex stacking fault (V_{CSF}) are shown. Also plotted are the Burgers vectors involved in eqns 5.1–4 with $\mathbf{b}_1 = \frac{1}{6}[\bar{2}11]$, $\mathbf{b}_2 = \frac{1}{6}[\bar{1}2\bar{1}]$, $\mathbf{b}_3 = \frac{1}{6}[11\bar{2}]$, $\mathbf{b}_4 = \frac{1}{6}[\bar{1}\bar{1}2]$, $\mathbf{b}_5 = \frac{1}{2}[\bar{1}10]$, $\mathbf{b}_6 = \frac{1}{2}[\bar{1}10]$, $\mathbf{b}_7 = \frac{1}{3}[\bar{1}2\bar{1}]$, $\mathbf{b}_8 = \frac{1}{3}[\bar{2}11]$ and $\mathbf{PQ} = [\bar{1}10]$].

Dislocations in the $L1_2$ compound A_3B

On a {111} plane, a perfect superdislocation with $\mathbf{b}$ = <110> may be dissociated into partial dislocations with Burgers vectors of $\frac{1}{2}\langle\bar{1}10\rangle$ ($\boldsymbol{V_{APB}}$), $\frac{1}{6}\langle\bar{2}11\rangle$ ($\boldsymbol{V_{CSF}}$) and $\frac{1}{3}\langle\bar{1}2\bar{1}\rangle$ ($\boldsymbol{V_{SISF}}$). The situation on the (111) plane is shown in Fig. 5.6 with the following possible configurations of the dissociation (Yamaguchi and Umakoshi, 1990).

$$[\bar{1}10] = \frac{1}{6}[\bar{2}11] + \text{CSF} + \frac{1}{6}[\bar{1}2\bar{1}] + \text{APB} + \frac{1}{6}[11\bar{2}] + \text{SISF}$$

$$+ \frac{1}{6}[\bar{1}\bar{1}2] + \text{APB} + \frac{1}{6}[\bar{2}11] + \text{CSF} + \frac{1}{6}[\bar{1}2\bar{1}] \qquad 5.1$$

$$[\bar{1}10] = \frac{1}{6}[\bar{2}11] + \text{CSF} + \frac{1}{6}[\bar{1}2\bar{1}] + \text{APB} + \frac{1}{6}[\bar{2}11] + \text{CSF} + \frac{1}{6}[\bar{1}2\bar{1}] \quad 5.2$$

$$[\bar{1}10] = \frac{1}{2}[\bar{1}10] + \text{APB} + \frac{1}{2}[\bar{1}10] \quad 5.3$$

$$[\bar{1}10] = \frac{1}{3}[\bar{1}2\bar{1}] + \text{SISF} + \frac{1}{3}[\bar{2}11] \quad 5.4$$

Reaction (5.1) is shown to be energetically not favoured and thus is unlikely to occur (Yamaguchi *et al.*, 1982). When the APB and CSF are stable, reactions (5.3) and (5.2) are preferred. Otherwise, reaction (5.4) is more likely to occur. On a {100} plane, a superdislocation with **b** = <110> may be dissociated into two $\frac{1}{2}$<110> partials, as in reaction (5.3), with an APB of a lower energy than that on a {111} plane (Yamaguchi and Umakoshi, 1990).

Dislocations in the $L1_0$ compound AB

As $L1_0$ is a face-centred tetragonal (FCT) structure with $a = b \neq c$, {*uvw*) and <*hkl*] are used for planes and directions, respectively, to indicate that the last index in each case, (i.e. *w* or *l*), is not equivalent to the first two and thus may not be permuted (Hug *et al.*, 1988). The perfect translation vectors are $\frac{1}{2}$ <110], <100], $\frac{1}{2}$ <112] and <011]. The atomic arrangement on the (111) plane is shown in Fig. 5.7 and the shifting vectors corresponding to APB, CSF and SISF are $V_{APB} = \frac{1}{2}$ <$\bar{1}$01], $V_{CSF} = \frac{1}{6}$ <$\bar{2}$11] and $V_{SISF} = \frac{1}{6}$ <11$\bar{2}$]. The following full dissociations are possible on a (111) plane (Fig. 5.7) (Hug *et al.*, 1988, Appel and Wagner, 1998),

$$\frac{1}{2}[\bar{1}10] = \frac{1}{6}[\bar{2}11] + \text{CSF} + \frac{1}{6}[\bar{1}2\bar{1}] \quad 5.5$$

$$\frac{1}{2}[11\bar{2}] = \frac{1}{6}[11\bar{2}] + \text{SISF} + \frac{1}{6}[2\bar{1}\bar{1}] + \text{APB} + \frac{1}{6}[11\bar{2}] + \text{CSF} + \frac{1}{6}[\bar{1}2\bar{1}] \quad 5.6$$

$$[01\bar{1}] = \frac{1}{6}[11\bar{2}] + \text{SISF} + \frac{1}{6}[\bar{1}2\bar{1}] + \text{APB} + \frac{1}{6}[11\bar{2}] + \text{CSF} + \frac{1}{6}[\bar{1}2\bar{1}] \quad 5.7$$

although the values of the respective energies of CSF, APB and SISF determine which sub-dissociations would actually take place. Similar to the case in $L1_2$, an <101] dislocation may dissociate into two $\frac{1}{2}$<101] on a {100) plane

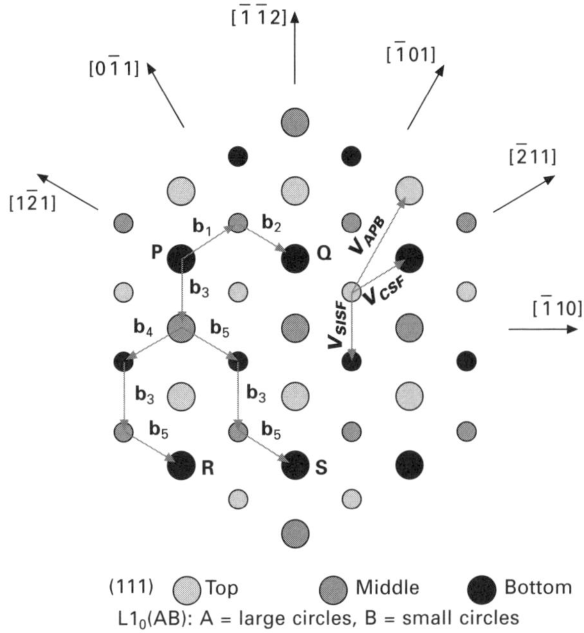

5.7 The atomic arrangement of three consecutive (111) planes in an $L1_0$ structured AB compound. The top layer is represented by light grey circles, the middle layer by dark grey ones and the bottom layer by black ones. The larger circles are A atoms and the smaller circles B atoms. The shifting vectors corresponding to the creation of an antiphase boundary ($\boldsymbol{V_{APB}}$), a superlattice intrinsic stacking fault ($\boldsymbol{V_{SISF}}$) and a complex stacking fault ($\boldsymbol{V_{CSF}}$) are shown. Also plotted are the Burgers vectors involved in eqns 5.5–7 with $\mathbf{b}_1 = \frac{1}{6}[\bar{2}11]$, $\mathbf{b}_2 = \frac{1}{6}[\bar{1}2\bar{1}]$, $\mathbf{b}_3 = \frac{1}{6}[11\bar{2}]$, $\mathbf{b}_4 = \frac{1}{6}[2\bar{1}\bar{1}]$, $\mathbf{b}_5 = \frac{1}{6}[\bar{1}2\bar{1}]$, $\mathbf{PQ} = \frac{1}{2}[\bar{1}10]$, $\mathbf{PR} = \frac{1}{2}[11\bar{2}]$ and $\mathbf{PS} = [01\bar{1}]$.

with an APB of lower energy than that on a {111) plane (Yamaguchi and Umakoshi, 1990). In addition, the decomposition of <011] into $\frac{1}{2}$<$\bar{1}$10] and $\frac{1}{2}$<112] with no planar defects between them and little energy reduction has been proposed and observed although further dissociations as in eqn 5.7 may follow (Greenberg, 1973, Hug *et al.*, 1988).

Dislocations in the B2 compound AB

The only stable planar defect in B2 is APB with $V_{APB} = \frac{1}{2}\langle 111\rangle$ (Yamaguchi and Umakoshi, 1990). On the (110) plane or the (211) plane, a [$\bar{1}$11] dislocation may be dissociated into

$$[\bar{1}11] = \frac{1}{2}[\bar{1}11] + \text{APB} + \frac{1}{2}[\bar{1}11] \qquad 5.8$$

SISF and CSF are not stable and no dissociation of dislocations forming SISF or CSF is observed (Yamaguchi *et al.*, 1981), as predicted in (Vitek, 1974) for BCC structures.

Dislocations in the DO_{19} compound A_3B

The shifting vectors are $V_{APB} = \frac{1}{3} <\bar{1}2\bar{1}0>$, $V_{CSF} = \frac{1}{3} <01\bar{1}0>$ and $V_{SISF} = \frac{2}{3} <\bar{1}100>$, respectively. The atomic arrangement on the basal plane (0001) is shown in Fig. 5.8 and possible dissociations of the superdislocation are

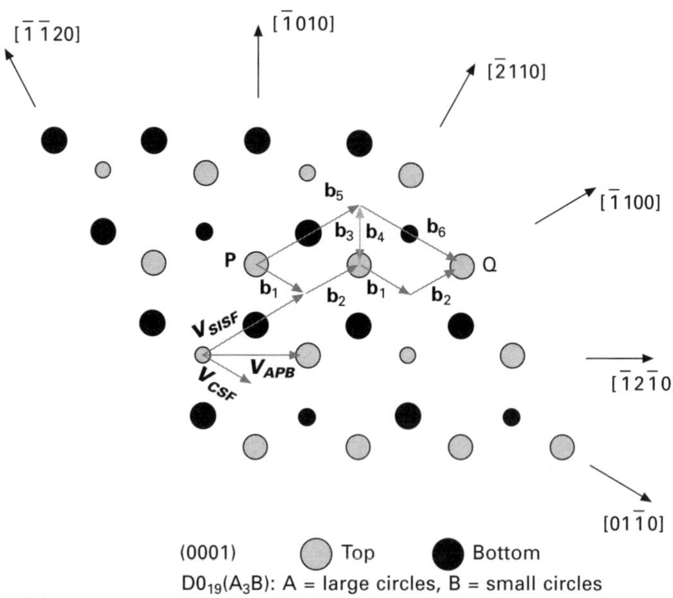

5.8 The atomic arrangement of two consecutive (0001) planes in a $D0_{19}$ structured A_3B compound. The top layer is represented by light grey circles and the bottom layer by black ones. The larger circles are A atoms and the smaller circles B atoms. The shifting vectors corresponding to the creation of an antiphase boundary ($\boldsymbol{V_{APB}}$), a superlattice intrinsic stacking fault ($\boldsymbol{V_{SISF}}$) and a complex stacking fault ($\boldsymbol{V_{CSF}}$) are shown. Also plotted are the Burgers vectors involved in equs 5.9–11 with $\mathbf{b}_1 = \frac{1}{3}[01\bar{1}0]$, $\mathbf{b}_2 = \frac{1}{3}[\bar{1}100]$, $\mathbf{b}_3 = \frac{1}{3}[\bar{1}010]$, $\mathbf{b}_4 = \frac{1}{3}[10\bar{1}0]$, $\mathbf{b}_5 = \frac{2}{3}[\bar{1}100]$, $\mathbf{b}_6 = \frac{2}{3}[01\bar{1}0]$, and $\mathbf{PQ} = \frac{2}{3}[\bar{1}2\bar{1}0]$.

$$\frac{2}{3}[\bar{1}2\bar{1}0] = \frac{1}{3}[01\bar{1}0] + \text{CSF} + \frac{1}{3}[\bar{1}100] + \text{APB}$$
$$+ \frac{1}{3}[01\bar{1}0] + \text{CSF} + \frac{1}{3}[\bar{1}100] \qquad 5.9$$

$$\frac{2}{3}[\bar{1}2\bar{1}0] = \frac{1}{3}[01\bar{1}0] + \text{CSF} + \frac{1}{3}[\bar{1}100] + \text{APB} + \frac{1}{3}[\bar{1}010]$$
$$+ \text{SISF} + \frac{1}{3}[10\bar{1}0] + \text{APB} + \frac{1}{3}[01\bar{1}0]$$
$$+ \text{CSF} + \frac{1}{3}[\bar{1}100] \qquad 5.10$$

$$\frac{2}{3}[\bar{1}2\bar{1}0] = \frac{2}{3}[\bar{1}100] + \text{SISF} + \frac{2}{3}[01\bar{1}0] \qquad 5.11$$

The SISF is always stable (Umakoshi and Yamaguchi, 1981b) whereas the APB and CSF may not be stable. An APB may also be created using $\boldsymbol{V_{APB}} = \frac{1}{3}[\bar{1}2\bar{1}0]$ on a prismatic plane, e.g., $(10\bar{1}0)$, according to (Yamaguchi and Umakoshi, 1990)

$$\frac{2}{3}[\bar{1}2\bar{1}0] = \frac{1}{3}[\bar{1}2\bar{1}0] + \text{APB} + \frac{1}{3}[\bar{1}2\bar{1}0] \qquad 5.12$$

If the APB energy on (0001) is much higher than that on $(10\bar{1}0)$, eqn 5.12 is preferred; otherwise, eqn 5.9 may be observed (Cserti *et al.*, 1992).

5.3 Plastic deformation

Plastic deformation in crystalline materials is primarily carried out by dislocation movement, gliding of dislocations on slip planes at lower temperatures and gliding and climbing at higher temperatures. When slip is restricted, twinning may play an important role. Other mechanisms such as diffusion and grain boundary sliding may become significant if deformation is slow, at very high temperatures and/or in materials with very fine grains. Since bonding, crystal structures and defects are different and more complicated in ordered intermetallics, some unique mechanical behaviours are observed. In this section, basic elements of plastic deformation (i.e. dislocation slip and twinning) in selected structures (single phase with stoichiometric composition) are described. This is followed by analysis of deformation behaviours at room temperature. Finally, the effects of temperature, composition and microstructures are discussed.

5.3.1 Dislocation slip

In monolithic metals of FCC, HCP and BCC structures, the easiest slip system consists usually of the close packed direction (the slip direction) on the close packed plane (the slip plane). Owing to the complex atomic structures and bonding, dislocation core structures are often complicated in ordered intermetallics (Vitek, 1998). Consequently, the actual slip systems are often not as straightforward. Typical slip behaviours in selected structures are analysed below and this forms the basis for further discussion about deformation behaviours in particular compounds.

Slip in the $L1_2$ compound A_3B

Although dislocations with **b** = <100> have the lowest energy, the resistance to their movement is high owing to a sessile core structure (Veyssière and Douin, 1985; Pasianot *et al.*, 1991). The mobile dislocations almost always have a total Burgers vector of <110> (usually dissociated into partials as described under 'Dislocations' above. At lower temperatures, the dominant slip systems are {111}<110>, similar to those in the disordered FCC structure, since the <110> superdislocations can dissociate on {111} planes into partials with completely planar core structures and slip is easy (Yamaguchi *et al.*, 1982; Paidar *et al.*, 1982). However, if the screw section of a $\frac{1}{2}$<110> superpartial moves onto a {001} plane (cross slip) and spreads its core on an intersecting {111} plane (i.e. the core of the partial is not on the same plane as the slip plane), the whole assembly becomes sessile (Yamaguchi *et al.*, 1982; Tichy *et al.*, 1986; Vitek, 1998). This non-planar configuration may be changed at high temperatures as a result of thermal activation and the {001}<110> slip may become active at high temperatures (Staton-Bevan and Rawlings, 1975). The core structure is also non-planar when <110> dissociates into two $\frac{1}{3}$<112> partials with a SISF, as in eqn 5.4 (Tichy *et al.*, 1986; Vitek, 1998) and their movement is expected to be difficult at lower temperatures as well.

Slip in the $L1_0$ compound AB

Slip takes place on {111) planes with **b** = $\frac{1}{2}$<110], $\frac{1}{2}$<112] and <101] at most temperatures. These perfect dislocations may dissociate into partials with APB, SISF and CSF, as described above. The $\frac{1}{2}$<110] dislocation may split into partials according to eqn 5.5 if the CSF energy is low, and the core structure is planar; when the CSF energy is high, the core of the screw component of the dislocation is non-planar and the dislocation is sessile (Mahapatra *et al.*, 1995; Simmons *et al.*, 1997). The $\frac{1}{2}$<112] dislocation may be dissociated on two parallel {111) planes forming an SESF together with an SISF (Mahapatra *et al.*, 1995; Inkson, 1998); depending on the configuration of the dissociation, the dislocation may be glissile or sessile (Inkson, 1998).

The dissociation of the $\frac{1}{2}$<112] dislocation may also be planar according to eqn 5.6 although the split forming CSF may be suppressed (Hug *et al.*, 1988). The <101] superdislocation may dissociate according to eqn 5.7 although the CSF forming split might be absent if the CSF energy is very high, and the core is planar (Hug *et al.*, 1988; Mahapatra *et al.*, 1995); however, at elevated temperatures, the APB associated with the screw part of the dislocation may lie on a {100) plane and the whole configuration becomes sessile (Hug *et al.*, 1988; Hemker *et al.*, 1993). In the case of low SISF energies, a non-planar dissociation of the screw component of a <101] dislocation

$$[10\bar{1}]_{\text{screw}} = \frac{1}{6}[11\bar{2}] + \text{SISF}_{(111)} + \frac{1}{3}[20\bar{1}] + \text{SISF}_{(\bar{1}1\bar{1})} + \frac{1}{6}[1\bar{1}\bar{2}] \qquad 5.13$$

has been proposed with SISF on two {111) planes, resulting in a sessile configuration (Mahapatra *et al.*, 1995).

Slip in the B2 compound AB

The possible slip systems in B2 involve dislocations with **b** = <100>, <110> and <111> on {110} and/or {112} planes (Yamaguchi and Umakoshi, 1990). At room temperature, the <100>{110} slip is favoured in strongly ordered compounds such as NiAl and CoAl although the <111> slip would become active if the resolved shear stress on the <100> systems is very low (Yamaguchi and Umakoshi, 1990; Field *et al.*, 1991; Baker, 1995). On the other hand, the <111>{110} slip dominates at room temperature in weakly ordered compounds like FeAl (Yamaguchi and Umakoshi, 1990; Veyssière and Douin, 1995) and the deformation behaviour is close to that observed in the disordered BCC structure. The <111> slip, however, would take place on {112} with increasing ordering energy and lowering temperature (Yamaguchi and Umakoshi, 1990). At elevated temperatures, however, slip activities on <110>{110} in strongly ordered compounds (Mills and Miracle, 1993) and on <100>{110} in weakly ordered alloys (Umakoshi and Yamaguchi, 1981a) are observed, respectively. The operation of the <100> dislocations on {110} planes which does not occur in the disordered BCC structure is attributed to their planar core structure in B2 (Yamaguchi and Umakoshi, 1975, Gumbsch and Schroll, 1999). The configuration of the <111> dislocations is more complicated: Whereas the core structure of the edge and mixed dislocations consist of planar dissociations, the screw component does not show a split and its core is non-planar (Gumbsch and Schroll, 1999; Umakoshi, 1993).

Slip in the DO_{19} compound A_3B

Although always along the $<\bar{1}2\bar{1}0>$ directions, slip may take place on the basal as well as prismatic planes. For example, the principal slip system in Mn_3Sn is the basal $<\bar{1}2\bar{1}0>$ (0001), but that in Ti_3Al is the prismatic $<\bar{1}2\bar{1}0>$ $\{10\bar{1}0\}$, with the prismatic and basal systems as secondary slip modes respectively (Bacon and Vitek, 2002). In addition, pyramidal slip systems may also operate (Yoo *et al.*, 2002). In the case of the superdislocation $\frac{2}{3}[\bar{1}2\bar{1}0]$ splitting into two $\frac{1}{3}[\bar{1}2\bar{1}0]$ superpartials on (0001), the superpartials may dissociate further into Shockley partials according to eqn 5.9. The core of the Shockley partial is shown to spread into the prismatic plane, i.e., the whole configuration is non-planar and thus sessile (Cserti *et al.*, 1992). When the dissociation of the superdislocation is on a $\{10\bar{1}0\}$ prism plane according to eqn 5.12, the superpartials would split into smaller partials with cores spreading into the basal plane, again raising the Peierls stress for their movement (Cserti *et al.*, 1992, Vitek, 1998).

5.3.2 Twinning

Twinning is considered an important plastic deformation mechanism in alloys with limited slip systems such as in metals with the hexagonal close packed structure (e.g. Zn and Mg). It may also be observed at low temperatures and/or high strain rates when dislocation slip becomes more difficult even in metals with FCC and BCC structures where plenty of slip systems are available otherwise. In ordered intermetallics, dislocation slip is often more restricted compared with that in disordered alloys. Consequently, twinning is more visible and considered more important in plastic deformation, and the major twinning systems have been documented for a number of important structures (Yoo, 2002). Among the structures focused on in this chapter (i.e. $L1_2$, $L1_0$, B2 and $D0_{19}$), significant deformation twinning has been widely reported only in the $L1_0$ structured TiAl and related materials (Shechtman *et al.*, 1974; Feng *et al.*, 1989; Jin and Bieler, 1992; Morris and Leboeuf, 1997; Yoo, 1998). Although twinning has been observed in TiAl based alloys at elevated temperatures under creep conditions (Jin and Bieler, 1992, Morris and Leboeuf, 1997), it is shown that during creep of fully lamellar TiAl extensive twinning occurs mainly in hard orientated grains especially in compression whereas dislocation slip dominates in softly orientated grains (Wu *et al.*, 2001, 2002, Xia *et al.*, 2004). It appears that twinning is still a secondary deformation mechanism in ordered intermetallics although its importance is increased owing to more restricted dislocation slip.

5.3.3 Deformation behaviour

As discussed so far, ordered intermetallic alloys generally possess stronger bonds and more complicated crystal structures and, consequently, they are stronger and stiffer especially at elevated temperatures. However, the same traits also cause reduction in ductility and toughness, as observed in all types of materials. It is a perennial effort to balance strength and ductility. In the following, the room temperature deformation behaviour of single phase, stoichiometric intermetallics will be described first. This will be followed by discussions of the effects of temperature, composition and microstructure on strength and ductility with a view to grasping the peculiar attributes shown by ordered intermetallics. The focus is mostly on intrinsic strength and ductility and wherever possible, extrinsic factors such as environment and impurity will be excluded except when their effects are discussed. The creep behaviour of ordered intermetallics, although very important for many applications, will be discussed when relevant but will not be addressed here as a topic.

Strength and ductility at room temperature in single phase, near-stoichiometric alloys

Strength

The strength of a material depends not only on its composition and crystal structure, but more importantly on its microstructure. In order to assess the inherent strength of a stoichiometric compound, the critical resolved shear stress (CRSS) of a single crystalline material in its fully annealed state is often used. It is useful to compare CRSS of the operative slip systems at room temperature between the ordered intermetallic compounds discussed earlier (Ni_3Al, TiAl, NiAl, FeAl and Ti_3Al) and the ingredient monolithic metals (Ni, Al, Ti and Fe). The values of CRSS and corresponding slip systems are presented in Table 5.2. Among the disordered pure metals, the FCC structured Ni and Al have very low CRSS (< 10 MPa), the BCC structured Fe has higher CRSS (~25–30 MPa), and the HCP structured Ti (with c/a = 1.588 < the ideal 1.632) has the highest CRSS of up to 90 MPa for the prismatic slip and over 100 MPa for the basal slip. In contrast, the ordered alloys show much higher CRSS. The FCC based Ni_3Al possess CRSS of over 150 MPa, much higher than pure Ni. Since Ni and Ni_3Al have comparable melting points, the bonding strength should also be of the same order of magnitude. The greatly enhanced CRSS in Ni_3Al may be attributed to its ordered structure and bonding character which is not purely metallic. Similar conclusions can be made between the BCC based Fe and FeAl/NiAl and between the HCP based Ti and Ti_3Al. In addition, CRSS may depend on the orientation of the crystals (Umakoshi *et al.*, 1984, 1993; Heredia and Pope, 1991b; Nitz and Nembach, 1998; Kawabata *et al.*, 1985) and the loading directions (Umakoshi *et al.*, 1984; Paidar *et al.*, 1984; Heredia and Pope,

Table 5.2 CRSS at room temperature corresponding to various slip systems in selected ordered intermetallic compounds and disordered monolithic metals

Material	Slip system	Loading direction	CRSS (MPa)	Reference
Ni_3Al	{111}<110>	Various loading directions	150–250	(Umakoshi *et al.*, 1984)
TiAl	{111)<110]	Various loading directions	60–170	(Inui *et al.*, 1997, Stucke *et al.*, 1995, Kawabata *et al.*, 1985)
NiAl	{110}<100>	Non <100> (soft orientation)	40–100	(Miracle, 1993)
	{112}<111>	<100> (hard orientation)	500–700	(Miracle, 1993)
FeAl	{110}<111>	Various loading directions	100–140	(Yoshimi *et al.*, 1995)
Ti_3Al	$(0001)\ <\bar{1}2\bar{1}0>$	Various loading directions	~330 (off stoichiometric)	(Inui *et al.*, 1993)
	$\{10\bar{1}0\}\ <\bar{1}2\bar{1}0>$	Various loading directions	60–70	(Minonishi, 1995) (Umakoshi *et al.*, 1993)
			90–110 (off stoichiometric)	(Inui *et al.*, 1993)
	$\{11\bar{2}1\}\ <\bar{1}\bar{1}26>$	Various loading directions	~450	(Umakoshi *et al.*, 1993)
			910 (off stoichiometric)	(Minonishi, 1995); Inui *et al.*, 1993)
Ni	{111}<110>		~10	(Andrade and Henderson, 1951)
Al	{111}<110>		~1	(Rosi and Mathewson, 1950; Noggle and Koehler, 1957)
Ti	$(0001)\ <\bar{1}2\bar{1}0>$		105–110	(Anderson *et al.*, 1953; Churchman 1954)
	$\{10\bar{1}0\}\ <\bar{1}2\bar{1}0>$		15–90 (depending on purity)	(Anderson *et al.*, 1953; Churchman, 1954)
Fe	{110}<111>		~25–30	(Cox *et al.*, 1957)

1991b; Zupan and Hemker, 2003) although a recent investigation disputes some of the findings in Ni_3Al (Hirano *et al.*, 1999). In general, the ordered intermetallics are inherently stronger in the absence of other strengthening mechanisms.

Ductility

Although the inherently higher resistance to dislocation movement may make those intermetallic compounds with relatively low cleavage strength susceptible to brittle fracture, many single crystalline compounds show considerable ductility at room temperature. For example, single crystalline Ni_3Al displays substantial tensile elongations up to 95% (Hirano *et al.*, 1999; Heredia and Pope, 1991a; Copley and Kear, 1967); NiAl (Darolia *et al.*, 1999) and FeAl (Wu and Baker, 2001) single crystals may reach elongations as high as 13–16%; at the orientations favouring prismatic slip, single crystal Ti_3Al shows a dramatic tensile elongation of over 200% at room temperature (Inui *et al.*, 1993). On the other hand, single crystalline TiAl is invariably brittle at room temperature even in compression (Kawabata *et al.*, 1985); this may be attributed to its relatively low cohesive energy of {111} planes, causing easy cleavage fracture (Appel *et al.*, 1995, Yoo and Fu, 1998).

However, good ductility may be lost in polycrystalline materials. Despite its FCC based structure and availability of sufficient slip systems, polycrystalline Ni_3Al was considered brittle until the famous discovery by Aoki and Izumi (1979) that the addition of boron could significantly improve its ductility. Much research followed to establish the status of boron as a magic bullet in the case of Ni_3Al and to find an explanation (e.g. (Liu *et al.*, 1985, Liu, 1991)). Considerable tensile ductility (as high as 50%) has been observed in polycrystalline Ni_3Al tested in oxygen (Liu, 1992, George *et al.*, 1992), leading to the conclusion that environmental embrittlement is responsible for the lack of plasticity and the role of B is to block the passage of hydrogen along grain boundaries (Liu and Pope, 1995). Nevertheless, efforts continue to investigate the chemistry and physics of grain boundaries for new insight (Masahashi, 1997). As described above, different slip systems are operative in NiAl and FeAl at room temperature despite the same B2 structure. In NiAl, the <100> slip does not provide a sufficient number of systems and polycrystalline NiAl is brittle despite limited improvement by adding alloying elements (George and Liu, 1990; Hahn and Vedula, 1989, Bowman *et al.*, 1992). In FeAl, the slip systems in operation at room temperature are <111>{110} providing a sufficient number of deformation modes and it also has a high cleavage strength compared to NiAl (Fu and Yoo, 1992); polycrystalline FeAl therefore shows considerable ductility (>10%) when environmental embrittlement is prevented (Liu *et al.*, 1989). Similar to NiAl, Ti_3Al does not have sufficient slip systems and its polycrystalline form behaves in a completely brittle manner (Lipsitt *et al.*, 1980). As may be expected from the behaviour of its single crystalline counterpart, polycrystalline, single phase γ-TiAl does not show much ductility either (Huang and Hall, 1991). A recent investigation reports high ductility at room temperature in ambient atmosphere in some B2 structured stoichiometric compounds of a rare earth metal and another metal, although the materials have very fine grains (<0.5 μm) (Gschneidner Jr *et al.*, 2003).

Effects of temperature

It is generally known that the strength of materials decreases with increasing temperature because atoms are more activated at high temperatures. However, quite a number of ordered intermetallic compounds display the so-called anomalous temperature dependence (Veyssière, 2001), i.e., their yield strengths increase with increasing temperature within certain ranges. The significance of this anomalous behaviour is obvious when the strength at elevated temperatures is concerned. Among the alloys considered here, anomalous temperature dependence has been observed in Ni_3Al (Westbrook, 1957; Copley and Kear, 1967; Umakoshi *et al.*, 1984; Nitz and Nembach, 1998); TiAl (Kawabata *et al.*, 1985; Inui *et al.*, 1997; Zupan and Hemker, 2003); Ti_3Al (Umakoshi *et al.*, 1993; Minonishi, 1991) and under certain conditions, FeAl (George and Baker, 1998).

Some of the most important observations in materials with $L1_2$ structure include: (i) the yield strength or CRSS increases with temperature until a peak temperature above which the strength decreases; (ii) the Schmid's law breaks down (i.e. the CRSS is orientation dependent) in a range of temperatures near the peak temperature (Takeuchi and Kuramoto, 1973) and there exists a tension/compression asymmetry (Ezz *et al.*, 1982) although this has been shown to be not necessarily the case in single crystals of stoichiometric Ni_3Al (Hirano *et al.*, 1999); (iii) there is a transition of slip from {111} to {100} planes at high temperatures (Copley and Kear, 1967); (iv) the anomalous temperature dependence is influenced by the strain at which the yield stress is taken (Thornton *et al.*, 1970); (v) the yield stress is considerably strain rate sensitive only above the peak temperature (Thornton *et al.*, 1970, Takeuchi and Kuramoto, 1973); (vi) the rate of work hardening increases with increasing temperature up to the peak temperature (Takeuchi and Kuramoto, 1973); and (vii) long, straight screw dislocations are typical of the structure deformed below the peak temperature while more complex structures appear at higher temperatures (Kear and Wilsdorf, 1962).

In the case of the $L1_0$ structured TiAl, the CRSS decreases with increasing temperature up to about room temperature and, after a relatively flat region, increases to reach a peak at around 600–800 °C (Zupan and Hemker, 2003). Many other features observed in the $L1_2$ structured materials are also observed in TiAl, such as the orientation dependence and tension/compression asymmetry, high work hardening rate and low strain rate sensitivity below the peak temperature and the straight dislocation configuration (Kawabata *et al.*, 1985, Inui *et al.*, 1997, Zupan and Hemker, 2003).

The models developed to explain the yield stress anomaly rely, to various degrees, on the famous Kear-Wilsdorf (K-W) lock. The formation of the K-W locks in the $L1_2$ structure involves cross slip of segments of the <110> screw dislocations from their original slip planes of {111} onto the {100} planes on which the APB energy is lower (Flinn, 1960; Kear and Wilsdorf, 1962; Takeuchi and Kuramoto, 1973) since the cross-slipped segments are no longer mobile on {111} planes, they become sessile 'locks'. The cross slip, however, needs thermal activation to produce constricted segments along the superdislocation line, and as temperature increases, more locks are formed and resistance to slip increases. In the first comprehensive model (Takeuchi and Kuramoto, 1973), a quantity called the cross slip probability is introduced and related to an activation energy. The activation energy is further divided into two parts, one related to the change in the core structure, the activation enthalpy, and the other to the stress-assisted activity. The allowance for the stress related part leads to the explanation of the dependence of the CRSS on orientations. When the resolved shear stress on {100} is increased (e.g. as a result of orientation change), the activation energy is reduced and the cross slip probability increases, resulting in more obstacles to the motion of the dislocation and thus greater CRSS. However, no analysis of the activation enthalpy other than an estimation of its value is possible in this model without a good knowledge of the core structure of dislocations, the configuration of the planar defects due to dislocation dissociation and the details of the cross slip process.

Based on atomistic simulation, Paidar *et al.* (1984) were able to present a calculation of the activation enthalpy. Two points were emphasised by them. First, in the $L1_2$ structured alloys a dislocation on a {111} plane would dissociate into two superpartials of the type $\frac{1}{2}$<110> with an APB between them (eqn 5.3); the superpartial would further dissociate into two Shockley partials of the type $\frac{1}{6}$<112> separated by a CSF on the {111} plane (eqn 5.2). Second, even if the leading superpartial had cross slipped to lie on a {100} plane, the dissociation of it into two Shockley partials would still occur on a {111} plane. Therefore, before the cross slip the whole configuration is mobile whereas the cross-slipped segments are sessile for the core of the superpartial does not lie on the same plane. The cross slip process could then be described as follows: the superpartial constricts itself on the original {111} plane and moves onto a {100} plane until it re-dissociates into Shockley partials on another {111} plane intersecting the {100} plane. Considering all these, they calculated the activation enthalpy which is related to the APB energy on {111} planes, the resolved shear stress on {100} planes and the width of the core dissociation on {111} planes. This so-called PPV model is used successfully to explain the tension/compression asymmetry observed.

In order for the deformation process to continue, dislocations have to be able to break away from these locks. Although Takeuchi and Kuramoto (1973) included a dynamic break-away process in their model, their analysis was rough and inconsistent (Xia, 1986). A thermally assisted unlocking process was proposed to modify the PPV model (Khantha *et al.*, 1992). A comprehensive treatment involving the creation, movement, pinning and unpinning of the so-called super kinks can be found in a later model (Hirsch, 1992). Despite the enormous amount of effort put into the study of the anomalous temperature dependence shown by many ordered intermetallics, some details of the mechanism responsible for the phenomenon are still not entirely clear. Nevertheless, this serves as a good example of unique behaviours in ordered intermetallics dictated by the fine structures of dislocations and planar defects.

Effects of composition

Here the discussion is restricted to compositions which do not result in the occurrence of new phases. In disordered alloys, solid solutions are formed with the solute atoms either substituting for the host atoms (substitutional) or staying between the host atoms (interstitial). In binary ordered intermetallics, the off-stoichiometric host atoms may substitute in the 'wrong' sites to form antisites or be accommodated by adjusting the number of vacancies in either site; foreign atoms may substitute for either or both of the host species or remain in interstitial sites. Consequently, their effects on mechanical properties are more complicated and less predictable. In one example of Ni_3Al, the peak yield strength of ~400 MPa at ~600 °C of the stoichiometric composition is increased to ~700 MPa with 26.5 at% Al and reduced to just over 300 MPa with 23.5 at% Al (Noguchi *et al.*, 1981). In another example, the peak yield strength of ~550 MPa of the binary Ni_3Al is increased to ~800, > 900, ~1000 and ~1200 MPa by additions of 5 at% Nb, 5 at% Ta, 14 at% Si and 4 at% Hf, respectively (Mishima *et al.*, 1986; Ochiai *et al.*, 1986). It is noticed that these very effective solution strengthening elements (Nb, Ta, Si and Hf) occupy the Al sites (Ochiai *et al.*, 1984) although Hf may occupy both Al and Ni sites (Stoloff, 1989). However, it is not clear why these elements are so effective in increasing the yield strength of Ni_3Al. Perhaps the most well-known example is the addition of B to polycrystalline Ni_3Al to improve its room temperature ductility (Aoki and Izumi, 1979), as already discussed earlier.

Effects of microstructure

Although pure composition manipulation appears to be fairly effective in improving strength and ductility in such alloys as Ni_3Al which can dissolve a reasonably large amount of solute atoms without introducing new phases, this may not be achievable or effective for many other alloys. Another, perhaps more versatile and effective, way of achieving a good combination of strength and ductility is through controlling microstructures, in particular microstructures consisting of multiple phases. Indeed, disordered alloys such as those of aluminium, titanium and iron rely on secondary phases for their high strength as pure metals are typically soft and solid solution can only bring about a moderate gain in strength. The main objective for ordered intermetallics is, however, quite the opposite, i.e., to manipulate the microstructure to gain ductility. This approach has been found to be more effective and flexible (Cahn, 1995; Kimura and Pope, 1998). Two examples will be described here.

The monolithic γ-TiAl has been shown to be brittle, as discussed earlier. However, the Ti-rich alloys containing ~40–49 at% Al possess two phases, the γ phase and the α_2 phase of Ti_3Al. The relevant section of the Ti-Al phase diagram is shown in Fig. 5.9, following the version used in (Huang and Chesnutt, 1995). Although a wide variety of microstructures may be obtained through heat treatment (Kim, 1994), the most important dual-phase microstructures include the duplex and the fully lamellar structures. When an alloy is solution treated in the single α region above the α transus temperature, T_α, equiaxed α grains will be formed. If this single phase

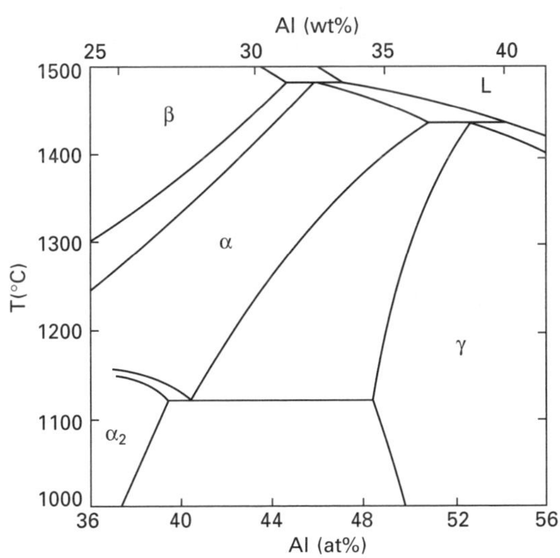

5.9 A portion of the Ti-Al phase diagram adapted from (Huang and Chesnutt, 1995).

microstructure is cooled slowly, γ lamellae will form first in the matrix of the α grains in the $\alpha + \gamma$ two phase field, and the α phase will become the ordered α_2 phase below the eutectoid temperature (Fig. 5.9). The resulting microstructure consists completely of lamellar grains of alternating γ and α_2 plates as shown in Fig. 5.10 and is referred to as the fully lamellar microstructure. The orientation relationship between the γ and α_2 lamellae has been identified to be: $(0001)_{\alpha_2} // \{111\}_\gamma$ and $<1210]_{\alpha_2} // <110]_\gamma$ (Blackburn, 1970). On the other hand, if the alloy is heated only to the $\alpha + \gamma$ two-phase region before being cooled, a duplex microstructure of equiaxed γ grains and

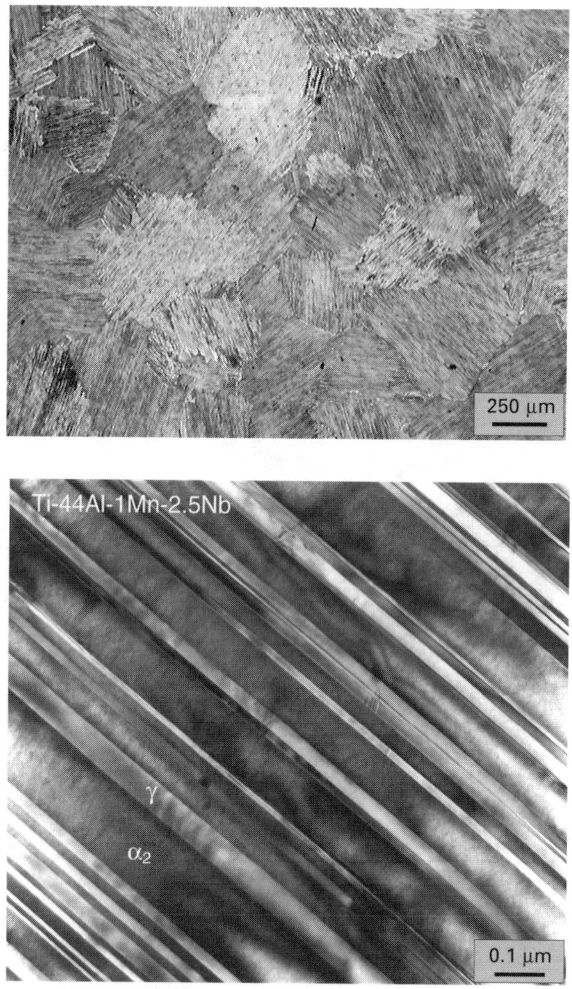

5.10 (a) Optical and (b) TEM microstructures showing the full lamellar structure of alternating γ and α_2 laths in a Ti-44Al-1Mn-2.5Nb (at%) alloy.

lamellar grains of various proportions (more or less equal proportions for technologically important alloys) will be formed.

The most important microstructural parameters are the grain sizes and the lamellar thicknesses as they determine the mechanical properties. With conventional processing, the fully lamellar microstructure consists of coarse lamellar grains of several hundred microns whereas the duplex microstructure is much finer with grains smaller than 50 μm. The lamellae are of the order of 1 μm thick. It is found that the duplex microstructure gives rise to better ductility and strength at room temperature (Huang and Hall, 1991) and the fully lamellar microstructure leads to better toughness and creep resistance (Huang, 1992; Chen *et al.*, 1999). However, when the grain sizes of the fully lamellar materials are refined to < 10 μm and the lamellar thicknesses to the order of 0.1 μm, tensile ductility of up to 5% is obtained which is better than that achieved in the duplex structured materials (~3%) with good strength (>800 MPa) and fracture toughness (>25 MPa $m^{1/2}$) (Liu *et al.*, 1996; Maziasz and Liu, 1998). At high temperatures, the creep resistance can be reduced significantly by decreasing the lamellar thicknesses (Maruyama *et al.*, 1997; Chen *et al.*, 1999; Xia *et al.*, 2004). It is noted, however, that the fine lamellar structure is not thermally stable and the steady state creep observed in the alloys with coarse lamellae is lost in the materials with fine lamellae, as shown in Fig. 5.11 (Xia *et al.*, 2004).

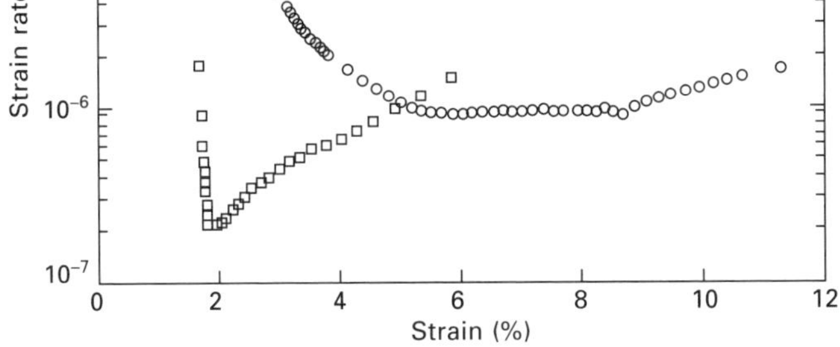

5.11 Strain rate versus strain curves corresponding to a coarse lamellar material with lamellar thicknesses of ~260 nm and a fine lamellar material with lamellar thicknesses of ~20 nm for a Ti-44Al-1Mn-2.5Nb alloy undergoing tensile creep at 750 °C and 220 MPa.

Another prominent example is Ti_3Al. It is only after the addition of a large amount of Nb (> ~10 at%) that the alloys show some reasonable room temperature ductility thanks to the presence of a significant amount of the B2 phase (Ti_2AlNb) (Banerjee, 1995). This ordered structure results from a high-temperature BCC structured β-Ti phase that has been stabilised by the addition of Nb. The B2 phase is shown to increase the strain needed to cause cleavage fracture in α_2. In addition, an ordered orthorhombic structure, the O phase, could introduce more varieties of microstructures (Banerjee, 1995) and may play important roles in improving the ductility and strength of Ti_3Al-Nb alloys (Wu *et al.*, 2004).

5.4 Deformation processing

Deformation processing has two primary purposes, to change the shape and to change the microstructure. Large plastic deformation is required and the material has to be ductile and possess relatively low flow stress at the processing temperature. For most intermetallics, this is possible only at relatively high temperatures although boron-containing Ni_3Al has been cold worked (Ball and Gottstein, 1993a,b; Escher *et al.*, 1998; Escher and Gottstein, 1998). The final microstructure depends not only on the deformation process, but also on dynamic recovery and recrystallisation. An important recent development is to use very large plastic deformation to bring about ultrafine structured materials. These will be discussed briefly whereas superplastic deformation, which is related to fine grain sizes and low strain rates, will be entirely omitted (interested readers may go to Chapter 8 for a general understanding and to the review by Nieh and Wadsworth (1999) for information related to intermetallics).

5.4.1 Deformation at high temperatures and resulting microstructures

Isothermal compression tests are most commonly used to assess hot formability of intermetallics although tensile tests are occasionally employed (Seetharaman and Semiatin, 1997). The stress versus strain curve may show work hardening, a steady state or work softening, depending on the temperature and strain rate. The behaviour is also closely related to the initial microstructure which affects deformation mechanisms and dynamic recovery and recrystallisation processes. In the example of TiAl, the starting microstructure may be nearly lamellar (Semiatin *et al.*, 1994; Sabinash *et al.*, 1995; Seetharaman and Semiatin, 2002), duplex (Semiatin *et al.*, 1992a; Sabinash *et al.*, 1995; Seetharaman and Semiatin, 1997; Kim *et al.*, 2003), or near-gamma (Sabinash *et al.*, 1995; Seetharaman and Semiatin, 1997; Millett *et al.*, 1999; Zhang *et al.*, 2000).

For a lamellar structured material deformed at 1093 °C, following an initial work hardening to reach a peak stress, there is significant work softening leading to an apparent steady state flow (Seetharaman and Semiatin, 2002). The deformation is thought to be controlled by dislocation glide and climb although a strong grain size dependence is observed. The softening is typical of such materials with lamellar structures and attributed to the breakdown of the lamellar structure through micro-buckling and dynamic spheroidisation (Seetharaman and Semiatin, 1996), leading to a partially or completely spheroidised structure depending on grain sizes (Seetharaman and Semiatin, 2002; Semiatin *et al.*, 1994). Flow softening is also considerable in duplex structured materials at relatively lower temperatures and higher strain rates although it is attributed primarily to deformation heating (Semiatin *et al.*, 1992a). At a much slower strain rate (0.001 s^{-1}) a steady state is observed at both a higher (1250 °C) and a lower (800 °C) temperature, but strain softening occurs at the intermediate temperatures (Kim *et al.*, 2003).

In materials with a near gamma starting microstructure, the extent of strain softening is reported to be mild (Millett *et al.*, 1999) and the peak and flow stresses are much lower than those in a comparable alloy (Zhang *et al.*, 2000). Dynamic recrystallisation is observed in materials with a variety of starting microstructures at lower temperatures and higher strain rates whereas dynamic recovery is believed to prevail at higher temperatures and slower strain rates (Sabinash *et al.*, 1995). In all the cases, the microstructure after hot deformation is finer and more uniform. A summary of the hot deformation behaviour and workability of γ-TiAl based alloys can be found in Semiatin *et al.* (1998) and Appel *et al.* (2002).

Some more limited investigations have been carried out into other intermetallic alloys such as Ti_3Al (Semiatin *et al.*, 1992b; Millett *et al.*, 2000a), Ni_3Al (Millett *et al.*, 2000b), NiAl (Millett *et al.*, 2001) and FeAl (Kupka, 2004). The unique bonding and structures of ordered intermetallics have not manifested themselves in high-temperature deformation as prominently as in low-temperature deformation. This is because the temperature is often high enough to enable sufficient dislocation movement although the softening processes may be more complicated (Baker, 2000). The more important factor may be the actual microstructure which often consists of several phases including disordered high-temperature phases. From a practical point of view, the flow stress is likely to be greater and the processing temperature higher, making their wrought processing more expensive and difficult.

5.4.2 Severe plastic deformation

Severe plastic deformation (SPD) has emerged as a major approach to producing bulk ultrafine structured materials (Valiev *et al.*, 2000). Equal channel angular pressing (ECAP) (Langdon *et al.*, 2000) and high-pressure torsion (HPT)

(Zhilyaev *et al.*, 2003) are among the most promising methods. In particular, ECAP has advantages over other processes since there is no cross-section reduction, plastic deformation is largely uniform and in the form of simple shear, and the total strain can be accumulated through repeated pressing in a number of different ways (Xia and Wang, 2001, Furukawa *et al.*, 1998). The resulting microstructure consists of ultrafine grains as small as 100 nm with high-angle grain boundaries. High-speed and low-temperature superplasticity (Langdon *et al.*, 1998) and simultaneously high strength and ductility at room temperature (Valiev *et al.*, 2002) have been realised in disordered metals. It is of great interest to see whether the same benefits can be obtained in ordered intermetallics.

ECAP has been carried out on TiAl alloys in a limited number of investigations (Semiatin *et al.*, 1995; Cornwell *et al.*, 1996; Sastry *et al.*, 2000; Sastry and Mahapatra, 2001). In ECAP of a nearly lamellar TiAl alloy, severe strain localisation leading to shear bands and cracking is observed at a lower processing temperature of 1150 °C whereas deformation is more uniform at 1250 °C. Within the deformed region, the microstructure is very much refined (Semiatin *et al.*, 1995). The formability is apparently much improved under isothermal conditions and several TiAl alloys with different microstructures are ECA pressed at an even lower temperature of 1000 °C (Sastry and Mahapatra, 2001); the starting nearly lamellar (average grain size 300 μm), duplex (33 μm) and equiaxed γ (53 μm) structures are transformed into much refined structures of partially recrystallised lamellar (< 5 μm in the recrystallised region), partially recrystallised duplex (< 2 μm in the recrystallised region) and mostly recrystallised γ (1–2 μm), respectively. However, the uniformity of deformation at a macroscopic level is not shown. The refined microstructures after ECAP are also shown to be stronger and more ductile. Although the degree of success overall is fairly limited, ECAP of such alloys is apparently feasible if the optimum starting microstructure and deformation conditions are used, and the results could be very attractive.

HPT, on the other hand, has been used to deform Ni_3Al (Languillaume *et al.*, 1993) and TiAl (Korznikov *et al.*, 1999b) at room temperature to very large strains. Grain sizes less than 50 nm are produced in both cases. The ordered structure of Ni_3Al is completely lost (Korznikov *et al.*, 1999a, 2001b) while the ordered structure of TiAl is transformed only at increased pressure and strain (Korznikov *et al.*, 1999b). Subsequent annealing can restore order in Ni_3Al to various degrees although its influence on strength and ductility does not show a clear trend (Korznikov *et al.*, 2001a). Superplasticity in Ni_3Al has been observed in the nanometer-sized materials at elevated temperatures (Valiev *et al.*, 2001).

In summary, SPD of ordered intermetallics is still in its infancy. The high temperatures and pressures often necessary in processing have limited the number of investigations. However, it is an important method in producing ultrafine and nano-structured bulk materials, and potential gains in much improved strength and ductility warrant more efforts.

5.5 Applications

Although functional intermetallics such as NiTi shape-memory alloys, FeCo magnetic materials and GaAs semiconductor materials have been known for years, it is their potential high-temperature performance under stress that attracts most recent developments for applications. In general, ordered intermetallics possess high melting points, low density, good strength and stiffness at elevated temperatures, and often better oxidation resistance. However, they are more brittle and less tough. Consequently, their reliability in service and amenability to manufacturing need much improvement. This section will focus on high-temperature structural applications in jet engines and turbochargers, although less critical applications appear to enjoy more success at the present time (Stoloff *et al.*, 2000).

5.5.1 Applications in jet engines

High-pressure turbine

The hottest part of a jet engine is in the high-pressure turbine where the temperature is well over 1000 °C (the hotter the more efficient the engine). Ni based superalloys, the work-horse materials, have been developed for over 50 years and the most advanced Ni superalloys can withstand temperatures over 1100 °C. In an effort to improve the efficiency of the jet engine further, an innovative combination of coating and cooling has raised the turbine rotor inlet temperature to > 1400 °C although this has caused more deviation from the ideal performance (Dimiduk and Perepezko, 2003). There appears to be very limited scope for any further improvement in performance without resorting to a new generation of materials. Among the most promising are intermetallic alloys and composites based on intermetallics or ceramics (Zhao and Westbrook, 2003). The new materials need to offer higher temperature capacity and/or lower weight.

Among the aluminides discussed in this chapter, only the nickel aluminides may be used at higher temperatures. Indeed, Ni_3Al is the reinforcing phase in the conventional Ni superalloys. However, Ni_3Al based alloys do not seem to possess sufficient creep strength at high temperatures compared to the Ni superalloys (Naka *et al.*, 1992) and improvements on other properties are not significant. Their uses are mostly limited to non-aerospace applications such as furnace components and hot dies (Sikka *et al.*, 2000). On the other hand, NiAl possesses considerably lower density, higher melting temperature, greater thermal conductivity, good oxidation resistance and high stiffness compared to Ni superalloys and is thus very attractive (Miracle, 1993).

However, brittleness and low toughness at room temperature and lower than desirable creep resistance at high temperatures have prevented NiAl alloys from being used in real applications although some engine parts have been tested successfully (Pope and Darolia, 1996; Zhao and Westbrook, 2003).

More exotic materials such as those based on molybdenum silicides (Pope and Darolia, 1996; Petrovic and Vasudevan, 1999; Dimiduk and Perepezko, 2003), niobium silicides (Subramanian *et al.*, 1996, 1997; Bewlay *et al.*, 2003) and the so-called platinum-group-metals (Cornish *et al.*, 2003) have received relatively less attention as they are more difficult to process and test. These alloys promise to offer significantly higher temperature capability and many are oxidation resistant. However, the structures and bonding of the phases involved are more complex, giving rise to mechanical behaviours more akin to ceramics. The high strength is more likely to be observed in compression and the ductility and toughness are generally low. Potentially viable materials often have complicated, multiphase microstructures. Applications of these materials in jet engines are not likely in the near future.

Low-pressure turbine and compressor

The temperature in the low-pressure turbine is below 850 °C and that in the compressor below 650 °C (Smarsly *et al.*, 2001; Clemens and Kestler, 2000). TiAl based alloys have clear advantages over conventional Ni superalloys, Ti alloys and other competing intermetallic alloys since they have low density and thus high specific strength and stiffness up to at least 800 °C (Dimiduk, 1999). Low-pressure turbine blades have been successfully designed, produced and tested (Loria, 2000; Gilchrist and Pollock, 2001). Casting (Loria, 2000), wrought processing (Appel *et al.*, 2000a) and powder metallurgical processing (Gerling *et al.*, 2004) can all be used to fabricate TiAl alloy parts. Other than considerations related to cost and manufacturing, concerns over reliability of the TiAl based alloys owing to their low ductility and fracture toughness remain to be completely removed. It is also important that creep resistance, especially that related to primary and long-term creep deformation, be improved (Appel *et al.*, 2000b) and the use temperature be raised to >850 °C to meet the requirement of new designs (Loria, 2000). Nevertheless, TiAl based alloys appear to be the sole contender in this type of application.

5.5.2 Automotive applications

Ni superalloys are also used in turbochargers. Considerable gain in performance can be achieved by replacing Ni alloys with lighter ones. The turbine inlet temperature may be as high as 1000 °C and the gas temperature inside the turbine exceeds 850 °C during running (Tetsui, 2001). However, the temperature at the root of the turbine wheel blade is lower due to the cooling effect from the mounting shaft. This makes TiAl based alloys the leading candidate material for turbine wheels. Tests show that a TiAl turbine wheel gives rise to a shorter response time (critical to the acceleration performance), allows higher rotation speeds (improved efficiency) and exhibits excellent oxidation resistance (Tetsui, 2001; Tetsui and Ono, 1999; Tetsui, 2002). TiAl turbine wheels have been manufactured by precision casting process (Noda, 1998, McQuay *et al.*, 1999) and commercial uses in one case have been realised (Tetsui, 2001).

Another important application that has been developed vigorously is the exhaust engine valve. Since the valve undergoes reciprocal movement in running, use of a lightweight material will increase efficiency and improve performance. It requires good strength at around 800 °C, good thermal shock resistance as well as oxidation and wear resistance (Clemens and Kestler, 2000; Smarsly *et al.*, 2001). TiAl based alloys are again very suitable for this application. Both casting (Noda, 1998; Eylon *et al.*, 1998) and wrought processing (Baur *et al.*, 2000) have been used and good testing results have been obtained (Clemens and Kestler, 2000).

5.6 Conclusion and further trends

Thanks to a mixture of metallic, covalent and ionic bonds and ordered structures, intermetallic alloys belong indeed to a unique class of materials. On the one hand, they are strong and stiff, especially at elevated temperatures, although relatively brittle and not tough, compared with disordered metallic alloys. On the other hand, they possess higher ductility and toughness and are deformable at high temperatures, compared with ceramic materials. Their microstructures can be manipulated using conventional heat and thermomechanical treatments. Other conventional processes such as liquid casting, solid forming, powder metallurgy, welding and machining may be used in their manufacture. It is thus expected that ordered intermetallic alloys will play significant roles in structural applications especially at elevated temperatures and in hostile environments.

Although much effort has been spent on improving the inherent brittleness of many single-phase intermetallics through alloying and phase selection (Fleischer and Taub, 1989), this approach is apparently not ideal. It may work if alloying is to improve fracture resistance. The best example here is perhaps the addition of boron in Ni_3Al based alloys. If, on the other hand, plasticity is increased at the expense of yield strength, the purpose is defeated. After all, brittleness is associated, to some extent, with the strong and often directional bonding which gives them the unique properties in the first place. It is also a huge effort to try to find the right alloying combinations through experiments. Although theories and atomistic modelling have made remarkable advances in the last decade and have contributed considerably to our understanding, they are still far away from providing direct engineering benefits.

The solution, it seems, is through the design and realisation of microstructures consisting of two or more phases. This conclusion should not have come as a surprise. Within conventional materials, the most prominent examples can be found in steels where a variety of microstructures from fully lamellar pearlite to tempered martensite provides solutions to very different engineering applications. Most light alloys (i.e. Al, Ti and Mg based) rely on precipitates (intermetallic phases in most cases) for greatest gain in strength. With intermetallics, significant progress has been made as in the examples of two-phase γ-TiAl alloys. In principle, it is necessary to have a relatively ductile matrix phase that is also environment resistant and to have strengthening phases that are stable and retain their strength at the designated service temperature.

The brittleness of ordered intermetallics results from either a low cohesive strength or the lack of a sufficient number of deformation elements such as mobile dislocations and slip systems. The former will lead directly to cleavage fracture, and the latter will cause intergranular fracture or strain localisation owing to insufficient work hardening. In both cases, a microstructure of ultrafine grains will enhance ductility considerably. The fine grains, however, may not be stable and may facilitate deformation at elevated temperatures. It is critical to have one or more strengthening phases of appropriate size, shape and distribution in order to stabilise the microstructure and increase strength at high temperatures.

The challenge is to actually produce such an ideal combination of microstructural features. Conventional thermomechanical treatment may work in some systems such as Ti-Al where a variety of microstructures can be obtained. Such an approach is limited by thermodynamics and by the ability of materials to undergo plastic deformation. It is envisioned here that future approaches will bring desirable combinations of phases in a more artificial way with more controls over their sizes and distribution as well as the interfaces between them. New developments in nano-materials processing and severe plastic deformation are predicted to play an important role. On the other hand, atomistic models making use of ever increasing computing power will continue to make a significant contribution to our fundamental understanding of these alloys. A new generation of structural materials based on ordered intermetallics is finally to bring the promised engineering applications into reality.

5.7 Sources of further information

Limited by space and scope, this chapter gives only a brief discussion of the important issues related to plastic deformation of ordered intermetallic alloys.

No attempt has been made to review thoroughly all the topics. In addition to the articles cited in each section and the conference proceedings mentioned in the Introduction, the following books offer concentrated information and data.

Westbrook J.H. and Fleischer R.L., eds, *Intermetallic Compounds: Vol. 1 – Principles, Vol. 2 – Practice*, and *Vol. 3 – Progress*, John Wiley, Chicheter, England, 1995 (Vols. 1 and 2) and 2002 (Vol. 3).

Stoloff N.S. and Sikka V.K., eds, *Physical Metallurgy and Processing of Intermetallic Compounds*, Chapman & Hall, New York, NY, 1996.

Villars P. and Calvert L.D., *Pearson's Handbook of Crystallographic Data for Intermetallic Phases*, vols. 1 to 4, ASM International, Materials Park, OH, 1991.

5.8 References

Anderson E.A., Jillson D.C. and Dunbar S.R. (1953), *Trans. AIME,* **197**, 1191–1197.

Andrade E.NdaC. and Henderson C. (1951), *Phil. Trans. R. Soc. Lond. A,* **244**, 177–203.

Aoki K. and Izumi O. (1979), *J. Jap. Inst. Met.,* **43**, 1190–1196.

Appel F. and Wagner R. (1998), *Mater. Sci. Eng.,* **R22**, 187–268.

Appel F., Christoph U. and Wagner R. (1995), *Phil. Mag. A,* **72**, 341–360.

Appel F., Brossmann U., Christoph U., Eggert S., Janschek P., Lorenz U., Müllauer J., Oehring M. and Paul J.D.H. (2000a), *Adv. Eng. Mater.,* **2**, 699–720.

Appel F., Oehring M. and Wagner R. (2000b), *Intermetallics,* **8**, 1283–1312.

Appel F., Kestler H. and Clemens H. (2002), In *Intermetallic Compounds – Principles and Practice*, Vol. 3 (eds, Westbrook, J.H. and Fleischer, R.L.) John Wiley, Chichester, England, pp. 617–642.

Bacon D.J. and Vitek V. (2002), *Metall. Mater. Trans. A,* **33A**, 721–733.

Baker I. (1995), *Mater. Sci. Eng.,* **A192/193**, 1–13.

Baker I. (2000), *Intermetallics,* **8**, 1183–1196.

Ball J. and Gottstein G. (1993a), *Intermetallics*, **1**, 171–185.

Ball J. and Gottstein G. (1993b), *Intermetallics*, **1**, 191–208.

Banerjee D. (1995), In *Intermetallic Compounds – Principles and Practice*, Vol. 2 (eds, Westbrook, J.H. and Fleischer, R.L.) John Wiley, Chichester, England, pp. 91–131.

Baur H., Joos R. and Clemens H. (2000), In *Intermetallics and Superalloys* (eds, Morris D.G., Naka S. and Caron P.) Wiley-VCH, Weinheim, Germany, pp. 384–390.

Bewlay B.P., Jackson M.R., Zhao J.-C., Subramanian P.R., Mendiratta M.G. and Lewandowski J.J. (2003), *MRS Bulletin,* **28**, 646–653.

Blackburn M.J. (1970), In *The Science, Technology, and Application of Titanium* (eds, Jaffee R.I. and Promisel N.E.) Pergamon Press, Oxford, UK, pp. 633–643.

Botton G.A. and Humphreys C.J. (1999), *Intermetallics,* **7**, 829–833.

Botton G.A., Guo G.Y., Temmerman W.M. and Humphreys C.J. (1996), *Phys. Rev. B,* **54**, 1682–1691.

Bowman R.R., Noebe R.D., Raj S.V. and Locci I.E. (1992), *Metall. Trans. A,* **23A**, 1493–1508.

Cahn R.W. (1995), *Phil. Trans. R. Soc. Lond. A,* **351**, 497–509.

Chen W.R., Triantafillou J., Beddoes J. and Zhao L. (1999) *Intermetallics,* **7**, 171–178.

Churchman A.T. (1954), *Proc. R. Soc. Lond. A,* **226**, 216–226.

Clemens H. and Kestler H. (2000), *Adv. Eng. Mater.*, **2**, 551–570.

Copley S.M. and Kear B.H. (1967), *Trans. Metall. Soc. AIME,* **239**, 977–984.

Cornish L.A., Fischer B. and Völkl R. (2003), *MRS Bulletin,* **28**, 632–638.
Cornwell L.R., Hartwig K.T., Goforth R.E. and Semiatin S.L. (1996), *Mater. Characterization,* **37**, 295–300.
Cox J.J., Horne G.T. and Mehl R.F. (1957), *Trans. ASM,* **49**, 118–131.
Cserti J., Khantha M., Vitek V. and Pope D.P. (1992), *Mater. Sci. Eng.,* **A152**, 95–102.
Darolia R., Walston W.S., Noebe R., Garg A. and Oliver B.F. (1999), *Intermetallics,* **7**, 1195–1202.
Dimiduk D.M. (1999), *Mater. Sci. Eng.,* **A263**, 281–288.
Dimiduk D.M. and Perepezko J.H. (2003), *MRS Bulletin,* **28**, 639–645.
Eberhart M.E., Clougherty D.P. and MacLaren J.M. (1993), *J. Mater. Res.,* **8**, 438–448.
Escher C. and Gottstein G. (1998), *Acta Mater.,* **46**, 525–539.
Escher C., Neves S. and Gottstein G. (1998), *Acta Mater.,* **46**, 441–450.
Eylon D., Keller M.M. and Jones P.E. (1998), *Intermetallics,* **6**, 703–708.
Ezz S.S., Pope D.P. and Paidar V. (1982), *Acta Metall.,* **30**, 921–926.
Feng C.R., Michel D.J. and Crowe C.R. (1989), *Scr. Metall.,* **23**, 1135–1140.
Field R.D., Lahrman D.F. and Darolia R. (1991), *Acta Metall. Mater.,* **39**, 2951–2959.
Fleischer R.L. and Taub A.I. (1989), *JOM,* **41**, 8–11.
Fleischer R.L., Field R.D. and Briant C.L. (1991), *Metall. Trans. A,* **22A**, 403–414.
Flinn P.A. (1960), *Trans. AIME,* **218**, 145–154.
Fox A.G. and Menon E.S.K. (1997), *J. Phase Equilibria,* **18**, 509–515.
Fox A.G. and Tabbernor M.A. (1991), *Acta Metall. Mater.*, **39**, 669–678.
Fu C.L. and Yoo M.H. (1992), *Acta Metall. Mater.,* **40**, 703–711.
Fu C.L. and Yoo M.H. (1993), *Intermetallics*, **1**, 59–63.
Fu C.L., Zou J. and Yoo M.H. (1995), *Scr. Metall. Mater.,* **33**, 885–891.
Furukawa M., Iwahashi Y., Horita Z., Nemoto M. and Langdon T.G. (1998), *Mater. Sci. Eng.,* **A257**, 328–332.
George E.P. and Baker I. (1998), *Phil. Mag. A,* **77**, 737–750.
George E.P. and Liu C.T. (1990), *J. Mater. Res.,* **5**, 754–762.
George E.P., Liu C.T. and Pope D.P. (1992), *Scr. Metall. Mater.*, **27**, 365–370.
Gerling R., Clemens H. and Schimansky F.P. (2004), *Adv. Eng. Mater.*, **6**, 23–38.
Gilchrist A. and Pollock T.M. (2001), In *Structural Intermetallics 2001* (eds, Hemker, K.J., Dimiduk D.M., Clemens H., Darolia R., Inui H., Larsen J.M., Sikka V.K., Thomas M. and Whittenberger J.D.) TMS, Warrendale, PA, pp. 3–12.
Greenberg B.F. (1973), *Phys. Status Solidi (b),* **55**, 59–67.
Gschneidner Jr, K., Russell A., Pecharsky A., Morris J., Zhang Z., Lograsso T., Hsu D., Chester Lo C.H., Ye Y., Slager A. and Kesse D. (2003), *Nature Mater.*, **2**, 587–590.
Gumbsch P. and Schroll R. (1999) *Intermetallics,* **7**, 447–454.
Hahn K.H. and Vedula K. (1989), *Scr. Metall.*, **23**, 7–12.
Harrison W.A. (1989), *Electron Structure and the Properties of Solids: The Physics of the Chemical Bond,* Dover Publications, New York.
Hemker K.J., Viguier B. and Mills M.J. (1993), *Mater. Sci. Eng.,* **A164**, 391–394.
Heredia F.E. and Pope D.P. (1991a), *Acta Metall. Mater.,* **39**, 2017–2026.
Heredia F.E. and Pope D.P. (1991b), *Acta Metall. Mater.,* **39**, 2027–2036.
Hirano T., Demura M. and Golberg D. (1999) *Acta Mater.,* **47**, 3441–3446.
Hirsch P.B. (1992), *Prog. Mater. Sci.*, **36**, 63–88.
Huang S.C. (1992), *Metall. Trans. A*, **23A**, 375–377.
Huang S.C. and Chesnutt J.C. (1995), In *Intermetallic Compounds – Principles and Practice*, Vol. 2 (eds, Westbrook J.H. and Fleischer R.L.) John Wiley, Chichester, England, pp. 73–90.

Huang S.C. and Hall E.L. (1991), *Metall. Trans. A,* **22A**, 427–439.
Hug G., Loiseau A. and Veyssiere P. (1988), *Philos. Mag. A*, **57**, 499–523.
Inkson B.J. (1998), *Phil. Mag. A,* **77**, 715–736.
Inui H., Toda Y. and Yamaguchi M. (1993), *Phil. Mag. A,* **67**, 1315–1332.
Inui H., Matsumuro M., Wu D.-H. and Yamaguchi M. (1997), *Phil. Mag. A,* **75**, 395–423.
Jin Z. and Bieler T.R. (1992), *Scr. Metall. Mater.*, **27**, 1301–1306.
Kawabata T., Kanai T. and Izumi O. (1985), *Acta Metall.*, **33**, 1355–1366.
Kear B.H. and Wilsdorf H.G.F. (1962), *Trans. Metall. Soc. AIME,* **224**, 382–386.
Kelly A. and Macmillan N.H. (1986), *Strong Solids,* Oxford University Press, Oxford, UK.
Khantha M., Cserti J. and Vitek V. (1992), *Scr. Metall. Mater.*, **27**, 481–486.
Kim Y.-W. (1994), *JOM*, **46**, 30–39.
Kim J.H., Ha T.K., Chang Y.W. and Lee C.S. (2003), *Metall. Mater. Trans. A*, **34A**, 2165–2176.
Kimura Y. and Pope D.P. (1998), *Intermetallics,* **6**, 567–571.
Korznikov A.V., Dimitrov O., Korznikova G.F., Dallas J.P., Idrisova S.R., Valiev R.Z. and Faudot F. (1999a), *Acta Mater.*, **47**, 3301–3311.
Korznikov A.V., Dimitrov O., Korznikova G.F., Dallas J.P., Quivy A., Valiev R.Z. and Mukherjee A.K. (1999b), *Nanostructured Mater.*, **11**, 17–23.
Korznikov A.V., Pakiela Z. and Kurzydlowski K.J. (2001a), *Scr. Mater.*, **45**, 309–315.
Korznikov A.V., Tram G., Dimitrov O., Korznikova G.F., Idrisova S.R. and Pakiela Z. (2001b), *Acta Mater.*, **49**, 663–671.
Kupka M. (2004), *Intermetallics*, **12**, 295–302.
Langdon T.G., Furukawa M., Horita Z. and Nemoto M. (1998), *JOM,* **50**, 41–45.
Langdon T.G., Furukawa M., Nemoto M. and Horita Z. (2000), *JOM*, **52**, 30–33.
Languillaume J., Chmelik F., Kapelski G., Bordeaux F., Nazarov A.A., Ganova G., Esling C., Valiev R.Z. and Baudelet B. (1993), *Acta Metall. Mater.*, **41**, 2953–2962.
Legros M., Couret A. and Caillard D. (1996), *Phil. Mag. A,* **73**, 61–80.
Lin W., Xu J.-H. and Freeman A.J. (1992), *J. Mater. Res.*, **7**, 592–604.
Lipsitt H.A., Shechtman D. and Schafrik R.E. (1980), *Metall. Trans. A,* **11A**, 1369–1375.
Liu C.T. (1991), *Scr. Metall. Mater.*, **25**, 1231–1236.
Liu C.T. (1992), *Scr. Metall. Mater.*, **27**, 25–28.
Liu C.T. and Pope D.P. (1995), In *Intermetallic Compounds – Principles and Practice*, Vol. 2 (eds, Westbrook J.H. and Fleischer R.L.) John Wiley, Chichester, England, pp. 17–51.
Liu C.T., Lee E.H. and McKamey C.G. (1989), *Scr. Metall.*, **23**, 875–880.
Liu C.T., Schneibel J.H., Maziasz P.J., Wright J.L. and Easton D.S. (1996), *Intermetallics,* **4**, 429–440.
Liu C.T., White C.L. and Horton J.A. (1985), *Acta Metall.*, **33**, 213–229.
Loria E.A. (2000), *Intermetallics*, **8**, 1339–1345.
Lu Z.W., Zunger A. and Fox A.G. (1994), *Acta Metall. Mater.,* **42**, 3929–3943.
Mahapatra R., Girshick A., Pope D.P. and Vitek V. (1995), *Scr. Metall. Mater.,* **33**, 1921–1927.
Maruyama K., Yamamoto R., Nakakuki H. and Fujitsuna N. (1997), *Mater. Sci. Eng.,* **A239–240**, 419–428.
Masahashi N. (1997), *Mater. Sci. Eng.,* **A223**, 42–53.
Masumoto K. and McGahan W.A. (1996), *MRS Bulletin*, **21**, 44–49.
Maziasz P.J. and Liu C.T. (1998), *Metall. Mater. Trans. A*, **29A**, 105–117.
McQuay P.A., Simpkins R., Seo D.Y. and Bishop T.R. (1999), In *Gamma Titanium Aluminides 1999* (ed., Kim, Y.-W.) TMS, Warrendale, PA, pp. 197–207.

Millett J.C.F., Brooks J.W. and Jones I.P. (1999), *Mater. Sci. Technol.,* **15**, 697–704.
Millett J.C.F., Brooks J.W. and Jones I.P. (2000a), *Mater. Sci. Technol.*, **16**, 617–624.
Millett J.C.F., Brooks J.W. and Jones I.P. (2000b), *Mater. Sci. Technol.*, **16**, 1041–1048.
Millett J.C.F., Brooks J.W. and Jones I.P. (2001), *Mater. Sci. Technol.*, **17**, 795–801.
Mills M.J. and Miracle D.B. (1993), *Acta Metall. Mater.*, **41**, 85–95.
Minonishi Y. (1991) *Phil. Mag. A*, **63**, 1085–1093.
Minonishi Y. (1995) *Mater. Sci. Eng.*, **A192/193**, 830–836.
Miracle D.B. (1993), *Acta Metall. Mater.*, **41**, 649–684.
Mishima Y., Ochiai S., Yodogawa M. and Suzuki T. (1986), *Trans. JIM,* **27**, 41–50.
Morinaga M., Saito J., Yukawa N. and Adachi H. (1990), *Acta Metall. Mater.*, **38**, 25–29.
Morris M.A. and Leboeuf M. (1997), *Intermetallics*, **5**, 339–354.
Naka S., Thomas M. and Khan T. (1992), *Mater. Sci. Technol.*, **8**, 291–298.
Nieh T.G. and Wadsworth J. (1999), *Inter. Mater. Rev.*, **44**, 59–75.
Nitz A. and Nembach E. (1998), *Metall. Mater. Trans. A,* **29A**, 799–807.
Noda T. (1998), *Intermetallics*, **6**, 709–713.
Noggle T.S. and Koehler J.S. (1957), *J. Appl. Phys.*, **28**, 53–62.
Noguchi O., Oya Y. and Suzuki T. (1981), *Metall. Trans. A*, **12A**, 1647–1653.
Ochiai S., Oya Y. and Suzuki T. (1984), *Acta Metall.*, **32**, 289–298.
Ochiai S., Mishima Y., Yodogawa M. and Suzuki T. (1986), *Trans. JIM*, **27**, 32–40.
Paidar V. and Vitek V. (2002), In *Intermetallic Compounds – Principles and Practice*, Vol. 3 (eds, Westbrook J.H. and Fleischer R.L.) John Wiley, Chichester, England, pp. 437–467.
Paidar V., Yamaguchi M., Pope D.P. and Vitek V. (1982), *Phil. Mag. A*, **45**, 883–894.
Paidar V., Pope D.P. and Vitek V. (1984), *Acta Metall.*, **32**, 435–448.
Pasianot R., Farkas D. and Savino E.J. (1991), *J. Physique III,* **1**, 997–1014.
Petrovic J.J. and Vasudevan A.K. (1999), *Mater. Sci. Eng.,* **A261**, 1–5.
Pope D.P. and Darolia R. (1996), *MRS Bulletin,* **21**, 30–36.
Pope D.P. and Ezz S.S. (1984), *Inter. Met. Rev.*, **29**, 136–167.
Rice J.R. and Thomson R. (1974), *Phil. Mag.*, **29**, 73–97.
Rohrer G.S. (2001), *Structure and Bonding in Crystalline Materials*, Cambridge University Press, Cambridge, UK.
Rosi F.D. and Mathewson C.H. (1950), *Trans. AIME,* **188**, 1159–1167.
Sabinash C.M., Sastry S.M.L. and Jerina K.L. (1995), *Mater. Sci. Eng.,* **A192–193**, 837–847.
Sastry S.M.L. and Mahapatra R.N. (2001), *Mater. Sci. Eng.,* **A329–331**, 872–877.
Sastry S.M.L., Mahapatra R.N. and Hasson D.F. (2000), *Scr. Mater.*, **42**, 731–736.
Sauthoff G. (1995), *Intermetallics*, VCH, Weinheim, Germany.
Schetky L.M. (1996), *MRS Bulletin*, **21**, 50–55.
Seetharaman V. and Semiatin S.L. (1996), *Metall. Mater. Trans. A*, **27A**, 1987–2004.
Seetharaman V. and Semiatin S.L. (1997), *Metall. Mater. Trans. A*, **28A**, 2309–2321.
Seetharaman V. and Semiatin S.L. (2002), *Metall. Mater. Trans. A*, **33A**, 3817–3830.
Semiatin S.L., Frey N., El-Soudani S.M. and Bryant J.D. (1992a), *Metall. Trans. A,* **23A**, 1719–1735.
Semiatin S.L., Lark K.A., Barker D.R., Seetharaman V. and Marquardt B. (1992b) *Metall. Trans. A,* **23A**, 295–305.
Semiatin S.L., Seetharaman V. and Jain V.K. (1994), *Metall. Mater. Trans. A*, **25A**, 2753–2768.
Semiatin S.L., Segal V.M., Goetz R.L., Goforth R.E. and Hartwig T. (1995), *Scr. Metall. Mater.*, **33**, 535–540.

Semiatin S.L., Seetharaman V. and Weiss I. (1998), *Mater. Sci. Eng.*, **A243**, 1–24.
Shechtman D., Blackburn M.J. and Lipsitt H.A. (1974), *Metall. Trans.*, **5**, 1373–1381.
Siegl R., Vitek V., Inui H., Kishida K. and Yamaguchi M. (1997), *Phil. Mag. A*, **75**, 1447–1459.
Sikka V.K., Deevi S.C., Viswanathan S., Swindeman R.W. and Santella M.L. (2000), *Intermetallics,* **8**, 1329–1337.
Sikora T., Jaouen M. and Hug G. (1999), *Phil. Mag. A*, **79**, 2157–2172.
Simmons J.P., Rao S.I. and Dimiduk D.M. (1997), *Phil. Mag. A*, **75**, 1299–1328.
Smarsly W., Baur H., Glitz G., Clemens H., Khan T. and Thomas M. (2001), In *Structural Intermetallics 2001* (eds, Hemker K.J., Dimiduk D.M., Clemens H., Darolia R., Inui H., Larsen J.M., Sikka V.K., Thomas M. and Whittenberger J.D.) TMS, Warrendale, PA, pp. 25–34.
Staton-Bevan A.E. and Rawlings R.D. (1975), *Phil. Mag.*, **32**, 787–800.
Stoloff N.S. (1989), *Inter. Mater. Rev.*, **34**, 153–183.
Stoloff N.S., Liu C.T. and Deevi S.C. (2000), *Intermetallics*, **8**, 1313–1320.
Stucke M.A., Vasudevan V.K. and Dimiduk D.M. (1995), *Mater. Sci. Eng.*, **A192/193**, 111–119.
Subramanian P.R., Mendiratta M.G. and Dimiduk D.M. (1996), *JOM,* **48**, 33–38.
Subramanian P.R., Mendiratta M.G., Dimiduk D.M. and Stucke M.A. (1997), *Mater. Sci. Eng.*, **A239–240**, 1–13.
Sun Y.-Q. (1995), In *Intermetallic Compounds – Principles and Practice*, Vol. 1 (eds, Westbrook, J.H. and Fleischer, R.L.) John Wiley, Chichester, England, pp. 495–517.
Takeuchi S. and Kuramoto E. (1973), *Acta Metall.*, **21**, 415–425.
Tetsui T. (2001), *Adv. Eng. Mater.*, **3**, 307–310.
Tetsui T. (2002), *Mater. Sci. Eng.*, **A329–331**, 582–588.
Tetsui T. and Ono S. (1999), *Intermetallics*, **7**, 689–697.
Thornton P.H., Davies R.G. and Johnston T.L. (1970), *Metall. Trans.,* **1**, 207–218.
Tichy G., Vitek V. and Pope D.P. (1986), *Phil. Mag. A*, **53**, 467–484.
Umakoshi Y. (1993), In *Plastic Deformation and Fracture of Materials*, Vol. 6 (ed., Mughrabi, H.) VCH, Weinheim, Germany, pp. 251–310.
Umakoshi Y., Pope D.P. and Vitek V. (1984), *Acta Metall.*, **32**, 449–456.
Umakoshi Y. and Yamaguchi M. (1981a), *Phil. Mag. A,* **44**, 711–715.
Umakoshi Y. and Yamaguchi M. (1981b), *Phys. Status Solidi (a),* **68**, 457–468.
Umakoshi Y., Nakano T., Takenaka T., Sumimoto K. and Yamane T. (1993), *Acta Metall. Mater.*, **41**, 1149–1154.
Valiev RZ., Islamgaliev R.K. and Alexandrov I.V. (2000), *Prog. Mater. Sci.*, **45**, 103–189.
Valiev R.Z., Song C., McFadden S.X., Mukherjee A.K. and Mishra R.S. (2001), *Phil. Mag. A,* **81**, 25–36.
Valiev R.Z., Alexandrov I.V., Zhu Y.T. and Lowe T.C. (2002), *J. Mater. Res.,* **17**, 5–8.
Van de Walle A. and Asta M. (2002), *Metall. Mater. Trans. A*, **33A**, 735–741.
Veyssière P. (2001), *Mater. Sci. Eng.,* **A309–310**, 44–48.
Veyssière P. and Douin J. (1985), *Phil. Mag. Lett.,* **51**, 1–4.
Veyssière P. and Douin J. (1995), In *Intermetallic Compounds – Principles and Practice*, Vol. 1 (eds, Westbrook, J.H. and Fleischer, R.L.) John Wiley, Chichester, England, pp. 519–558.
Villars P. and Calvert L.D. (1985), *Pearson's Handbook of Crystallographic Data for Intermetallic Phases,* ASM International, Materials Park, OH.
Vitek V. (1974), *Crystal Lattice Defects*, **5**, 1–34.
Vitek V. (1992), *Prog. Mater. Sci.*, **36**, 1–27.

Vitek V. (1998), *Intermetallics,* **6**, 579–585.
von Mises R. (1928), *Z. angew. Math. Mech.*, **8**, 161–185.
Westbrook J.H. (1957), *Trans. AIME,* **209**, 898–904.
Westbrook J.H. (1995), In *Intermetallic Compounds – Principles and Practice*, Vol. 1 (eds, Westbrook J.H. and Fleischer R.L.) John Wiley, Chichester, England, pp. 3–18.
Westbrook J.H. (1996), *MRS Bulletin,* **21**, 37–43.
Wu, D. and Baker, I. (2001) *Intermetallics*, **9**, 57–65.
Wu X., Song D. and Xia K. (2001), In *Structural Intermetallics 2001* (eds, Hemker K., Dimiduk D., Clemens H., Darolia R., Inui H., Larsen J.M., Sikka, V.K., Thomas M. and Whittenberger J.D.) TMS, Warrendale, PA, pp. 633–642.
Wu X., Song D. and Xia K. (2002), *Mater. Sci. Eng.,* **A329-331**, 821–827.
Wu Y., Zhen L., Yang D.Z., Kim M.S., Hwang S.K. and Umakoshi Y. (2004), *Intermetallics*, **12**, 43–53.
Xia K. (1986), *Research Report*, University of Southern California, pp. 1–37.
Xia K. and Wang J. (2001), *Metall. Mater. Trans. A*, **32A**, 2639–2647.
Xia K., Wu X. and Song D. (2004), *Acta Mater.,* **52**, 841–849.
Yamaguchi M. and Umakoshi Y. (1975), *Phys. Status Solidi (a),* **31**, 101–106.
Yamaguchi M. and Umakoshi Y. (1990), *Prog. Mater. Sci.*, **34**, 1–148.
Yamaguchi M., Pope D.P., Vitek V. and Umakoshi Y. (1981), *Phil. Mag. A,* **43**, 1265–1275.
Yamaguchi M., Paidar V., Pope D.P. and Vitek V. (1982), *Phil. Mag. A,* **45**, 867–882.
Yamaguchi M., Inui H. and Ito K. (2000), *Acta Mater.*, **48**, 307–322.
Yoo M.H. (1998), *Intermetallics*, **6**, 597–602.
Yoo M.H. (2002), In *Intermetallic Compounds – Principles and Practice*, Vol. 3 (eds, Westbrook J.H. and Fleischer R.L.) John Wiley, Chichester, England, pp. 403–436.
Yoo M.H. and Fu C.L. (1998), *Metall. Mater. Trans. A*, **29A**, 49–63.
Yoo M.H., Morris J.R., Ho K.M. and Agnew S.R. (2002), *Metall. Mater. Trans. A*, **A33**, 813–822.
Yoshimi K., Hanada S. and Yoo M.H. (1995), *Acta Metall. Mater.*, **43**, 4141–4151.
Zhang J., Zhang Z., Su X., Zou D., Zhong Z. and Li C. (2000), *Intermetallics,* **8**, 321–326.
Zhao J.-C. and Westbrook J.H. (2003), *MRS Bulletin,* **28**, 622–627.
Zhilyaev A.P., Nurislamova G.V., Kim B.K., Baro M.D., Szpunar J.A. and Langdon T.G. (2003), *Acta Mater.*, **51**, 753–765.
Zupan M. and Hemker K.J. (2003), *Acta Mater.*, **51**, 6277–6290.

6
Discontinuously reinforced metal matrix composites

M FERRY, University of New South Wales, Australia

6.1 Introduction

Metal matrix composites (MMCs) have received considerable attention over a number of years because of their potentially high elastic modulus and specific strength (Taya and Arsenault, 1989; Clyne and Withers, 1993; Lloyd, 1994; Clyne, 2001). MMCs are usually classified into three types depending on the morphology of the reinforcement (Nair *et al.*, 1985; Clyne and Withers, 1993). A metal matrix may be reinforced with ceramic or metallic materials in the form of:

- long, continuous or unbroken fibres (Fig. 6.1a)
- whiskers of a substantially defect-free material which are discontinuous in nature, that is, they have a finite length/diameter ratio (Fig. 6.1b)
- particulates or equiaxed particles (Fig. 6.1c).

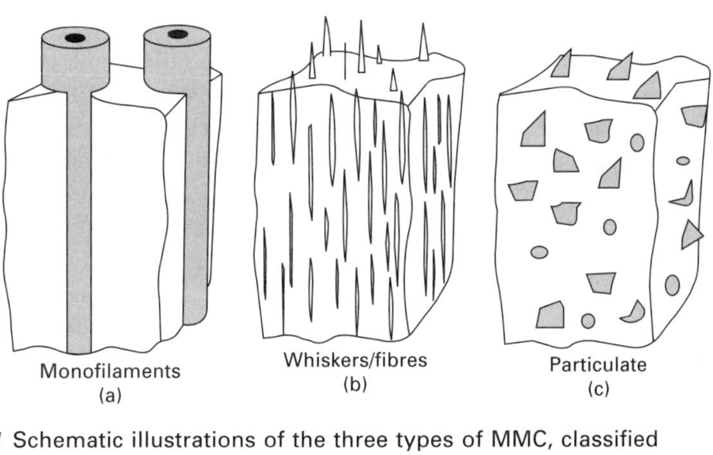

6.1 Schematic illustrations of the three types of MMC, classified according to the type of reinforcement: (a) monofilaments; (b) whiskers or fibres, and (c) particulate, after Clyne and Withers (1993), with kind permission of Cambridge University Press.

A substantial amount of the earlier research on MMCs was concentrated on continuous-fibre reinforcement, see e.g. Clyne and Withers (1993), but these materials have restricted application because of their limited secondary formability as well as their anisotropic properties and high overall production cost. Consequently, considerable effort has been directed towards understanding the thermomechanical behaviour of discontinuously reinforced MMCs (hereafter termed DRCs) containing either whiskers, platelets or particulates of the type SiC, Al_2O_3, TiC, TiB_2, TiN or B_4C etc., since these materials often have mechanical properties superior to their unreinforced alloy counterparts and approaching that of continuous-fibre reinforced MMCs (Clyne and Withers, 1993). Since DRCs are also more expensive to produce than monolithic alloys, their use will be restricted to cost-effective applications, where the strength-stiffness savings outweigh the increased cost.

One of the prime advantages of DRCs is their ability to be fabricated into billets then mechanically processed using technologies developed for monolithic alloys; for example, hot and cold rolling, extrusion, forging and superplastic forming. The basic structural difference between DRCs and their alloys is that the volume fraction of the reinforcement phase is usually considerably higher (up to 30 vol. %) than the particle content of the unreinforced alloys (~<5 vol. %). Subsequently, the thermomechanical behaviour of these composites is expected to be different from that of the unreinforced alloy. The aim of this chapter is to provide an up-to-date overview of the influence of secondary processing on the microstructure and mechanical properties of DRCs.

6.1.1 Primary processing routes and reinforcement distribution

Most of the current processing routes produce DRCs with a heterogeneous dispersion of the reinforcing phase (Clyne and Withers, 1993; Clyne, 2001). The mechanical properties of the final product are critically dependent on the integrity and distribution of the reinforcement since reinforcement fracture and clustering may give rise to initiation sites for premature fracture (Lloyd, 1991a,b; 1994; Boyd and Lloyd, 2001). Where clustering is less severe, inhomogeneous matrix deformation may also give rise to strain localisation and premature failure (Arsenault *et al.*, 1989; Barlow and Hansen, 1991). For the reinforcement to be used to its full potential, it should be resistant to comminution and must be reasonably homogeneously distributed throughout the metal matrix by the careful choice both of the initial fabrication method and secondary working operations. Considering the importance of the initial composite microstructure on final properties (Lloyd, 1991a; Clyne, 2001), the more notable primary fabrication routes will be described. For a more extensive treatment of fabrication methods, the reader is referred to several books and reviews (Humphreys, 1988; Mortensen and Jin, 1991; Clyne and Withers, 1993; Lloyd, 1994; Clyne, 2001).

Powder metallurgy

This fabrication method has been used extensively both for whisker reinforced MMCs (WRCs) and particulate reinforced MMCs (PRCs). In this method, alloy powder and reinforcement are mixed and subsequently compacted by a combination of temperature and pressure then processed further by either hot-extrusion or hot rolling. This method is very flexible since the type, size and volume fraction of reinforcement and type of matrix can be tailored to satisfy the intended final application. For example, particulates in the size range 1–100 μm and volume fractions up to 0.5 have been blended with metal powder to produce PRCs (Miller and Humphreys, 1989). The success of this processing route will be dependent on the ability of both the metal particles and reinforcement to be blended to produce a uniform mix, without substantial clustering of the reinforcement. Both alloys and composites produced by this method usually contain fine oxide particles that originate from the surface of the atomised powder (Hansen, 1969). These particles are generally difficult to avoid and affect the development of microstructure and texture during further processing.

Liquid metal infiltration

This process has been used most extensively for producing MMCs containing discontinuous fibres with volume fractions of up to 0.2 (Chi *et al.*, 1987). In this technique, a preform is made to the desired shape, then placed in a die and infiltrated by molten metal. Infiltration is either achieved by evacuating the die, or more usually, by applying a pressure as for example in squeeze casting. To obtain a sound product, it is necessary to control carefully the preform temperature, pressure cycle, melt superheat and solidification conditions (Caron and Masounave, 1990). Principal advantages of molten metal infiltration include reduced matrix/reinforcement chemical interaction and attractive matrix microstructures when cold dies and reinforcement preforms are used, few or no microstructural defects when the process is well-controlled, great versatility in the allowable reinforcements, and the potential for rapid net or near-net shape production of components (Caron and Masounave, 1990).

Spray deposition

In this process, the metal is atomised and sprayed with, or onto the reinforcement and since the reinforcement is only in contact for a brief time with the molten metal, chemical reactions are usually minimal. However, there are inherent difficulties in homogeneously mixing the sprayed components and, subsequently, the reinforcement is limited to particulates with sizes greater than 10 μm and volume fractions less than 0.15–0.2. Full densification of the as-sprayed composite is achieved only by further mechanical processing such as hot rolling, forging or extrusion.

Molten metal mixing

The molten metal route is, in principle, the most direct and least expensive method of producing PRCs, and is currently used on a commercial scale (Hoover, 1990). In this process, the reinforcement is stirred into the molten alloy, and the resultant mixture is cast by conventional techniques. The type of reinforcement is generally restricted to those ceramics that do not react appreciably with the molten metal and Al-based composites reinforced with SiC, Al_2O_3 or B_4C particulates have been produced successfully with volume fractions up to 0.2 and in the size range 10–15 μm. This upper limit of the volume fraction arises because of problems of inhomogeneity of the particle distribution at higher reinforcement contents due to effects such as dendrite pushing of the particulate during solidification. Lloyd (1991a) has shown that a uniform particle distribution in an as-cast billet can be achieved by: (i) a high mixing efficiency of the melt to uniformly distribute the particulate and (ii) restricting the degree of particulate sedimentation after mixing.

The influence of solidification parameters on the distribution of particulate in Al-based PRCs has been considered in detail by Lloyd (1989; 1991a; 1994). It was shown that commercial solidification rates generate a cell size that is not significantly affected by the addition of the particulates (Lloyd, 1991a). However, the reinforcement will be rejected at the meniscus of the growing solid during normal solidification conditions, which results in particle entrapment within interdendritic regions during final solidification (Fig. 6.2). While particle clustering is difficult to avoid, higher solidification rates will minimise the tendency for particle pushing and reduce the degree of clustering (Lloyd, 1991a).

6.1.2 Dislocation density in as-fabricated composites

The coefficient of thermal expansion (CTE) of a metal matrix is usually substantially greater than the reinforcing phase (Taya and Arsenault, 1989), so cooling from an elevated temperature after either initial fabrication or secondary processing will generate matrix dislocations near the matrix/reinforcement interface (Fig. 6.3). This dislocation density may be as high as 10^{13}–10^{14} m^{-2} (Arsenault and Fisher, 1983) with the actual density dependent on the temperature interval of cooling, the difference in CTE between each

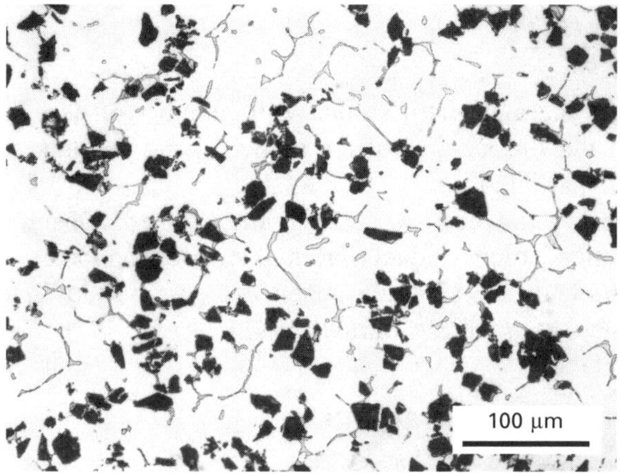

6.2 Optical micrograph of the as-cast microstructure of a slowly solidified Al-20%-Al_2O_{3p} composite showing clustering of particulate at interdendritic regions.

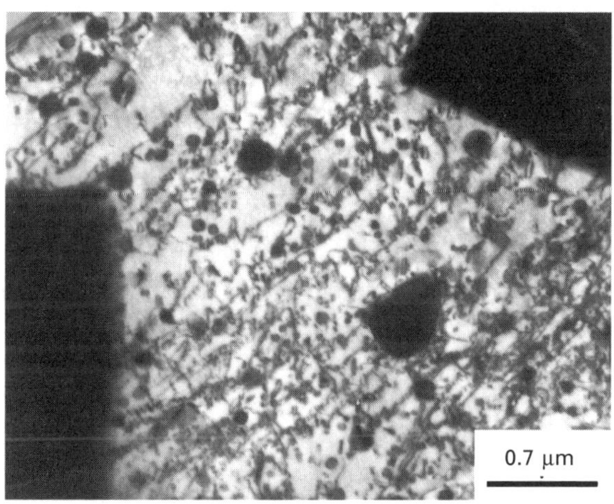

6.3 Bright field TEM micrograph showing a high density of dislocations in aluminium generated in the Al matrix in the vicinity of Al_2O_3 particles by cold water quenching from a temperature of 500 °C.

phase and the shape of the reinforcement (Humphreys *et al.*, 1991; Taya *et al.*, 1991). Arsenault and Shi (1986) predicted an increase in dislocation density ($\Delta\rho$) due to spherical particles given by:

$$\Delta\rho = \frac{\Delta\alpha\Delta TNA}{b} \qquad 6.1$$

Miller and Humphreys (1989) obtained a similar expression assuming cube-shaped inclusions:

$$\Delta\rho = \frac{12\Delta\alpha\Delta T f}{bd} \qquad 6.2$$

where $\Delta\alpha\Delta T$ is the thermal misfit strain, N the number of particles, b the Burgers vector, A the surface area of the particle, f the volume fraction of particles and d the particle size.

While the stresses induced by cooling are relieved considerably by dislocation generation, some residual (internal) stresses can locally surround the reinforcement, with a magnitude dependent on its shape and spatial distribution (Withers *et al.*, 1989; Barlow, 1991).

6.2 Deformation behaviour

During deformation of a polycrystalline metal and metal-base composite, the original grains change shape and an internal substructure forms, texture changes take place, precipitation may occur at sub-boundaries and other heterogeneities, dynamic recovery or recrystallization processes are possible and the constituent particles and the reinforcement phase may fracture and redistribute. These changes are affected by the processing method, temperature of deformation, the degree and variation in strain rate and the degree of strain. The most notable material factors that affect both the deformation microstructure and texture include stacking fault energy (SFE) of the matrix, matrix crystal structure, initial grain size and the size, shape and volume fraction of a second phase (Humphreys and Hatherly, 2004). The nature of the deformed state in both single-phase and two-phase alloys has been documented in detail over many years but DRCs are more recent materials and much less is known about their behaviour. The following sections will therefore illustrate the influence of a considerable volume fraction of whiskers, platelets or particulates on the deformation microstructure and texture of DRCs.

6.2.1 Cold deformation

Cold deformation may be carried out by rolling, extrusion, forging or drawing and is generally restricted to low homologous temperatures ($T_{def} < 0.5T_m$) where T_m is the absolute melting temperature of the matrix. During cold deformation, there is a marked increase in dislocation density of the metal matrix resulting in considerable work hardening and grains subdivide in a complex way to produce a myriad of possible features such as a cellular substructure, microbands, deformation twins, deformation bands and larger-scale heterogeneities such as shear bands (Humphreys and Hatherly, 2004).

Influence of reinforcement on deformation microstructure

Matrix dislocation density and deformation zone formation

The addition of a ceramic reinforcement to an alloy may produce an enhanced matrix dislocation density prior to deformation ($\Delta\rho$) (section 6.1.3) with cold deformation generating further dislocations in the vicinity of the reinforcement. An upper limit of these geometrically necessary dislocations (ρ_g) has been shown to be (Ashby, 1966, 1970):

$$\rho_g = \frac{k\varepsilon f}{rb} \qquad 6.3$$

where k is a constant, f the volume fraction of particles, b the Burgers vector, r the particle radius and ε the true plastic strain. Consequently, the presence of the reinforcement may result in a greater dislocation density ($\Delta\rho + \rho_g$) in DRCs compared with their unreinforced alloys.

During deformation, the local stresses that build up at a particle can be relaxed by one or more mechanisms (Humphreys, 1983): (i) the generation of new dislocations at the particle/matrix interface by local dislocation glide; (ii) diffusional relaxation; (iii) interfacial decohesion; (iv) cracking or (v) deformation of the particles. Plastic relaxation in the vicinity of particles may be achieved by any of these mechanisms and is dependent both on the level of strain and particle size. At high strains ($\varepsilon > 0.2$) very large rotations (>30°) can be produced in the vicinity of non-deformable particles greater than ~1 μm (Humphreys, 1979). These highly rotated and distorted regions are usually termed 'deformation zones' and generally exhibit a high density of dislocations rearranged into a very fine substructure. Figure 6.4a shows a schematic diagram of a deformation zone around a large angular particle.

Similar to particle-containing alloys, high strain deformation of DRCs will generate deformation zones in the vicinity of the reinforcement (Liu *et al.* 1989; Humphreys *et al.*, 1990; Ferry *et al.*, 1992). It has been found that the shape of the reinforcement strongly affects the uniformity of the matrix flow with fibres exhibiting large aspect ratios (Liu *et al.*, 1989) and angular particles (Ferry *et al.*, 1992) generating complex regions with large orientation gradients at the ends of the reinforcement. If DRCs contain a high volume fraction of particles or whiskers, the particle/whisker spacing may be small enough to allow the interaction between deformation zones of adjacent particles/whiskers (Fig. 6.5). This may result in a substantial fraction of the matrix having a complex microstructure, characterised by a fine cellular substructure with localised lattice rotations greater than 30° (Liu *et al.*, 1989; 1991a, b; Humphreys *et al.*, 1990; Liu and Barlow, 1995).

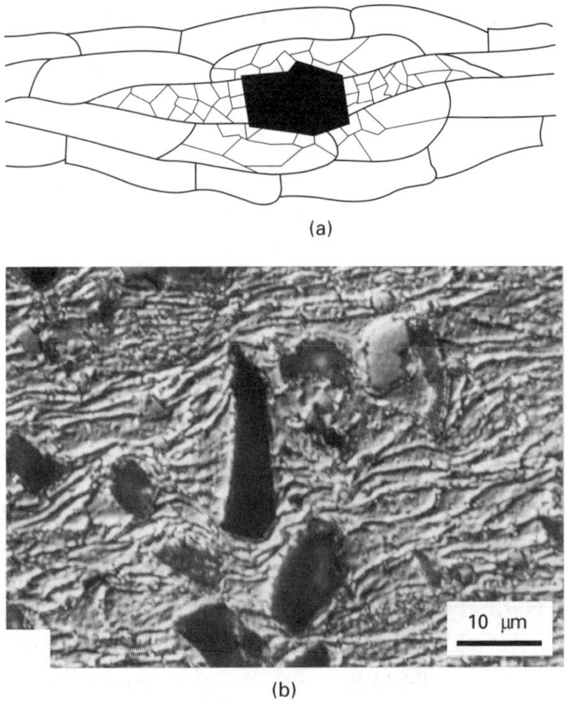

6.4 Schematic illustration of the deformation zones that form around: (a) an angular, non-deformable particle, after Clyne and Withers (1993) with kind permission of Cambridge University Press; and (b) optical micrograph showing tortuous matrix flow around an Al_2O_3 particle in an Al-base DRC.

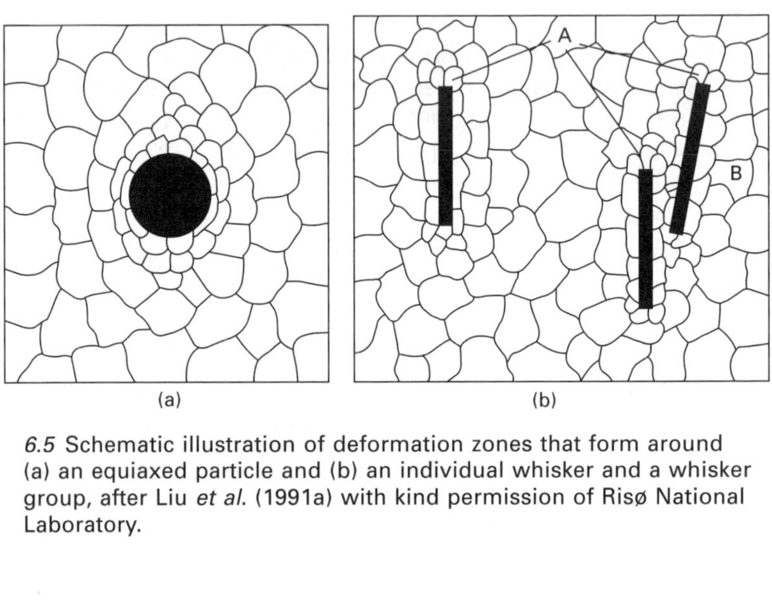

6.5 Schematic illustration of deformation zones that form around (a) an equiaxed particle and (b) an individual whisker and a whisker group, after Liu *et al.* (1991a) with kind permission of Risø National Laboratory.

Local strain gradients

The matrix strain distribution in DRCs after cold working is affected by the volume fraction, size, shape and distribution of the reinforcement (Clyne and Withers, 1993). Large angular particles or fibres generate tortuous matrix flow (Fig. 6.4b) with local matrix strains often considerably greater than the macroscopic strain (Humphreys, 1988). However, it has been reported that the matrix within particle clusters may also be shielded from deformation which causes the cluster to behave like a larger particle (Shahani, 1991). There has been extensive interest in understanding the nature and magnitude of the local strain gradients that develop near non-deformable inclusions during deformation, see e.g. Christman *et al.* (1989), Shen *et al.* (1995), Sorensen *et al.* (1995), Bassani *et al.* (2001), Shu and Barlow (2000). Both modelling and experimental studies are generally concerned with the stress and strain distribution near the reinforcement during low-strain tensile deformation and show clearly that large strain and orientation gradients develop readily. However, only very few modelling studies have been carried out on the influence of reinforcement shape and size on the strain and orientation gradients that develop during high strain ($\varepsilon > 0.2$) processes such as rolling or forging (Humphreys and Bate, 2003).

Other deformation heterogeneities

In moderately cold-rolled Al-10/20%-SiC_p (Humphreys *et al.*, 1990), Al-20%-Al_2O_{3p} (Ferry *et al.*, 1992) and cold-drawn Al-10%-Al_2O_{3p} (Shahani and Clyne, 1991a) composites, shear bands may form with characteristics similar to those observed in more highly strained unreinforced alloys. The transmission electron microscope (TEM) micrograph in Fig. 6.6 shows a matrix shear band in the vicinity of Al_2O particulates after 50% cold work. It has been argued that during deformation, the particulates, particularly those in clusters, may slide relative to each other and to accommodate this sliding motion, the matrix must deform in a shear-like manner otherwise voids will form at the particulate/matrix interface (Zhao *et al.*, 1992). Clearly, the reinforcement may contribute to premature plastic instability in DRCs during cold working.

In some DRCs, the general deformation microstructure has been shown to be more homogeneous than the unreinforced alloy (Liu *et al.*, 1991b). For example, the occurrence of inhomogeneities such as microbands and transition bands in Al-2/4%-SiC_w were lower than the alloy as the SiC whiskers tended to break up these features. For an Al-10%-SiC_w composite, Liu and Barlow (1995) reported that a rolling reduction of 93% ($\varepsilon = 2.66$) generated a uniform subgrain size throughout the entire microstructure with a large fraction of the subgrains exhibiting high angle boundaries. They attributed this effect to

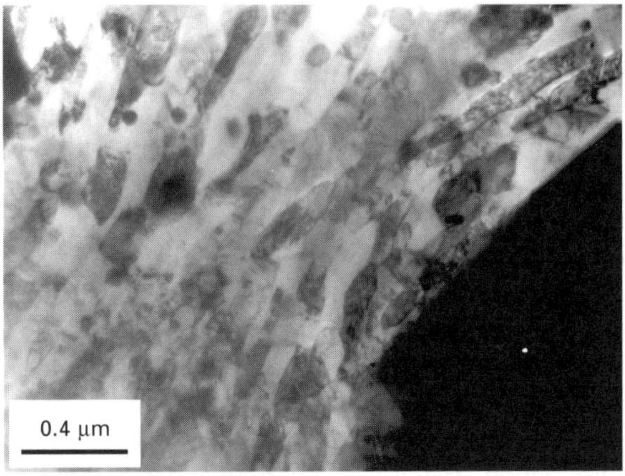

6.6 Bright field TEM micrograph of a 50% cold rolled Al-Al_2O_{3p} composite showing a shear band traversing diagonally through the microstructure near Al_2O_3 particles.

the substantial strain gradients that develop as a consequence of the very close spacing (2–3 μm) between whiskers.

Internal stresses

The change in internal stress during loading and unloading has been studied in Al-SiC_w composites (Hansen *et al.*, 1992), and the results show there is build up of stresses, both in the matrix and reinforcement during loading, and that internal stresses remain after the external load is removed. These internal stresses may then eventually relax at room temperature. The internal stresses generated in an Al-SiC_w composite during loading were found to be generally non-uniform, and that a larger concentration of stress and strain is in the vicinity of sharp corners of the fibres (Shu and Barlow, 2000). For spherical particles, the distribution of matrix stresses and strains were significantly less than for whiskers (Hansen *et al.*, 1992). It is expected that internal stresses produced by loading will contribute to the total stored energy of the composite and, therefore, will have a small influence on recovery and recrystallization.

6.2.2 Hot deformation

Hot working is usually carried out by the same processes as cold working but generally occurs at $T_{def} > 0.5T_m$ and is distinguished from cold working by the absence or near absence of strain hardening and lower matrix dislocation content (Zaidi and Wert, 1989). Furthermore, deformation becomes more homogeneous with increasing temperature which reduces the frequency of microstructural inhomogeneities such as microbands, transition bands and shear bands (Humphreys and Hatherly, 2004).

Hot worked microstructure of DRCs

The microstructural development during hot extrusion of Al-SiC_p and Al-SiC_w composites has been studied by Juul Jensen *et al.* (1991a). In these materials, a well-defined subgrain structure developed and individual subgrains contained a high internal dislocation density, and, for a given extrusion temperature, the subgrain size decreased with increasing concentration of reinforcement. Extrusion at a true strain rate of ~$0.1 s^{-1}$ at 450 °C generated complete or incomplete deformation zones at whiskers and coarse (3 μm) particles but fine (0.8 μm) particles had a negligible effect. For both composites, the frequency of formation of deformation zones decreased with increasing extrusion temperature.

The microstructural development during multi-pass hot rolling of a range of Al-SiC_p (Humphreys *et al.*, 1990) and Al-Al_2O_{3p} (Kalu and McNelley, 1991) composites has shown that the reinforcement affects the deformation microstructure in a complex manner, since recrystallization may occur between passes. For low volume fraction composites ($f < 0.05$), grain boundary nucleation is predominant which produces a coarse, final grain size, in a manner similar to conventional Al alloys (Humphreys *et al.*, 1990). For intermediate volume fractions (and small particle sizes), recrystallization is generally inhibited due to particle pinning effects (Zener, 1948). However, high volume fractions ($f > 0.1$) of coarse particulates results in particle stimulated nucleation (PSN) between rolling passes and results in continual refinement of the grain size (Kalu and McNelley, 1991).

Dynamic restoration mechanisms

The principal dynamic restoration mechanisms that operate to alleviate the effect of strain hardening are either dynamic recovery (DRV) or dynamic recrystallization (DRX) (Sellars, 1986). In high SFE face centred cubic (fcc) metals such as Al, Ni, etc., DRV is the predominant restoration mechanism and occurs via dislocation climb and cross slip. However, extensive DRV does not readily occur in low SFE fcc alloys or those where the number of slip systems is limited and DRX is the predominant restoration mechanism characterised by the presence in the microstructure of fine, equiaxed grains with a high internal dislocation density, bounded by high-angle boundaries (Humphreys and Hatherly, 2004).

Dynamic recrystallization has been observed, as expected, in Mg-SiC_p composites during hot extrusion (Lee *et al.*, 1995), but there is uncertainty as to its occurrence in Al-based composites. While DRV is often found to be the sole restoration mechanism in these latter composites (Almas, 1989; Juul Jensen *et al.*, 1991a; Chandra and Yu, 1993; Ferry and Munroe, 1995a), DRX has been argued to occur under certain deformation conditions (Tuler *et al.*, 1988; Bhat *et al.*, 1992; Zhao *et al.*, 1992; Ko and Yoo, 1998; Mazen, 1999; Park *et al.*, 2002). In many of the studies supporting DRX, no TEM evidence was presented, so it is difficult to draw any firm conclusions. However, McQueen and co-workers (McQueen *et al.*, 1993; Xia *et al.*, 1996; Xia and McQueen, 1997; McQueen *et al.*, 1998) carried out TEM on a range of hot-worked Al-SiC_p and Al-Al_2O_{3p} composites and found that, while DRV was the principal restoration mechanism, DRX produced in the vicinity of the reinforcement, occasional large (2–3 μm) highly misorientated equiaxed grains. In summary, DRCs with a high-SFE matrix will generally undergo extensive DRV during hot working, and DRX may occur under certain processing conditions but will have only a small influence on the dynamic restoration behaviour.

6.2.3 Deformation textures

The orientation changes that take place during deformation are not random but are a consequence of the fact that deformation occurs on the most favourably orientated slip or twinning systems and it follows that the deformed metal acquires a preferred orientation or texture. The strength of the texture and the balance between the various texture components will depend on the texture of the starting material and the degree of deformation. There are a number of other factors which may affect the deformation texture such as deformation geometry, temperature of deformation, crystal structure, SFE, initial grain size and the presence of a second phase (Humphreys and Hatherly, 2004).

For a given level strain, the strength of the texture that develops in DRCs during rolling, extrusion or drawing is less than the unreinforced alloys (Bowen *et al.*, 1991; Bowen and Humphreys, 1991; Juul Jensen *et al.*, 1991a,b; Poudens *et al.*, 1991, 1996; Shahani and Clyne, 1991a; Chen *et al.*, 1998). Drawing or extrusion of Al-alloys generally develops a strong <111> fibre texture whereby grains align with <111> parallel to the extrusion axis. It has been shown that such a texture is considerably weakened as the concentration of reinforcement is increased (Juul Jensen *et al.*, 1991a; Poudens *et al.*, 1991, 1996; Shahani and Clyne, 1991a). Figure 6.7 shows the influence of size, morphology and volume fraction of SiC on the development of the <111> fibre texture during extrusion of Al where it can be seen that an increased concentration of fine (0.8 μm) particles has a far smaller influence on texture weakening than fibres and large (3 μm) particles. Bowen and co-workers (1991) showed that the typical copper-type texture that develops in rolled aluminium is weakened considerably when the volume fraction of

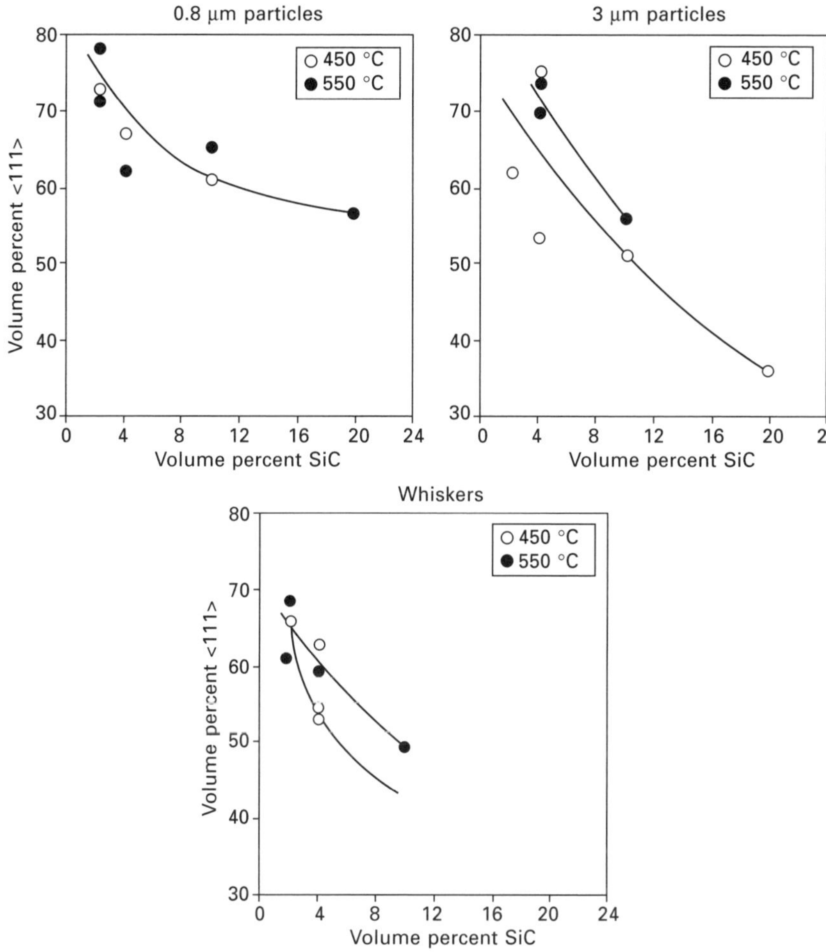

6.7 Influence of size, morphology and volume fraction of SiC in aluminium on the development of the <111> fibre texture during hot extrusion at 450 and 550 °C, adapted from Juul Jensen *et al.* (1991a) with kind permission of Risø National Laboratory.

SiC_p is greater than ~0.02, but particle size was found to have a minor influence on texture development (Fig. 6.8).

The weak textures that generally develop in DRCs during mechanical processing have been attributed to the following effects (Bowen and Humphreys, 1991): (i) during deformation, a zone of material near the reinforcement will be strongly rotated (~>15°) from the matrix orientation and, while the orientations in these zones are not random, the spread away from the matrix texture will cause a net weakening of the overall texture; and (ii) a zone extends further from the particle which is less strongly rotated (<15°), but

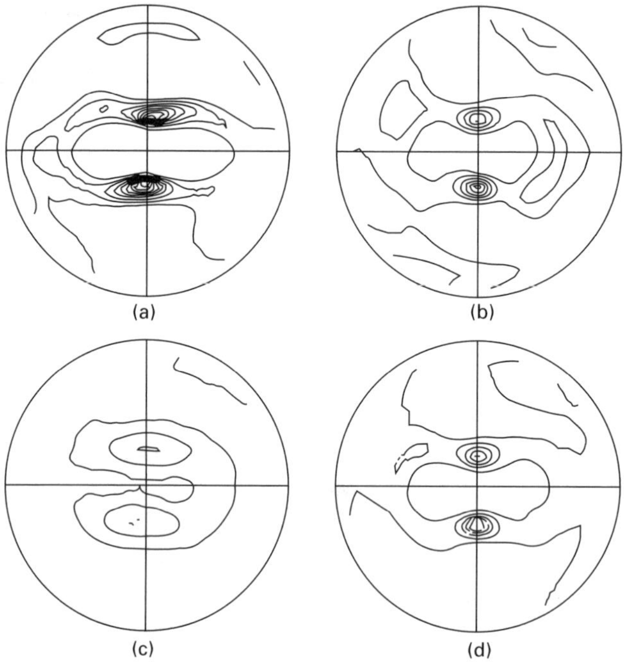

6.8 (111) pole figures of 1050Al-SiC$_p$ composites after rolling to 80% reduction. Contours are plotted for levels 1, 1.5, 2, 2.5, etc. x random. The rolling direction is vertical: (a) unreinforced alloy; (b) 2 vol.% 3 μm SiC$_p$; (c) 20 vol.% 3 μm SiC$_p$, and (d) 2 vol.% 20 μm SiC$_p$, after Bowen *et al.* (1991), with kind permission of Risø National Laboratory.

will contribute to further weakening of the texture. It was shown that for PRCs with $f > \sim 0.07$, very little of the matrix is available to deform independently of the particles, thereby resulting in a very weak texture. Following deformation of a Cu-W composite, Poole *et al.* (1992) observed that 1mm diameter tungsten wires generated an inhomogeneous matrix strain distribution and the texture, on a global scale, was weaker than the unreinforced alloy. A similar effect may occur in DRCs that contain a very high volume fraction of large particles or fibres.

6.3 Annealing processes following deformation

6.3.1 Static recovery

Static recovery (SRV) results in a release of stored energy and a change in microstructure in cold-worked metals and proceeds by the reduction in point defect concentration, dislocation annihilation and rearrangement into low-energy configurations in the form of sharp and ordered walls of subgrains through dislocation cross slip and climb. Studies of SRV in Al-SiC$_w$ (Liu *et al.*, 1992; Hansen *et al.*, 1992) and Al-Al$_2$O$_{3p}$ (Ferry and Munroe, 2003) showed a more rapid decline in hardness in the composites compared with the alloys during annealing at low temperatures (≤ 300°C). In addition, a more rapid rate of subgrain growth occurred in the composites, particularly in the vicinity of whisker or particulate groups (Fig. 6.9). The rapid subgrain growth was attributed to the higher matrix stored energy which originates from the higher dislocation density produced by deformation, and also from the generation of long-range internal stresses. In DRCs that contain a dispersion of fine (< 100 nm) oxide particles as a result of their method of production (section 6.1.2), there is a strong retarding influence on the rate of recovery due to particle pinning of individual dislocations and subgrain boundaries (Liu *et al.*, 1992).

6.3.2 Static recrystallization

Following cold working, static recrystallization (SRX) in DRCs may occur in a manner similar to conventional particle-containing alloys. However,

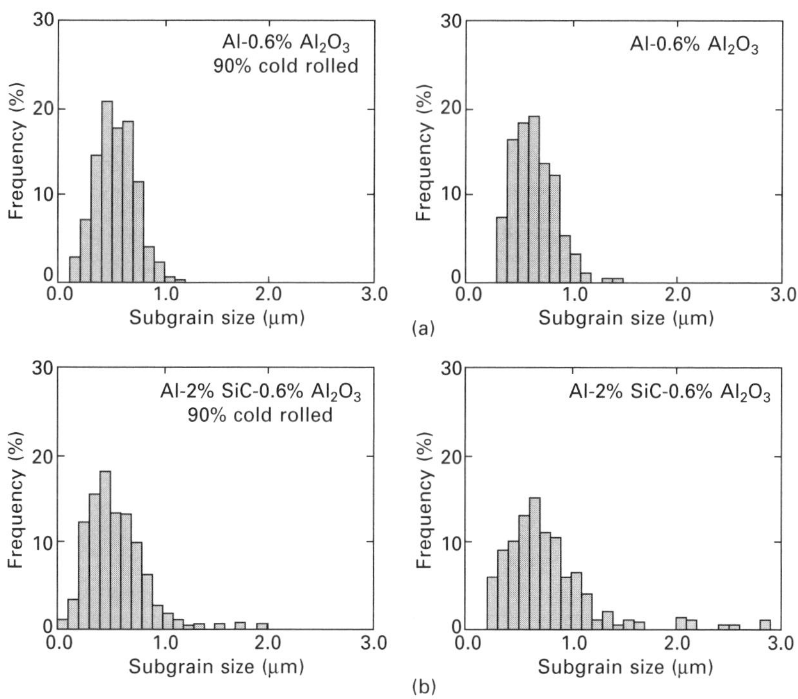

6.9 Distribution of subgrain sizes following cold rolling to 90% reduction and then annealing for 1 h at 200 °C: (a) Al-0.6%-Al$_2$O$_{3p}$ alloy and (b) Al-2%-SiC$_w$-0.6%Al$_2$O$_3$ composite, adapted from Liu *et al.* (1991a) with kind permission of Risø National Laboratory.

depending on the size, shape and volume fraction of the reinforcement, the principal nucleation sites, kinetics of recrystallization, recrystallized grain size and shape and recrystallization texture can be considerably different from the unreinforced alloys.

Nucleation of recrystallization

Large (> 1 μm) particles are known to accelerate recrystallization and decrease the recrystallized grain size by providing enhanced nucleation sites for recrystallization (Humphreys, 1977). For a particle to act as a nucleation site, a deformation zone must form in its vicinity and the overall driving force for recrystallization must be sufficient for a potential nucleus to grow into the surrounding less heavily deformed matrix (Humphreys, 1977). Humphreys showed that particle stimulated nucleation of recrystallization (PSN) following cold deformation was dependent on both the degree of strain and particle diameter, but particle shape can also strongly influence nucleation. For example, particles with angular (Es-Said and Morris, 1994) or plate-shaped morphologies (Hatherly and Dillamore, 1975; Herbst and Huber, 1978), exhibit very large lattice rotations at their ends and these regions act as preferential sites for nucleation. Similarly, Shahani and Clyne (1991a,b) found that nucleation in DRCs is dependent on the morphology of the reinforcement and showed that fibres have a much higher nucleation efficiency than particulates due to the large lattice rotations that develop at their ends. Several workers (Liu *et al.*, 1989, 1991a,b; Ferry *et al.*, 1992) have found that clusters of fibres/particles develop highly complex deformation zones in their vicinity and these have a greater propensity to nucleate recrystallized grains than individual fibres/particles. Figure 6.10a shows nucleation at a cluster of Al_2O_{3p} particles in an Al-base DRC with Fig. 6.10b showing schematically PSN and the progress of recrystallization in a PRC.

In addition to PSN effects, an inhomogeneous matrix strain distribution, as observed in a deformed Cu-W fibre composite (section 6.2.1), may strongly influence recrystallization with nucleation most likely in regions of high von Mises strain (i.e. regions containing the greatest stored energy). The work on Cu-W illustrates the importance of the reinforcement on the overall distribution in the matrix stored energy and, hence, the subsequent nucleation sites at which recrystallization will commence.

Recrystallization kinetics

When nucleation occurs predominantly near the reinforcement, an increase in the volume fraction usually results in a higher nucleation rate leading to a lower recrystallization temperature (Liu *et al.*, 1989, 1991a, 1992; Humphreys *et al.*, 1990; Shahani and Clyne, 1991b; Chandra and Yu, 1991; Ferry *et al.*,

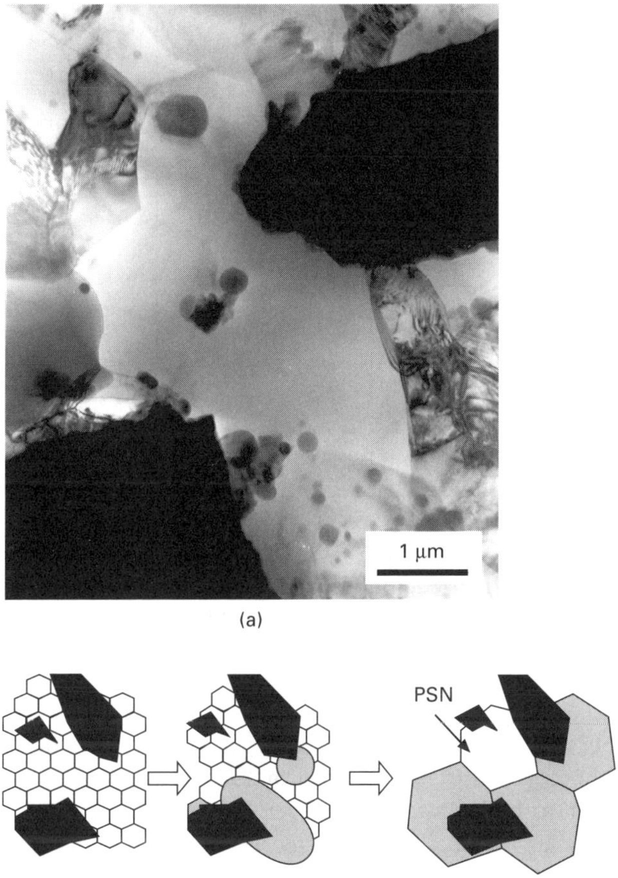

(a)

(b)

6.10 (a) Bright field TEM micrograph showing nucleation of recrystallization in the vicinity of Al_2O_3 particles, and (b) schematic diagram showing PSN and the progress of recrystallization in a PRC.

1992; Taha, 1996; Manna *et al.*, 1996; Yu *et al.*, 1994). Figure 6.11 illustrates the substantial decrease in recrystallization temperature in Al-SiC_p for an increase in volume fraction and a decrease in size of particulates. This effect was attributed to a larger stored energy, enhanced rate of nucleation, and the smaller distance by which grains must grow before mutual impingement.

Similar to particle-containing alloys, an increase in the level of cold work increases the rate of recrystallization in DRCs (Liu *et al.*, 1989, 1991b; Humphreys *et al.*, 1990; Shahani and Clyne, 1991b; Ferry and Munroe, 1993). It has been shown that, while the rate of recrystallization in an Al-20%-Al_2O_{3p} composite was more rapid than an unreinforced alloy at low strains ($\varepsilon < 50\%$), the rates were comparable after high strains (Ferry, 1994).

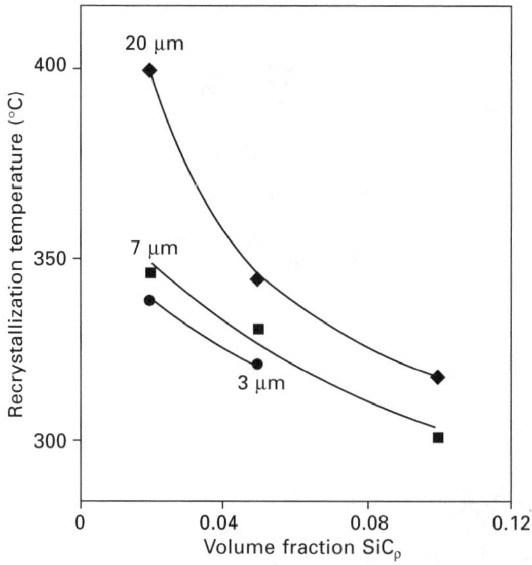

6.11 The recrystallization temperature (1 h to full recrystallization) of 60% cold rolled Al-SiC_p composites, after Humphreys *et al.* (1990), with kind permission of Institute of Materials, Minerals and Mining, UK.

Such behaviour may be a result of the large (15 μm) particles in the composite generating inhomogeneous deformation at low strains thereby promoting recrystallization. At higher strains, however, copious PSN at coarse (>2 μm) constituent particles in the alloy allows recrystallization to proceed at a rate approaching the composite.

For DRCs that contain a very high volume fraction of whiskers or small particles, recrystallization may be suppressed to such an extent that recovery is the sole restoration process (Humphreys, 1988; Liu *et al.*, 1991b; Liu and Barlow, 1995). Such an effect is often attributed to the relationship between the size and spacing of the reinforcement and their effect on grain boundary pinning. However, Liu and Barlow (1995) found, in a 93% cold-rolled Al-10%-SiC_w composite, subgrains of uniform size exhibiting a large fraction of high angle grain boundaries (HAGBs) and these undergo uniform subgrain growth throughout the material during annealing, in a similar manner to normal grain growth.

Since the kinetics of recrystallization is controlled both by the rate of nucleation and the growth rate of the nucleated grains, the accelerating effect of the reinforcement may be offset by the retarding influence of any fine particles present in the matrix (Liu *et al.*, 1989; Humphreys *et al.*, 1990). In those DRCs containing either fine oxide or precipitate particles, recrystallization will be influenced in a complex manner both by the reinforcement phase and fine particles (Ferry *et al.*, 1992; Liu and Barlow, 1995; Manna *et al.*, 1996; Yu *et al.*, 1994).

Recrystallization textures

The weak deformation textures that develop in DRCs during rolling, extrusion or drawing are generally weakened further by recrystallization (Bowen *et al.*, 1991; Shahani and Clyne, 1991a; Juul Jensen *et al.*, 1991a,b; Liu *et al.*, 1992; Poudens and Bacroix, 1996; Taha, 1996). This behaviour is generally attributed to extensive nucleation in the vicinity of the high volume fraction of large particles/fibres present in these materials. It was found that the recrystallization textures in a range of cold rolled Al-SiC_p composites were generally weak and relatively independent of both the volume fraction and particle size (for $d \geq 3$ μm). Nevertheless, the recrystallized composites where found to exhibit grains with orientations within the spread of orientations of the deformation zones that form at the particles (Bowen *et al.*, 1991) as expected for conventional particle-containing alloys (Ferry and Humphreys, 1996; Humphreys and Hatherly, 2004).

Recrystallized grain size

The addition of a discontinuous reinforcement to an alloy has, in all reported cases, resulted in a refinement of the recrystallized grain size (Almas, 1989; Liu *et al.*, 1989, 1991b; Miller and Humphreys, 1989; Shahani and Clyne, 1991a, b; Ferry *et al.*, 1993; Van Aken *et al.*, 1995). Figure 6.12 demonstrates the effect of reinforcement size and volume fraction on the recrystallized grain size in a range of Al-SiC_p composites cold rolled 80% and annealed at 600 °C.

The following relation was obtained for the recrystallized grain size, D_r, in PRCs by assuming that each particle acts to nucleate a single grain (Humphreys, 1988):

$$D_r = d(1 - f)^{1/3}/f^{1/3} \qquad 6.4$$

Figure 6.13 shows the influence of particle parameters on the final grain size, as predicted by eqn 6.4. It is important to note that this relation is an estimate only, since parameters such as the initial grain size, particle shape and their distribution, degree of cold work, recrystallization temperature, heating rate, etc., affect the recrystallized grain size in DRCs in a manner similar to conventional particle-containing alloys (Juul Jensen *et al.*, 1991b; Liu *et al.*, 1991b; Ferry and Munroe, 1993; Van Aken *et al.*, 1995; Taha, 1996). As an example, further grain refinement may be achieved via the higher nucleation efficiency associated with fibres compared with more regular shaped particles (Shahani and Clyne, 1991b).

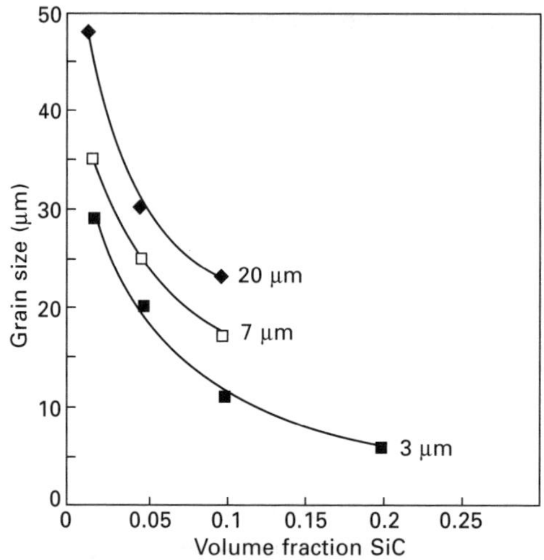

6.12 The recrystallized grain size in Al-SiC$_p$ composites after cold rolling to 80% reduction and annealed at 600 °C, after Humphreys and Hatherly (2004) with kind permission of Elsevier Science limited.

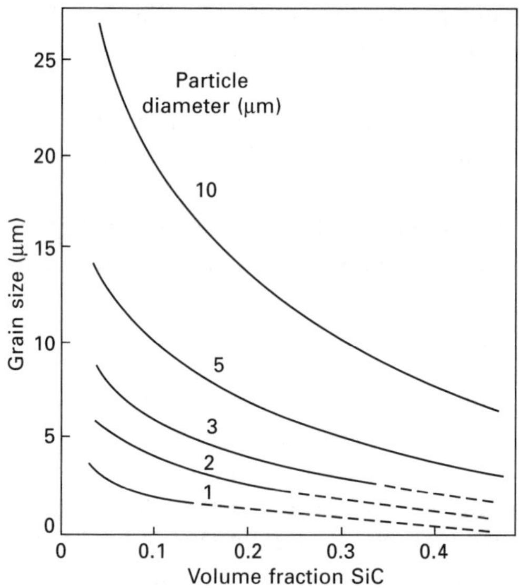

6.13 The predicted grain size for a PSN efficiency of unity using (eqn 6.4). Recrystallization is unlikely in the region indicated by the dashed lines, after Humphreys and Hatherly (2004) with kind permission of Elsevier Science Limited.

Equation 6.4 predicts a decrease in grain size as either the volume fraction increases or particle size decreases which has been verified experimentally for a range of Al-SiC_p composites with d = 3–100 μm and f = 0.05–0.3. Miller and Humphreys (1989) showed that large (≥ 20 μm) particles control the recrystallized grain size by nucleation (i.e. eqn 6.4), but a large volume fraction of small (~≤ 3 μm) particles produced a grain size corresponding more closely to a grain growth limited grain size, D_G, as predicted by Zener's pinning relation (Zener, 1948):

$$D_G = 2d/3f \qquad 6.5$$

Recrystallization after hot working

The annealing of Al-SiC_p composites following multi-pass hot rolling has shown that recrystallization during a high temperature anneal (> 500 °C) is more rapid than the unreinforced alloy (Humphreys *et al.*, 1990; Sparks and Sellars, 1992). Similar to cold working and annealing, the recrystallized grain size decreases with either increasing volume fraction of SiC particles or with decreasing particle size. In all noted cases (Humphreys *et al.*, 1990; Sparks and Sellars, 1992; Ferry and Munroe, 1995b), hot working generates a coarser recrystallized grain size than that produced after cold working. However, deformation of heat treatable Al-based DRCs at very high temperatures and low strain rates can suppress recrystallization completely during subsequent solution heat treatment (Ferry and Munroe, 1995c).

6.3.3 Grain size stability after recrystallization

The recrystallized microstructure of DRCs is highly resistant to grain growth (Ferry *et al.*, 1993). Grain growth inhibition may be due to Zener pinning by the reinforcement (Van Aken *et al.*, 1995; Ozdemir *et al.*, 2003) and from fine particles present in the matrix (Juul Jensen *et al.*, 1991b), but reinforcement spacing may also be important (Ferry *et al.*, 1993). In DRCs containing fine oxide or other particles aligned in the working direction by rolling or extrusion, Zener drag due to this non-random particle dispersion will result in grain growth along the working direction and an elongated grain structure, similar to conventional alloys, will be produced.

Despite the considerable effort devoted to predicting the limiting grain size in particle-containing materials, see e.g., Manohar *et al.* (1998), there are no models that can satisfactorily take into account the microstructural features typical of DRCs. These include the heterogeneity of the reinforcement distribution, irregular shapes and high volume fractions of reinforcement with a range of sizes. Figure 6.14 shows the close correlation between the particulate distribution in Al-20%-Al_2O_{3p} and the matrix grain size distribution during prolonged annealing at high temperature.

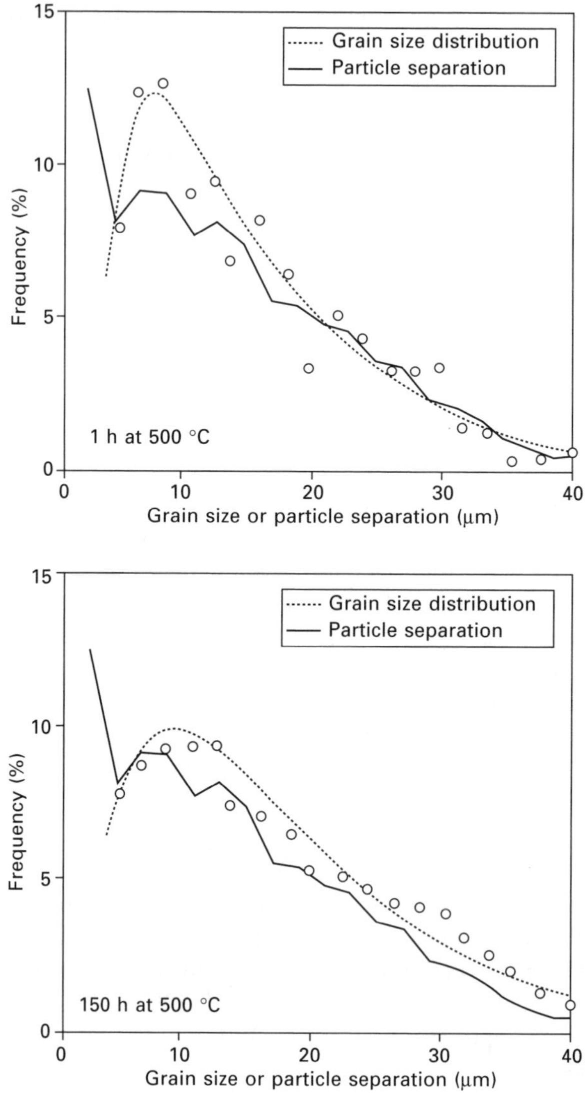

6.14 Frequency distribution of the interparticulate distribution in Al-20%-15 μm-Al_2O_{3p} composite (unbroken curve), together with linear intercept grain size distributions (dashed curves) following cold rolling 50% and annealing at 500 °C for 30 s (as-recrystallized) and 150 h (open symbols denote the percentage of grains below a given size for 2 μm intervals), after Ferry (1994).

6.4 Processing and processability of DRCs

6.4.1 Flow stress, ductility and workability

At low deformation temperatures, it is generally found that the initial work hardening rate (WHR) and flow stress (σ) of an alloy is significantly increased with the addition of a reinforcing phase (McDanels, 1985; Nair *et al.*, 1985; Jarry *et al.*, 1987; Tuler *et al.*, 1988; Chanda *et al.*, 1991; Zhao *et al.*, 1992; Sarkar and Surappa, 1993; Lloyd, 1994). As the temperature of deformation increases, the differential in WHR and flow stress ($\Delta\sigma$) becomes negligible (Ferry and Munroe, 1995a), and it has been shown that deformation of DRCs close to the matrix melting point may result in the flow stress falling below the monolithic alloy due to the onset of grain boundary sliding in the finer grained composites (Mahoney and Ghosh, 1987).

While the low-temperature strength of a DRC is higher than its unreinforced alloy, its ductility is usually considerably lower (Nieh, 1984; Pickens *et al.*, 1987; Humphreys, 1988; Clyne and Withers, 1993; Lloyd, 1991b, 1994; Boyd and Lloyd, 2001). Such limited ductility is problematic as it will restrict the potential end uses of DRCs as well as reducing their formability to such a degree that mechanical processing should be carried out under carefully chosen processing conditions. Nevertheless, Brechet *et al.* (1993) has demonstrated that an imposed hydrostatic pressure during tensile straining improves the low-temperature ductility of DRCs but the high pressures required will restrict this technique to more exotic processing routes.

Relationship between processing parameters – micromechanisms of deformation

For conditions normally encountered during industrial processing, the Sellars-Tegart constitutive relation has been found to be useful as it incorporates flow stress (σ), strain rate ($\dot{\varepsilon}$) and deformation temperature (T) in a single relation (Sellars and Tegart, 1966):

$$\dot{\varepsilon} = A(\sinh \alpha\sigma)^n \exp(-Q/RT) \quad 6.6$$

where A, α and n are material constants, R the universal gas constant and Q an activation energy term.

Calculation of Q using mechanical test data and eqn 6.6 can provide useful information on the micromechanisms of deformation at elevated temperature (Dieter, 1986). However, calculated Q-values are often not comparable with the activation energy of a particular atomic process, such as lattice, grain boundary or surface diffusion, etc., and problems often arise in its interpretation. Nevertheless, Q is incorporated in various hot working models (such as eqn 6.6) which are useful for furthering our understanding of the effect of processing parameters on the development of microstructure and properties of materials produced by rolling, forging, extrusion, superplastic forming, etc.

Equation 6.6 has been used to calculate Q for a range of Al-SiC and Al-Al_2O_3 composites deformed in torsion, compression and extrusion (Table 6.1). A few workers have shown that Q is similar in both the alloy and composites and close to the activation energy of self diffusion of Al (~142 kJ/mol). Most investigations, however, report Q-values much greater than the unreinforced alloy and considerably higher than the activation energy of self diffusion. McQueen *et al.* (1999) argued that high Q-values are a result of composites undergoing additional matrix strain hardening due to the constraints imposed by the rigid reinforcement. Table 6.1 shows that values of n and Q can vary considerably, even in composites with the same general matrix composition and type of reinforcement. A number of important factors may influence the flow behaviour of DRCs, such as the mode of deformation, initial processing route, starting grain size, degree of solute supersaturation (matrix temper) before testing, and reinforcement parameters (size, shape, volume fraction and distribution). Hence, the calculated constants in eqn 6.6 will be influenced to some degree by these factors. Similar disparities in the values of n and Q have been reported by Clyne and Withers (1993) on the creep behaviour of Al-based PRCs.

Models for optimising the workability of DRCs

Physical-based models

The high rate of work hardening in DRCs at low working temperatures may limit their formability due to stress accumulation at the reinforcement which can result in excessive fragmentation and void formation in their vicinity. It is well known that WHR in particle-containing alloys is strongly temperature dependent and decreases considerably over a small temperature interval (Hirsch and Humphreys, 1969; Stewart and Martin, 1975; Humphreys and Kalu, 1987). This decrease in WHR is argued to be a result of the removal of local stresses at particles via enhanced cross-slip and climb assisted dislocation motion and diffusional relaxation (vacancy transport) (Stewart and Martin, 1975). A number of models have been proposed for predicting the conditions under which stresses will not accumulate at particles during deformation (Ashby, 1970; Stobbs, 1973; Koeller and Raj, 1978; Mori *et al.*, 1980). Humphreys and Kalu (1987) have derived a useful stress relaxation model which takes into account the regime in which dislocation climb is affected by both volume diffusion and interfacial diffusion. The critical strain rate, $\dot{\varepsilon}$, below which dislocation climb is predominant, was shown to be a function of both the particle diameter and the deformation temperature:

$$\dot{\varepsilon} = K_1 \frac{\exp(-Q_b / RT)}{d^2 T} + K_2 \frac{\exp(-Q_i / RT)}{d^3 T} \qquad 6.7$$

Table 6.1 Experimental stress exponent (n) and activation energy (Q) data determined from eqn 6.6 for hot working of Al-based DRCs

Matrix alloy	Reinforcement (volume %)	Reinforcement (size) (μm)	Fabrication method	Mode of deformation	Temperature (°C)	Strain rate (s^{-1}	n	Q (kJ/mol)	Reference
AA6061	20% Al_2O_{3p}	15	Stir-cast	Compression	200–500	0.001–1.0	5.1	220	Chandra and Yu (1993)
AA6061	20% Al_2O_{3p}	15	Stir-cast	Compression	200–500	0.001–1.0	4.4	220	Chen *et al.* (1996)
AA6061	20% Al_2O_{3p}	–	Casting	Torsion	375–500	0.003–10	5.6	150	Spigarelli *et al.* (2001)
AA6061	20% Al_2O_{3p}	20	Stir-cast	Compression	350–500	0.001–0.1	6.0	156	Cerri *et al.* (2002)
AA2014	20% Al_2O_{3p}	15	Stir-cast	Compression	200–500	0.01–1.0	5.8	227	Ferry and Munroe (1995a)
AA2024	20% Al_2O_{3p}	15	PM route	Extrusion	475–575	0.001–1.7	2–3	180	Kim and Hong (1994)
AA2024	15% SiC*p*/SiC_w	8(p)	PM route	Torsion	300–500	0.001–2.4	–	254	Ko and Yoo (1998)
AA2618	10–20% Al_2O_{3p}	10–20	Stir-cast	Torsion	200–500	0.1–5.0	–	315–400	Zhao *et al.* (1996)
AA7075	10% Al_2O_{3p}	10–20	Stir-cast	Torsion	250–540	0.1–4.0	2.4	274	McQueen *et al.* (1999)
AA7075	15% Al_2O_{3p}	10–20	Stir-cast	Torsion	250–540	0.1–4.0	2.4	300	Qin and McQueen (2000)
A356	15% SiC_p	12	Stir-cast	Torsion	250–540	0.1–4.0	3.2	263	McQueen *et al.* (1998)

where K_1 and K_2 are derived constants and Q_b and Q_i is the activation energy for bulk diffusion and interface diffusion, respectively.

Based on eqn 6.7, Humphreys (1988) constructed a deformation map for PRCs which illustrates the critical strain rate at which stresses will be relaxed at particles, as a function of both particle size and temperature (Fig. 6.15). These and similar processing maps (see e.g. Raj, 1981) are useful since they may be used to predict the conditions under which DRCs may be successfully hot worked. For example, if deformation is carried out under conditions where deformation zones do not form at the reinforcement, the tendency for reinforcement fracture and matrix cavitation will be minimised (see section 6.4.2), thereby enhancing the workability of the composite. If we take a typical commercial Al-based PRC with an average particle size of 10 μm, Fig. 6.15 predicts that deformations zones will not form during typical conditions for extrusion ($\dot{\varepsilon} > 1\ s^{-1}$) and rolling ($\dot{\varepsilon} > 10\ s^{-1}$) for deformation temperatures greater than ~450 °C and ~520 °C, respectively. While this is feasible for extrusion, it is likely that multi-pass hot rolling will always generate deformation zones at the reinforcement unless the rolled plate is reheated between passes. Juul Jensen *et al.* (1991a) carried out high-temperature extrusion studies on Al-SiC_p and Al-SiC_w, composites and compared the observed conditions at which deformation zones do not form at the reinforcement with that predicted by eqn 6.7. While prediction was good for PRCs, it was necessary to adjust the model for WRCs by assuming that the diffusion distance for relaxation at whiskers is their length rather than diameter. Such a requirement was proposed initially by Stanford-Beale and Clyne (1988).

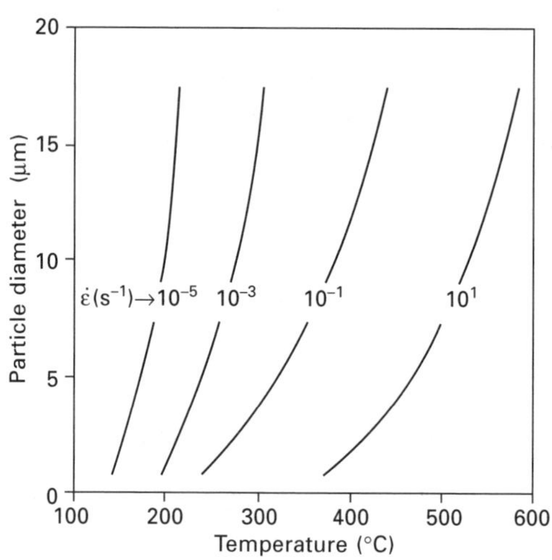

6.15 Processing map showing the critical temperature below which stresses will accumulate at particles for a range of strain rates, adapted from Humphreys (1988), with kind permission of Risø National Laboratory.

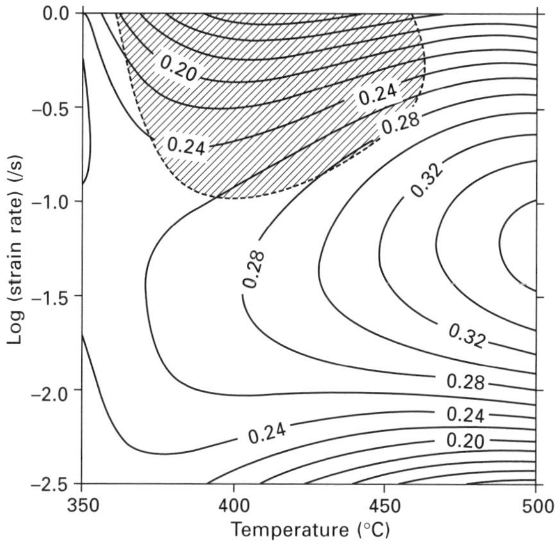

6.16 Power efficiency map for 6061Al-20%-Al_2O_{3p} deformed in compression to $\varepsilon = 0.5$ showing the predicted (shaded) region of unstable flow using eqn 6.9, after Spigarelli *et al.* (2002) with kind permission of Elsevier Science Limited.

Dynamic materials models

These models consider the workpiece to be a power dissipater whereby the power dissipated by processes such as recovery, recrystallization, cavity formation, phase transformations, etc., may be computed. On this basis, mechanical test data may be used to determine the variation in the efficiency of power dissipation (η) (Prasad, 1990):

$$\eta = \left.\frac{2m}{m+1}\right|_{\dot{\varepsilon},T} \qquad 6.8$$

where m is the strain rate sensitivity of flow stress defined as $m = \partial \log \sigma / \partial \log \dot{\varepsilon}$. The parameter η describes the constitutive response of the workpiece in terms of various microstructural mechanisms that operate in a given range of temperature and strain rate.

Equation 6.8 has been used extensively to construct power dissipation maps for a range of Al-based DRCs in an attempt to locate optimum conditions of workability (Bhat *et al.*, 1992, 1995, 2000; Ko and Yoo, 1999; Yoo and Ko, 2000; Spigarelli *et al.*, 2002). A typical deformation map is given in Fig. 6.16 for hot compressed 6061Al-20%-Al_2O_{3p} which is characterised by the presence of a region of high η at high deformation temperatures (> 450 °C) for strain rates between 3×10^{-2} and 3×10^{-1} s^{-1} and a region of low η at intermediate deformation temperatures (375–400 °C) for high strain rates

(0.6–1.0 s^{-1}). Regions of high η are often interpreted as zones of optimum workability where restoration processes such as DRV or DRX are claimed to be predominant and both matrix flow localisation and particulate damage is expected to be low.

Superimposed on the map in Fig. 6.16 is a cross-hatched region that is believed to represent metallurgical instability in the composite. This zone was determined from the following instability criterion based on the principles of irreversible thermodynamics of large plastic flows (Prasad, 1990):

$$\xi(\dot{\varepsilon}) = \frac{\partial \ln[m/(m+1)]}{\partial \ln \dot{\varepsilon}} + m < 0 \qquad 6.9$$

It is pertinent to note that the physical significance of eqn 6.8 and the associated processing maps have been critically reviewed by Montheillet *et al.* (1996) and several alternative treatments have recently been proposed for analysing the stability of plastic flow in metals and DRCs, see e.g., (Montheillet *et al.*, 1996; Murthy and Rao, 1998).

Matrix strain localisation

Avoiding matrix strain localisation in the form of shear banding during deformation is important because it leads to large strain and stress gradients which can produce premature failure of the material (Wilson, 1986). From a formability viewpoint, the onset of shear banding is the limiting factor in determining the strain obtainable in a formed part (Lloyd, 1986). For example, the local strains associated with shear banding can be sufficient to produce incipient fracture along these inhomogeneities (Ferry and Munroe, 1994), which manifests itself as edge cracking during rolling. Strain localisation in the form of shear bands has been observed in a number of moderately cold deformed Al-based DRCs (section 6.2.1). Since these composites are expected to undergo high levels of strain during rolling, drawing or forging, deformation at an elevated temperature is desirable.

6.4.2 Reinforcement integrity during processing

Fracture and decohesion of the reinforcement

Since the reinforcing phase in DRCs is usually a ceramic exhibiting irregular shapes and internal faults, the localised stresses built up during deformation may result in fracture or decohesion with the matrix. In general, comminution of the reinforcement should be minimised during processing, since fractured particles or whiskers will adversely affect, not only the workability, but also the mechanical properties of the final product (Lloyd, 1991a,b). For example, cracked particles will have a reduced ability to sustain load, and properties such as elastic modulus (E) will decrease as their number of cracked particles increases (Lloyd, 1991a,b; Mochida *et al.*, 1991a,b; Llorca *et al.*, 1993). The effect of the degree of particulate fracture on the degradation of the elastic modulus in PRCs has been given as (Mochida *et al.*, 1991a):

$$E_c/E_m = 1/[1 + \eta_1(1 + B)f + \eta_2 fB] \quad 6.10$$

where E_c and E_m is the modulus of the particle and matrix, respectively, f the volume fraction of particulates, B the volume fraction of broken particulates, and η_1 and η_2 are constants. Since the stiffness-to-weight ratio is an important property of these composites, secondary processing routes that minimise particulate cracking must be sought (see section 6.4.1).

A description of the processes leading to reinforcement fracture is complex, principally because they do not have a specific fracture stress. Furthermore, DRCs exhibit a high rate of work hardening in the initial stages of cold deformation and damage to the reinforcement begins at very low plastic strains (Lloyd, 1991b; Llorca *et al.*, 1993). Brechet and co-workers (1993) observed that the percentage of cracked particles increased approximately linearly with strain up to 75% cold-work with other workers noting a similar breakup of the reinforcement during processing (Lewandowski *et al.*, 1987). It has also been observed that particles with high aspect ratios (>2.0), and those larger than the average diameter, have the greatest tendency to fracture during cold deformation (Stanford-Beale and Clyne, 1989; Humphreys *et al.*, 1990; Lloyd, 1991a,b; Llorca *et al.*, 1993).

The principal criterion for particulate failure is argued to be one of stress rather than strain (Lloyd 1991b). Subsequently, the degree of cracking will be a function of the flow stress of the matrix which is dependent on alloying additions and matrix temper (Lloyd, 1991b) as well as temperature/strain rate combination (Stanford-Beale and Clyne, 1989). For example, Al-base DRCs show a reduced susceptibility to reinforcement fracture when they are extruded (Stanford-Beale and Clyne, 1989) or forged (Durrant *et al.*, 1991) under conditions where deformation zones do not form (see eqn 6.7 and Fig. 6.15). If there is a need to retain a very high aspect ratio of fibres in WRCs, semi-solid processing may be an attractive route since the flow stresses are extremely low (Clyne and Withers, 1993). It is generally expected that some degree of reinforcement fracture will occur during solid-state deformation. Subsequently, a viable processing route must be one that has the ability to redistribute any fragments of reinforcement and heal interfacial voids; factors that will improve the ductility and fracture toughness of the final component.

Redistribution of the reinforcement

During secondary processing of DRCs, the reinforcement not only fractures but, due to its irregular shape, will undergo alignment in the direction of working (Lewandowski *et al.*, 1987; Ehrstrom and Kool, 1988; McKimpson and Scott, 1990; Kalu and McNelley, 1991; Lloyd, 1994; Chen *et al.*, 1996; Kang *et al.*, 2000; Zhang *et al.*, 2001). It has been found that hot extrusion can largely break up agglomerations of reinforcement in the as-cast billet (McKimpson and Scott, 1990), thereby producing a more uniform distribution with additional hot rolling further improving microstructural uniformity (Lewandowski *et al.*, 1987; Kalu and McNelly, 1991). Surappa (1993) demonstrated that die shape is an important parameter during extrusion of Al-SiC_p composites whereby a circular cross-section minimised the dead zone and improves significantly the homogeneity of particulate distribution. Lloyd (1991a) showed that hot extrusion of as-cast 6061Al-SiC_p redistributes the particulates at low extrusion ratios (~10:1) and improves the ductility of the final component, but substantial clustering generally persists even after high ratios (i.e. 324:1) with little further improvement in ductility. Figure 6.17 shows the distribution of interparticulate spacing of the composite following both casting and extrusion where significant clustering, revealed as the peak at spacings below ~5 μm, persists after extrusion.

A further problem associated with extrusion of DRCs is the formation of ceramic rich bands parallel to the extrusion axis with short fibres found to be most susceptible to banding (Ehrstrom and Kool, 1988). To reduce banding, it has been argued that peak strain rates and strain rate gradients during extrusion need to be reduced through the use of tapered or streamlined dies (Rack and Niskanen, 1984; Stanford-Beale and Clyne, 1989; Goswami *et al.*, 1999). The general difficulty in eliminating clustering in DRCs highlights an inherent problem; clusters act as sites for premature failure during loading and will therefore affect adversely the mechanical properties of the finished

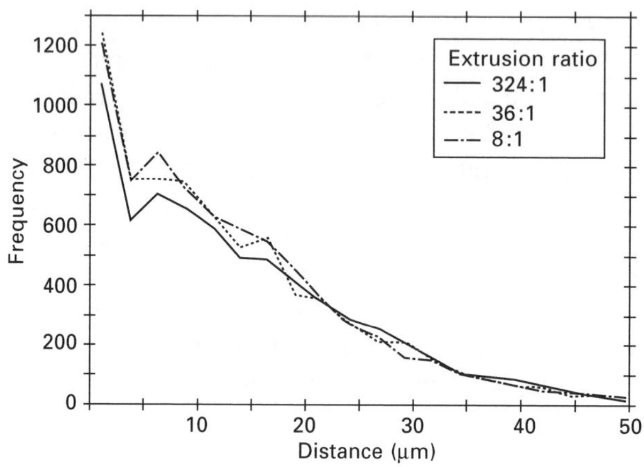

6.17 Influence of extrusion ratio on the distribution of particle separation in 6061Al-SiC_p, after Lloyd (1991a) with kind permission of Risø National Laboratory.

product (Lewandowski *et al.*, 1987; Lloyd, 1991a,b, 1994; Boyd and Lloyd, 2001).

6.4.3 Influence of processing parameters on final microstructure and properties

The Sellars-Tegart relation (eqn 6.6) may be written in terms of the Zener-Hollomon parameter, Z, (Zener and Hollomon, 1944), which is defined as:

$$Z = \dot{\varepsilon} \exp(Q/RT) \tag{6.11}$$

While it is often difficult to relate Q to any particular deformation mechanism (Table 6.1), Z is useful for discussions of hot working as it incorporates several key processing variables into a single relation. Since Z is related to the flow stress, it can therefore be related to dislocation density and subgrain size during deformation and post-deformation restoration processes such as the kinetics of recrystallization and final grain size. As an example, the influence of Z on the hot-worked subgrain size (D_{sg}) of 2014Al and 2014Al-Al_2O_{3p} was found to fit a relation of the form (Ferry and Munroe, 1995a):

$$1/D_{sg} = \alpha \ln Z - \beta \tag{6.12}$$

For the alloy $\alpha = 0.06$, $\beta - 1.65$ and for the composite $\alpha = 0.12$, $\beta = 3.73$. Close inspection of these values indicates, for a given Z-value, that the subgrain size in the composite is finer than in the alloy, with the difference becoming negligible at higher temperatures. It has also been shown for these materials that the nucleation rate ($\dot{N}$) of recrystallized grains during subsequent solution heat treatment is related to matrix subgrain size, D_{sg}, by a single relation (Fig. 6.18):

$$\log \dot{N} = \phi / D_{sg}^3 \tag{6.13}$$

For both materials, recrystallization will be suppressed if a hot-worked matrix subgrain size greater than ~2 μm is produced via the selection of a particular strain rate and temperature (Z); parameters which differ between materials (see constants in eqn 6.12). Such a correlation between recrystallization behaviour and deformation microstructure is industrially significant as, for example, hot extruded heat treatable Al-based alloys and composites often undergo a high temperature solution heat treatment prior to age hardening. Hence, a subgrain structure may be retained with high extrusion temperatures, low exit speeds and/or the use of streamlined dies to minimise strain rate gradients (Stanford-Beale and Clyne, 1989). There is, however, a range of flow stress (and Z) where the subsequent solution treatment results in nucleation, but at such a low rate that SRX is restricted to the outer rim of the extrudate to produce a coarse grained (> 200 μm) outer band (Sheppard and Vierod, 1987). Such a duplex microstructure is known to result in undesirable

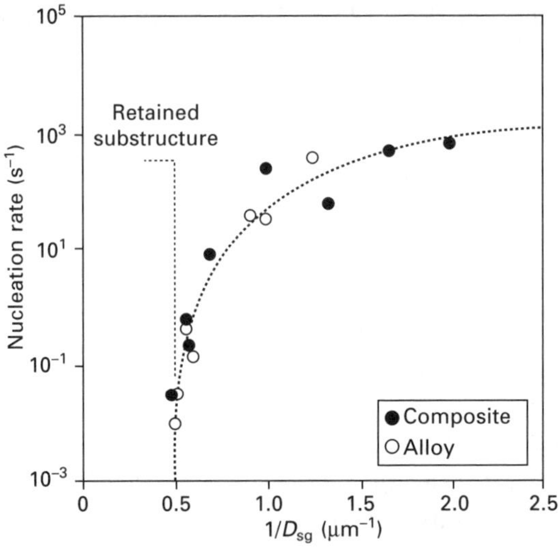

6.18 Influence of hot-worked subgrain size on the apparent nucleation rate during recrystallization in both 2014Al and 2014Al-20%-15 μm-Al_2O_{3p}, after Ferry (1994).

mechanical properties and poor corrosion resistance of the final component (Bryant, 1975).

Figure 6.19 shows processing maps for hot compressed 2014Al and 2014Al-Al_2O_{3p} showing the influence of strain rate and temperature on the solution-treated microstructure. Desirable microstructures are those that exhibit either a retained substructure (region A) or fine recrystallized grains (region B) whereas undesirable microstructures are found in region C (coarse grained and partially recrystallized zones). If we take normal extrusion conditions ($\dot{\varepsilon} > 1\ s^{-1}$), it is clear that a retained substructure in the composite will not be possible since high deformation temperatures will result in incipient melting of the aluminium matrix. Conversely, low deformation temperatures will easily generate a fine-grained microstructure but these processing conditions will also result in substantial particulate fracture (section 6.4.2).

6.4.4 Other factors affecting composite processability

Mill performance and billet surface integrity

The previous sections have shown that DRCs may be processed via rolling, extrusion, drawing and forging by conventional methods but with some necessary adjustment in processing parameters (Jeffrey and Holcomb, 1990; Watanabe *et al.*, 1990; Chen *et al.*, 1996; Goswami *et al.*, 1999). As an example, Brusethaug *et al.* (1990) has shown, for Al-15%–SiC_p, that the presence of the reinforcement necessitates a reduction in the maximum

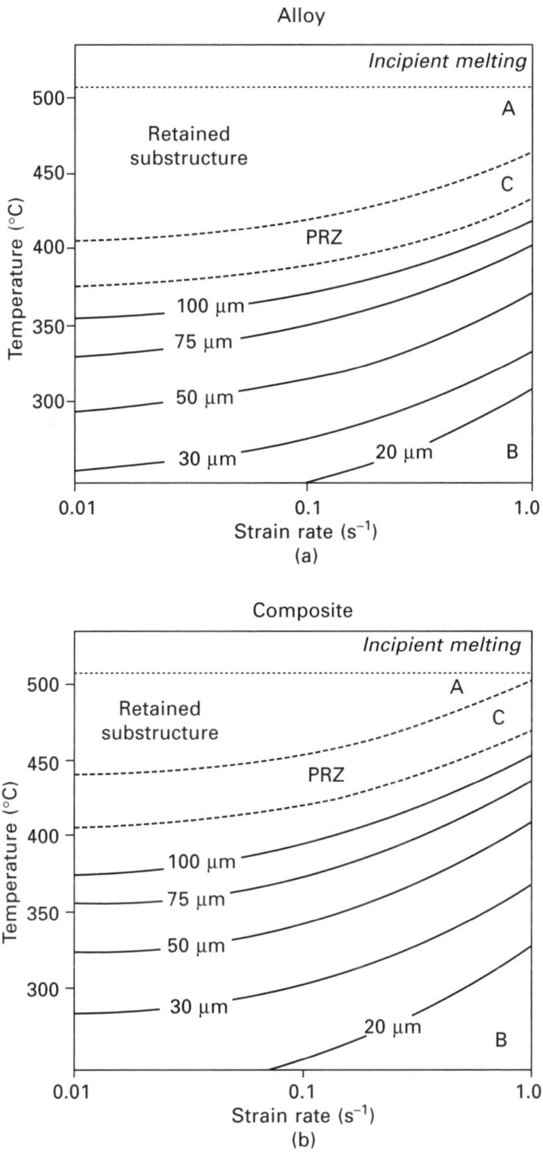

6.19 Processing maps of (a) 2014Al and (b) 2014Al-20%-15 μm-Al_2O_{3p} showing the effect of deformation temperature and true strain rate on post-deformation solution-treated microstructure (samples were deformed to 60% reduction and annealed for 1 h at 500 °C), after Ferry (1994).

extrusion (ram) speed by 30–40% and, for a given extrusion speed, both the billet surface temperature and mill pressure is increased. It has also been shown that rolling of Al-SiC_p composites generates an increase in mill load

(Sparks and Sellars, 1992). For extrusion, it is possible to lower the extrusion pressure with streamlined dies as they do not generate dead zones, but the disadvantage with this technique is the need to pre-machine billets to fit the die profile and this will generate a higher scrap level. Extrusion temperatures and pressures can also be reduced drastically by hydrostatic extrusion, which may ultimately allow extrusion to be carried out at room temperature (Grow and Lewandowski, 1995). Here, the extrusion ram does not contact the billet directly, but through a high-pressure fluid, which not only extrudes the composite, but isolates it from the container wall, hence lowering frictional effects. The superimposed hydrostatic pressure also reduces the tendency of void formation and growth during extrusion, which improves the overall quality of the extrudate.

During conventional extrusion of DRCs, the exit speed is limited by surface tearing of the extrudate due to the inherently low ductility of these materials and local liquation (Brusethaug *et al.*, 1990). Extrusion limit diagrams, as illustrated in Fig. 6.20 for 6061Al and 6061Al-Al_2O_{3p}, have been developed to define the processing window for satisfactory extrusion (Dixon and Lloyd,

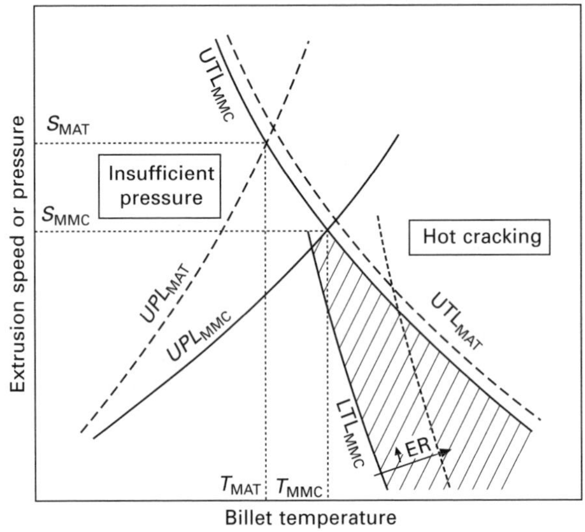

6.20 Extrusion limit diagrams for 6061Al and 6061Al-Al_2O_{3p} showing the optimum processing window (cross-hatched region) for satisfactory extrusion of the composite. LPL_{MMC} is the lower pressure limit for the composite, UPL_{MAT} and UPL_{MMC} is the upper pressure limit for the alloy and composite, respectively and UTL_{MAT} and UTL_{MMC} is the upper temperature limit for the alloy and composite, respectively. For the composite, LTL_{MMC} shifts to higher temperatures with an increase in extrusion ratio (ER) thereby further restricting the processing window. Adapted from Dixon and Lloyd (1996) with kind permission of the Metals, Minerals and Materials Society, USA.)

1996). Compared with the unreinforced alloy, extrusion of the DRC is restricted to a reduced region with the high temperature limit controlled by hot cracking of the matrix whereas the higher flow stress at low temperatures results in a higher extrusion pressure and a shift in the pressure limit line. At the optimum extrusion billet temperature, the maximum extrusion speed for the composite (S_{MMC}) is considerably lower than the equivalent maximum speed for the unreinforced alloy (S_{MAT}).

Rolling is regarded as one of the most difficult working processes for DRCs since it involves near plane strain conditions which result in unconstrained edges of the ingot that are susceptible to edge cracking. Nevertheless, edge cracking can be greatly reduced if the composite is first extruded or closed die forged to break up the as-cast structure before rolling and, hence increase its ductility (Lloyd and Jin, 2001). It is clear that extrusion and closed die forging are more attractive processing routes since both are constrained deformation processes which minimise the problem of edge cracking.

Die and roll wear

While conventional secondary processing facilities can adequately handle the slight adjustment in parameters needed for thermomechanical processing of DRCs, the hard reinforcement has a detrimental effect on die and roll life. The problem of die wear represents one of the greatest technical challenges in composite extrusion but considerable progress has been made in controlling wear via the development of specialised die shapes, wear-resistant materials and advanced surface treatments (Jeffrey and Holcomb, 1990; Dixon and Lloyd, 1996; Lefstad and Brusethaug, 1996). For reasonable die life, powder metallurgy tool steels, steel matrix TiC composites and bonded carbides have been used for simple profiles while more complex shapes were found to require chemical vapour deposition coatings of TiC/TiN on conventional tool steels.

6.4.5 Superplasticity and superplastic forming of DRCs

During deformation, a superplastic crystalline material is characterised by a low flow stress (< 10 MPa) and high resistance to non-uniform thinning, see, e.g., Chapter 8. These materials generally exhibit very high ductilities (> 200% in tension) which allows the near-net-shape forming of sheet material into complex shapes. In order to resist localised necking during forming, the material must have a high strain rate sensitivity of the flow stress, or *m*-value, where:

$$\sigma = K\dot{\varepsilon}^m \qquad 6.14$$

This relation may be described by a generalised power-law equation, see, e.g., Mukherjee (2002):

$$\dot{\varepsilon} = K' \frac{D_i E b}{kT} \left(\frac{b}{D} \right)^p \left(\frac{\sigma - \sigma_o}{E} \right)^{1/m} \tag{6.15}$$

where K and K' are constants, D_i an appropriate diffusivity (lattice, grain boundary, etc.), D the matrix grain size, p the grain size exponent (usually in the range 2–3) and σ_o a threshold stress.

For a material to exhibit superplastic behaviour, m-values greater than 0.3 are required (Pilling and Ridley, 1989). In DRCs, it has been found that an increase in deformation temperature can increase m to the level where superplastic deformation becomes possible and this has led to several hundred publications over the past few decades on the superplastic behaviour of these materials. These studies are based mainly on a matrix of **Al** (Jarry *et al.*, 1987; Mahoney and Ghosh, 1987; Wei *et al.*, 1994; Han and Chan, 1996; Wang *et al.*, 2001; Kaibyshev *et al.*, 2002), but DRCs based on **Ti** (Yamada *et al.*, 2000, 2001), **Zn** (Kim *et al.*, 1991; Nishimura *et al.*, 1994; Lu and Van Aken, 1996) and **Mg** (Nieh and Wadsworth, 1995; Sugumata, 1996; Watanabe *et al.*, 2000, 2001; Yan *et al.*, 2003) have received attention. There has been considerable recent interest in high strain rate superplasticity (HSRS) in DRCs where strain rates in excess of 10^{-2} s^{-1} are possible, see, e.g., Nieh and Wadsworth (1991); Nieh *et al.* (1994); Matsuki *et al.* (1997); Kim and Hong (2000); Chan and Tong (2000); Hikosaka *et al.* (2001a,b) and Lou *et al.* (2002).

The mechanisms of superplastic deformation of conventional metals involve grain boundary sliding, dislocation glide and climb and diffusional processes and the microstructural requirement is for a stable, fine equiaxed grain size containing high angle boundaries (Pilling and Ridley, 1989). Mishra *et al.* (1997) have shown that HSRS is possible in a range of Al-based DRCs and is characterised by a high Q-value (293–338 kJ/mol) and an inverse grain size and reinforcement size dependence, that is:

$$\dot{\varepsilon} = 3.6 \times 10^{18} \frac{D_o G b^3}{DdkT} \exp\left(-\frac{Q}{RT} \right) \left(\frac{\sigma - \sigma_o}{E} \right)^2 \tag{6.16}$$

Unlike conventional alloys, the likely mechanism of HSRS in DRCs is particle/matrix interface diffusion controlled superplasticity (Mishra *et al.*, 1997). These workers subsequently generated useful superplastic mechanism maps for DRCs by combining eqn 6.15 (Q = 84 kJ/mol, K' = 40, p = 2, m = 0.5) with the relaxation models of Koeller and Raj (1978) and Mori *et al.* (1980). Figure 6.21 is a typical map for PRCs where the region of interfacial sliding control (shaded region) denotes the optimum processing zone for superplastic forming. The upper limit for superplasticity is based on eqn

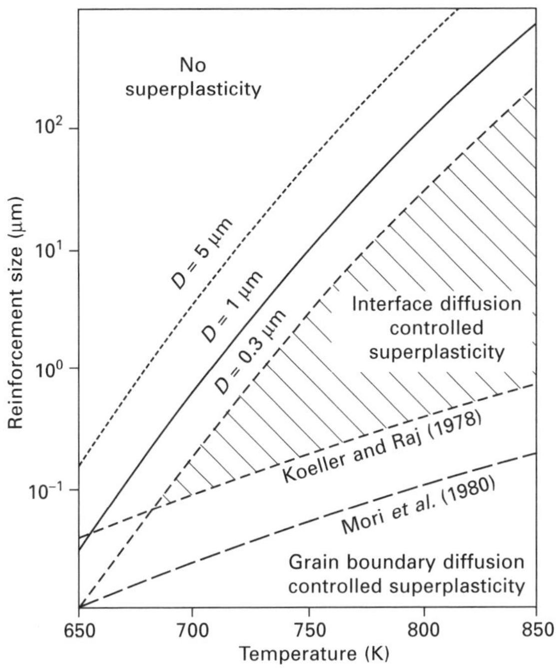

6.21 Superplastic mechanism map for PRCs showing the influence of reinforcement size, deformation temperature and matrix grain size on the propensity for superplastic deformation. Superimposed on the diagram is a cross-hatched region which indicates processing conditions for a PRC with a matrix grain size of 0.3 μm where interface diffusion controlled superplasticity is expected to be dominant. Adapted from Mishra *et al.* (1997) with kind permission of Elsevier Science Limited.)

6.15 and represents the unreinforced matrix behaviour. Figure 6.21 is useful as it shows that superplastic behaviour of PRCs is a function of reinforcement size, grain size and temperature and indicates that a large reinforcement size is to be avoided as this leads to cavitation and, hence, poor formability. Since most commercial PRCs exhibit average grain sizes of 5–10 μm and particle sizes of 5–20 μm, these materials are expected to undergo interface diffusion controlled HSRS at temperatures greater than 350–400 °C.

6.5 Final microstructure and mechanical properties

Certain mechanical properties of DRCs (such as elastic modulus and strength) are generally superior to the unreinforced alloys, although the ductility and fracture toughness are usually significantly lower, see, e.g., Nair *et al.* (1985), Humphreys (1988), Clyne (2001), Lloyd (1994). The mechanical properties of composites reinforced with continuous fibres are reasonably well predicted from fibre strengthening theories (continuum models), where the properties are largely determined from global parameters such as the volume fraction and mechanical properties of both the matrix and the reinforcement phase (Kelly and MacMillan, 1986; Taya and Arsenault, 1989; Clyne and Withers, 1993). For DRCs, however, the situation is more complex, and the prediction of strength on the basis of continuum modelling is not entirely satisfactory (Humphreys, 1988; Miller and Humphreys, 1989, 1991; Lloyd, 1991a).

Predicting the strength of DRCs from first principles is difficult since they invariably have complex matrix microstructures and the reinforcement is usually distributed non-uniformly (section 6.1.2). Despite these complications, composite strengthening can be understood on the basis of the current knowledge of the deformation processes that occur on an atomic level, that is, from theories based on dislocation micromechanisms (Humphreys, 1988; Clyne and Withers, 1993). The possible individual mechanisms contributing to the overall strength of DRCs will be discussed only briefly as more detailed discussions are given elsewhere (Miller and Humphreys, 1989; Clyne and Withers, 1993; Lloyd, 1994).

Quench strengthening

The significant difference in CTE between the reinforcement (SiC, Al_2O_3, TiC, TiB_2 etc.) and the metal matrix (Al, Mg, Ti etc.) results in the generation of dislocations during cooling from the initially high processing temperatures, see, e.g., eqns 6.1 and 6.2. Strengthening by dislocation generation has been argued to be one of the most significant contributing factors affecting the strength of these composites (Arsenault and Fisher, 1983; Vogelsang *et al.*, 1986; Arsenault *et al.*, 1991). Other workers have established that dislocation strengthening, to a lesser degree, is also important (Miller and Humphreys, 1989; Taya *et al.*, 1991).

Internal stresses

Although most of the stresses induced by quenching are relieved by dislocation generation, some residual (internal) stresses remain (Withers *et al.*, 1989). These internal stresses tend to surround the reinforcement locally, with a magnitude dependent on their actual shape and spatial distribution. In general, the residual stress distribution in WRCs is more anisotropic compared with PRCs (Barlow, 1991; Shu and Barlow, 2000). It is therefore likely that stresses in the vicinity of irregularly shaped particles or whiskers may make a small contribution to the overall strength of the composites.

Grain size strengthening

Both the subgrain and recrystallized grain size may contribute to the strengthening in DRCs on the basis of the well-established Hall-Petch relation (Hall, 1951; Petch, 1953). Thus, if the grain size can be refined via thermomechanical processing, a significant contribution to the total strength of the composite is possible (Parker, 1991).

Reinforcement strengthening

Since the reinforcement is in the form of whiskers or particles, some strengthening may result from Orowan bypassing of the reinforcement by dislocations (Humphreys, 1983). However, dispersion strengthening is not significant if the reinforcement size is greater than ~1 μm in diameter which is usually the case for PRCs.

Initial work hardening

The work hardening of DRCs will, to some extent, be influenced by the dislocation structure formed during quenching (Humphreys, 1988; Shu and Barlow, 2000). However, other factors will increase WHR of the composite compared with that of the matrix alloy. In the early stages of deformation there will be load transfer between the matrix and the reinforcement phase by such mechanisms as Orowan loop formation. As deformation proceeds, the stress induced at the particle/matrix interface leads to the relaxation of Orowan loops at true strains of $\sim 5 \times 10^{-4}$ (Humphreys, 1979). At higher strains, the particle will contribute to work hardening due to the creation of geometrically necessary dislocations (eqn 6.3). It is therefore likely that dislocations generated by work hardening will contribute substantially to the observed strength of these composites.

In summary, the enhanced strengthening in DRCs appears to be a result of a contribution of various mechanisms but it is unlikely that they can simply be summed to estimate the overall strength of the composite, since several mechanisms may interact synergistically (Miller and Humphreys, 1989, 1991). Therefore, to predict the strength of DRCs using a micro-mechanical approach, it is necessary to take into account the complex interactions between the various strengthening components.

6.6 DRC applications

A wide range of useful properties is possible in DRCs via the combination of a particular metallic matrix and reinforcement. Control of properties of a component is achieved by the careful manipulation of: (i) matrix alloy composition; (ii) reinforcement volume fraction, size and aspect ratio, and (iii) thermomechanical treatments. Table 6.2 provides a summary of some of the most useful properties of DRCs. Based on these properties, applications may be found in the aerospace, automotive, biomedical, electronic packaging and thermal management industries. A more extensive range of applications will depend on the development of cost-effective large-scale production processes that result in materials with improved combinations of properties over currently available materials. While there are numerous discussions of the applications and case studies of DRCs in various industries (see, e.g., Clyne and Withers, (1993), Chapter 6 of *Comprehensive Composite Materials* (2001), the following section provides a brief discussion of the potential applications of Al-based DRCs in the aircraft industry.

Table 6.2 Important properties affecting the performance of DRC components, tabulated from Hunt Jr (2001)

Property	Comments, after Hunt Jr (2001)
Specific stiffness	The addition of stiff metallic or ceramic reinforcement materials to the metal matrix results in an increase in elastic modulus. In the case of lightweight metals such as Al, Ti, and Mg, the increases can be very significant even at moderate levels of reinforcement addition. Many structural applications are designed primarily by stiffness, and the availability of materials with increased specific stiffness can enable thinner sections and hence weight saving.
Specific strength	In addition to the higher elastic modulus of reinforcement materials, many possess high strength. Gaining advantage from the strength of the reinforcement material is strongly dependent on the specific characteristics of the reinforcement material, its morphology (fibre, whisker, chopped fibre, or particulate), and the specifics of its bonding to the metal matrix, including the occurrence of reaction at the reinforcement-matrix interface.
Fatigue resistance	Another property that can be enhanced in metal matrix composites relative to the unreinforced matrix is fatigue resistance. The mechanisms of fatigue resistance enhancement differ depending on the morphology of the reinforcement and the reinforcement-matrix interface.
Wear resistance	The increased hardness of the typical ceramic reinforcement material can also affect the tribological properties of the metal matrix composite relative to the unreinforced matrix.
Coefficient of thermal expansion	The typical ceramic reinforcements for DRCs have significantly smaller values of the coefficient of thermal expansion (CTE) than the metal matrices into which they are incorporated. Thus, the addition of ceramic reinforcements to the high expansion metals such as Al, Mg, Cu, and Ti can result in substantial reductions in the CTE.

6.6.1 The use of DRCs as aircraft components (after Hunt Jr, (2001))

To be used as a structural component on an aircraft, a material generally requires a combination of properties, including adequate strength, damage tolerance (including ductility, fracture toughness and fatigue resistance) and corrosion resistance. DRCs typically have lower damage tolerance properties than their unreinforced alloy counterparts which is expected to limit the extent of their use in the primary structure of an aircraft. Nevertheless, DRCs may still be found in several parts of an aircraft structure due to their higher specific stiffness compared with conventional alloys. For example, an Al-based composite reinforced with SiC particulate (6092Al-17.5%-SiC_p) is now used in the ventral fins of the US Air Force F-16 aircraft due to its high lateral stability during sweeping manoeuvres. Sheet material of this composite has replaced the unreinforced aluminium skins in the honeycomb structure of the fin since its high specific stiffness significantly increases the service life of the component. The same composite is found as fuel access door covers of the F-16 due to not only its high specific stiffness but to alleviate overload-associated cracking problems of conventional Al alloys (Maruyama, 1999). Al-base DRCs have also been successfully incorporated in aircraft engines due to their high specific stiffness and specific strength, especially at elevated temperature, together with their fatigue, creep, and oxidation resistance. For example, 6092Al-17.5%-SiC_p produced by extrusion and artificial ageing (T6 temper) is found in fan exit guide vanes for the Pratt and Whitney 4000 series engines that power many of the Boeing 777 commercial aircraft. This composite is used in preference to graphite/epoxy materials as it is less expensive over the life of the component and has improved resistance both to impact damage and erosion from foreign objects (Maruyama, 1999).

6.7 Conclusion

Metal matrix composites (MMCs) are usually based on a matrix of Al, Mg, Cu, Ti, etc., reinforced with a hard second phase of the form of a ceramic such as SiC, Al_2O_3, TiC, TiB_2, TiN, B_4C, etc., an intermetallic compound such as TiAl, Ni_3Al, etc., or a metal such as W. These composites may be classified into three principal types depending on the morphology of the reinforcement (section 6.1.1). An MMC may contain: (i) long, continuous or unbroken fibres; (ii) discontinuous whiskers or fibres, or (iii) particulates or equiaxed particles. The combination of a metallic matrix with either of the last two phases is described as a discontinuously reinforced composite (DRC). These materials have received considerable attention over the past few decades since they can either be fabricated directly into a final or near-final shape or produced in billet form by methods such as powder metallurgy, liquid metal infiltration, spray deposition and molten metal mixing (section 6.1.2). Billets can be further processed using existing technologies such as hot and cold rolling, extrusion, forging and superplastic forming (section 6.4).

The basic structural difference between conventional alloys and DRCs is that the volume percentage of the reinforcing phase is usually considerably higher (up to 30 vol.%) than the particle content of the unreinforced alloys (~<5 vol.%). Subsequently, the thermomechanical behaviour of a DRC is often markedly different from that of its unreinforced alloy counterpart (section 6.2 and 6.3). For example, the deformation microstructure of DRCs is usually more complex than in conventional particle-containing alloys, principally due to the high volume concentration of the reinforcement present which is not regularly shaped or randomly dispersed, and has a considerable difference in CTE to the matrix. During mechanical working, the reinforcement may increase the matrix dislocation density and generate extensive matrix deformation zones with high local strain gradients that may lead to matrix shear banding and premature failure. Considering the complex deformation microstructures in DRCs, the operative dynamic and static restoration processes are often quite different from conventional alloys since recovery, recrystallization and grain growth are not only dependent on deformation and annealing conditions but also strongly affected by reinforcement parameters (section 6.3). For example, static recovery and recrystallization is usually accelerated in as-deformed DRCs but these softening processes may actually be retarded if the DRC contains fine oxide particles produced by initial fabrication or when the volume fraction of reinforcement is considerable. The overall annealing response of DRCs is therefore complex and governed both by material factors such as the type of metal matrix, and the reinforcement size, volume fraction, shape and distribution together with processing parameters such as the mode of deformation, degree and uniformity of both strain and strain rate and deformation temperature (section 6.3).

In addition to the influence of the reinforcing phase on the deformation microstructure (dislocation density, cell size, deformation heterogeneities and texture) and subsequent annealing response (kinetics of recovery and recrystallization, final grain size and texture), it has a strong influence on the flow stress, ductility and overall processability. For example, a given secondary processing route may have a deleterious effect on reinforcement integrity where it may undergo fracture, matrix decohesion and redistribution into bands (section 6.4). The presence of the reinforcing phase may result in increased mill loads, cause excessive roll and die wear and may generate components with poor surface integrity. Hence, both the processability and final properties of a DRC are also affected by various material factors and processing parameters (section 6.4 and 6.5). These differences in both the thermomechanical behaviour and processability between the composite and alloy will necessitate some modification of the processing schedules and plant equipment, as designed for conventional alloys (section 6.4).

An understanding of microstructural evolution of DRCs during the various processing stages is important for optimising the mechanical properties of the final component. Compared with unreinforced alloys, DRCs generally possess lower ductility and fracture toughness (section 6.4.1) but their strength is usually enhanced as a consequence of the various mechanisms outlined in section 6.5. It is pertinent to note, however, that prediction of mechanical properties (such as flow stress) of DRCs from first principles is a formidable task since the complex interactions between the various strengthening mechanisms must be taken into consideration.

On a final note, DRCs have found application in areas that can cost-effectively capitalise on improvements in specific stiffness, specific strength, fatigue resistance and wear resistance, etc. These materials have already made their presence felt in the aerospace, automotive, electronics and thermal management industries. However, a wider range of applications will depend not only on further improvements in the properties of these materials but also on the ability to reduce the cost of raw materials and manufacturing.

6.8 Sources of further information

6.8.1 Monographs on MMCs

Taya M. and Arsenault R.J. (1989), *Metal Matrix Composites: Thermomechanical Behaviour*, Pergamon Press, Oxford, UK.

Clyne T.W. and Withers P.J. (1993), *An Introduction to Metal Matrix Composites*, Cambridge University Press, Cambridge, UK.

6.8.2 Multi-author, edited compilations on MMCs

Bader M.G., Kedward K. and Sawada Y. (eds) (2001), *Comprehensive Composite Materials – Volume 6: Design and Applications of Composites*, Pergamon Press, Oxford, UK.

Clyne T.W. (ed.) (2001), *Comprehensive Composite Materials – Volume 3: Metal Matrix Composites*, Pergamon Press, Oxford, UK.

6.8.3 Proceedings of conferences on MMCs

International Conferences on Composite Materials (ICMM series held every 3 years), e.g.:

Matthews F.L., Hodgkinson J.M. and Morton J. (eds) (1987), *Proceedings of 6th International Conference on Composite Materials (ICMM VI)*, Elsevier Science, London, UK.

Fishman S.G. and Dhingra A.K. (eds) (1988), *Proceedings of Conference on Cast Reinforced Metal Composites*, Metals Park, OH, ASM International, USA.

Gungor M.N. and Liaw P.K. (eds) (1989), *Proceedings of Conference on Fundamental Relationships between Microstructure and Mechanical Properties of Metal Matrix Composites*, TMS, Warrendale, USA.

Andersen S.I., Lilholt H and Pedersen O.S. (eds) (1989), *Proceedings of 9th International Risø Symposium on Materials Science: Mechanical and Physical Behaviour of Metallic and Ceramic Composites*, Risø National Laboratory, Roskilde, Denmark.
Masounave J. and Hamel F.G. (eds) (1990), *Proceedings of Conference on Fabrication of Particulates Reinforced Metal Composites*, Metals Park, OH, ASM International, USA.
Hansen N., Juul Jensen D., Leffers T., Lilholt H., Lorentzen T., Pedersen A.S., Pedersen O.S. and Ralph B. (eds) (1991), *Proceedings of 12th International Risø Symposium on Materials Science: Metal Matrix Composites – Processing, Microstructure and Properties*, Risø National Laboratory, Roskilde, Denmark.
Rohatki P.K. (ed.) (1996), *Proceedings of Conference on Processing, Properties and Applications of Cast Metal Matrix Composites*, TMS, Warrendale, USA.

6.8.4 Review articles and books containing chapters on MMCs

Lloyd D.J. (1994), *Int. Mater. Rev.*, 39, 1.
Mortensen A. and Jin. I. (1992). *Int. Mater. Rev.*, 37, 101.
Nair S.V., Tien J.K. and Bates R.C. (1985), *Int. Met. Rev.*, 30, 257.

6.8.5 WWW homepages on MMCs

http://www.compcompmat.com/

6.9 References

Almas M. (1989), *Ph.D. Thesis*, Imperial College, London.
Arsenault R.J. and Fisher R.M. (1983), *Scripta Metall.*, 17, 67.
Arsenault R.J. and Shi N. (1986), *Mat. Sci. Eng.*, A81, 175.
Arsenault R.J., Shi N.N., Feng C.R. and Wang L. (1989), *Mat. Sci. Eng.*, A131, 55.
Arsenault R.J., Wang L. and Feng C.R. (1991), *Acta Metall. Mater.*, 39, 47.
Ashby M.F. (1966), *Phil. Mag.*, 14, 1157.
Ashby M.F. (1970), *Phil. Mag.*, 21, 399.
Barlow C.Y. (1991), in *Proc. 12th Int. Risø Symp.*, Risø National Laboratory, Roskilde, Denmark, 1.
Barlow C.Y. and Hansen N. (1991), *Acta Metall. Mater.*, 39, 1971.
Bassani J.L., Needleman A. and Van der Giessen E. (2001), *Int. J. Solids and Structures*, 38, 833.
Bhat B.V.R., Mahajan Y.R., Roshan H. and Prasad Y.V.R.K. (1992), *Metall. Trans.*, 23A, 2223.
Bhat B.V.R., Mahajan Y.R., Roshan H. and Prasad Y.V.R.K. (1995), *Mat. Sci. Tech.*, 11, 167.
Bhat B.V.R., Mahajan Y.R. and Prasad Y.V.R.K. (2000), *Metall. Mater. Trans.*, 31A, 629.
Bowen A.W. and Humphreys F.J. (1991), in *Proc. 9th Int. Conf. on Textures: Textures and Microstructures*, 14, 715.
Bowen A.W., Ardakani M. and Humphreys F.J. (1991), in *Proc. 12th Int. Risø Symp.*, Risø National Laboratory, Roskilde, Denmark, 241.
Boyd J.D. and Lloyd D.J. (2001), in *Comprehensive Composite Materials*. ed. T.W. Clyne, Pergamon Press, 3, 139.

Brechet Y., Newell J., Tao S. and Embury J.D. (1993), *Scripta Metall. Mater.*, 28, 47.
Brusethaug S., Reiso O. and Ruch W. (1990), in *Proc. Fabrication of Particulates Reinforced Metal Composites*, eds J. Masounave and F.G. Hamel, Metals Park, OH, ASM International, 173.
Bryant A.J. (1975), *Met. Tech.*, 2, 21.
Caron S. and Masounave J. (1990), in *Proc. Fabrication of Particulates Reinforced Metal Composites*, eds J. Masounave and F.G. Hamel, Metals Park, OH, ASM International, 79.
Cerri E., Spigarelli S., Evangelista E. and Cavaliere P. (2002), *Mat. Sci. Eng.* A324, 157.
Chan K.C. and Tong G.Q. (2000), *Mat. Sci. Eng.* A286, 218.
Chanda T., Lavernia E.J. and Wolfenstine J. (1991), *Scripta Metall. Mater.*, 25, 131.
Chandra T. and Yu D. (1991), *Mat. Forum*, 15, 343.
Chandra T. and Yu D. (1993), *Mat. Forum*, 17, 361.
Chen L.Q., Lu Y.X., Lee C.S., Bi J. and Li R.K.Y. (1998), *Textures and Microstructures*, 31, 43.
Chen W.C., Davies C.H.J., Samarasekera I.V., Brimacombe J.K. and Hawbolt E.B. (1996), *Metall. Mater. Trans.*, 27A, 4095.
Chi F.K., Maier R.D., Krucek T.W. and Boymel, P.M. (1987), in *Proc. 6th Int. Conf. on Composite Materials*, eds F.L. Matthews, J.M. Hodgkinson and J. Morton, Elsevier Science, London, 449.
Christman T. Needleman A. and Suresh S. (1989), *Acta Metall. Mater.*, 37, 3029.
Clyne T.W. (2001), in *Comprehensive Composite Materials.* ed. T.W. Clyne, Pergamon Press, 3, 1
Clyne T.W. and Withers P.J. (1993), *An Introduction to Metal Matrix Composites.* Cambridge University Press, Cambridge, UK.
Davies C.H.J., Chen W.C., Lloyd D.J., Hawbolt E.B., Samarasekera I.V. and Brimacombe J.K. (1996), *Metall. Mater. Trans.*, 27A, 4113.
Dieter G.E. (1986), *Mechanical Metallurgy*, 3rd edn, McGraw Hill, 358.
Dixon W. and Lloyd D.J. (1996), in *Processing, Properties and Applications of Cast Metal Matrix Composites*, TMS, Warrendale, USA, 259.
Durrant G., Scott V.D., Clift S.E. and Trumper R.L. (1991), in *Proc. 12th Int. Risø Symp., Roskilde, 1991*, Risø National Laboratory, Roskilde, Denmark, 311.
Ehrstrom J.C. and Kool W.H. (1988), *J. Mat. Sci. Lett.*, 7, 578.
Es-Said O.S. and Morris J.G. (1994), in *Aluminium Alloys: Their Physical and Mechanical Properties*, eds A. Starke and T.H. Sanders, Engineering Materials Advisory Service, Warley, 1, 451.
Ferry M. (1994), *Ph.D. Thesis*, University of New South Wales.
Ferry M. and Humphreys F.J. (1996), *Acta Mater.*, 44, 3089.
Ferry M. and Munroe P.R. (1993), *Scripta Metall. Mater.*, 29, 741.
Ferry M. and Munroe P.R. (1994), *Scripta Metall. Mater.*, 29, 123.
Ferry M. and Munroe P.R. (1995a), *Mat. Sci. Tech.*, 11, 663.
Ferry M. and Munroe P.R. (1995b), *Mat. Sci. Tech.*, 11, 734.
Ferry M. and Munroe P.R. (1995c), *Scripta Metall. Mater.*, 33, 857.
Ferry M. and Munroe P.R. (2003), *Mat. Sci. Eng.*, A358, 142.
Ferry M., Munroe P.R., Crosky A. and Chandra T. (1992), *Mat. Sci. Tech.*, 8, 43.
Ferry M., Munroe P.R. and Crosky A. (1993), *Scripta Metall. Mater.*, 28, 1235.
Goswami R.K., Sikand R., Dhar A., Gupta A.K., Grover O.P. and Jindal U.C. (1999), *Mat. Sci. Tech.*, 15, 443.
Grow A.L. and Lewandowski J. (1995), *SAE Paper 950260*, SAE, 1995.

Hall E.O. (1951), *Proc. Phys. Soc. London*, 643, 747.
Han B.Q. and Chan K.C. (1996), *Mat. Sci. Eng.*, A212, 256.
Hansen N. (1969), *Powder Met.,* 12, 23.
Hansen N., Juul Jensen D. and Liu Y.L. (1992), in *Proc. Recrystallization 92; Mat. Sci. Forum,* ed. M. Fuentes, 113, 55.
Hatherly M. and Dillamore I.L. (1975), *J. Inst. Metals*, 20, 71.
Herbst P. and Huber J. (1978), in *Proc. 5th Int. Conf. on Textures*, eds G. Gottstein and K. Lücke, Berlin, 1, 453.
Hikosaka T., Imai T., Kobayashi T. and Toda H. (2001a), *J. Japan. Inst. Light. Metals*, 51, 86.
Hikosaka T., Imai T., Kobayashi T. and Toda, H. (2001b), *J. Japan. Inst. Light. Metals*, 51, 169.
Hirsch P.B. and Humphreys F.J. (1969), in *Physics of Strength and Plasticity*, ed. A. Argon, MIT Press, 189.
Hoover W.R. (1990), in *Proc. Fabrication of Particulates Reinforced Metal Composites*, eds J. Masounave and F.G. Hamel, Metals Park, OH, ASM International, 115.
Humphreys F.J. (1977), *Acta Metall.*, 25, 1323.
Humphreys F.J. (1979), *Acta Metall.*, 27, 1801.
Humphreys F.J. (1983), in *Proc. 4th Int. Risø Symp.,* Risø National Laboratory, Roskilde, Denmark, 41.
Humphreys F.J. (1988), in *Proc. 9th Int. Risø Symp.,* Risø National Laboratory, Roskilde, Denmark, 51.
Humphreys F.J. and Bate P. (2004), *Scripta Mater.*, 48, 173.
Humphreys F.J. and Hatherly M. (2004), *Recrystallization and Related Annealing Phenomena*, 2nd edn, Pergamon Press, Oxford, UK.
Humphreys F.J. and Kalu P.N. (1987), *Acta Metall.*, 35, 2815.
Humphreys F.J., Miller W.S. and Djazeb M.R. (1990), *Mat. Sci. Tech.*, 6, 1139.
Humphreys F.J., Basu A. and Djazeb M.R. (1991), in *Proc. 12th Int. Risø Symp.,* Risø National Laboratory, Roskilde, Denmark, 51.
Hunt Jr W.H. (2001), in *Comprehensive Composite Materials.* eds Bader, M.G., Kedward K. and Sawada Y., Pergamon Press, 6, 57.
Jarry P., Loué W. and Bouvaist J. (1987), in *Proc. 6th Int. Conf. on Composite Materials*, eds. F.L. Matthews, J.M. Hodgkinson and J. Morton, Elsevier Science, London, 2, 350.
Jeffrey P.W. and Holcomb S. (1990), *in Proc. Fabrication of Particulates Reinforced Metal Composites*, eds J. Masounave and F.G. Hamel, Metals Park, OH, ASM International, 181.
Juul Jensen D., Liu Y.L. and Hansen N. (1991a), in *Proc. 12th Int. Risø Symp.,* Risø National Laboratory, Roskilde, Denmark, 417.
Juul Jensen D., Hansen N. and Liu Y.L. (1991b), *Mat. Sci. Tech.*, 7, 369.
Kaibyshev R., Kazyhanov V. and Musin F. (2002), *Mat. Sci. Tech.*, 18, 777.
Kalu P.N. and McNelley T.R. (1991), *Scripta Metall. Mater.*, 25, 853.
Kang G.C., Kim N.H. and Kim B.M. (2000), *J. Mater. Proc. Tech.*, 100, 53.
Kelly A. and Macmillan N.H. (1986), *Strong Solids*, 3rd edn, Oxford Science Publications.
Kim H.Y. and Hong S.H. (1994), *Scripta Metall. Mater.*, 30, 297.
Kim W.J. and Hong S.H. (2000), *J. Mat. Sci.*, 35, 2779.
Kim J.S., Sugamata M. and Kaneko J. (1991), *J. Japan Inst. Metals*, 55, 994.
Ko B.-C. and Yoo Y.-C. (1998), *Comp. Sci. Tech.*, 58, 479.
Ko B.-C. and Yoo Y.-C. (1999), *Mat. Sci. Tech.*, 14, 765.

Koeller R.C. and Raj R. (1978), *Acta Metall.*, 26, 1551.
Lee J.S., Lee D.M. and Lee C.H. (1995), *J. Korean Inst. Metall. Mater.*, 33, 214.
Lefstad M. and Brusethaug S. (1996), in *Processing, Properties and Applications of Cast Metal Matrix Composites*, TMS, Warrendale, USA, 689.
Lewandowski J.J., Liu C. and Hunt Jr, W.H. (1987), in *Processing and Properties of Powder Metallurgy Composites*, eds P. Kumar, K. Vedula and A. Ritter, TMS Warrendale, 117.
Liu Y.L. and Barlow C.Y. (1995), in *Proc. 16th Int. Risø Symp.,* Risø National Laboratory, Roskilde, Denmark, 151.
Liu Y.L., Hansen N. and Juul Jensen D. (1989), *Metall. Trans.*, 20A, 1743.
Liu Y.L., Hansen N. and Juul Jensen D. (1991a), in *Proc. 12th Int. Risø Symp.,* Risø National Laboratory, Roskilde, Denmark, 67.
Liu Y.L., Hansen N. and Juul Jensen D. (1991b), *Mat. Sci. Tech.*, 7, 270.
Liu Y.L., Juul Jensen D. and Hansen N. (1992), *Metall. Trans.*, 23A, 807.
Llorca J. (2001), in *Comprehensive Composite Materials.* ed. T.W. Clyne, Pergamon Press, 3, 91.
Llorca J., Martín A., Ruiz J. and Elices M. (1993), *Metall. Trans.*, 24A, 1575.
Lloyd D.J. (1986), in *Formability and Metallurgical Structure*, eds A.K. Sachdev and J.D. Embury, TMS Warrendale, 193.
Lloyd D.J. (1989), *Comp. Sci. Tech.*, 35, 159.
Lloyd D.J. (1991a), in *Proc. 12th Int. Risø Symp.*, Risø National Laboratory, Roskilde, Denmark 81.
Lloyd D.J. (1991b), *Acta Metall. Mater.*, 39, 59.
Lloyd D.J. (1994), *Int. Mater. Rev.*, 39, 1.
Lloyd D.J. and Jin I. (2001), in *Comprehensive Composite Materials.* ed. T.W. Clyne, Pergamon Press, 3, 555.
Lou B.Y., Huang J.C., Wang T.D. and Langdon T.G. (2002), *Trans. JIM*, 43, 501.
Lu J. and Van Aken D.C. (1996), *Metall. Mater. Trans.*, 27A, 2565.
Mahoney M.W. and Ghosh A.K. (1987), *Metall. Trans.*, 18A, 653.
Manna R., Sarkar S. and Surappa M.K. (1996), *J. Mater. Sci.*, 31, 1625.
Manohar P.A., Ferry M. and Chandra T. (1998), *ISIJ Int.*, 38, 913.
Maruyama B. (1999), *Advanced Materials and Processes*, 156, 47.
Matsuki K., Tokizawa M., Kawakami H. and Murukami S. (1997), *Mat. Sci. Tech.*, 13, 1039.
Mazen A.A. (1999), *J. Mater. Eng. Perform.*, 8, 487.
McDanels D.L. (1985), *Metall. Trans*, 16A, 1105.
McKimpson M.G. and Scott T.E. (1990), *Mat. Sci. Eng.*, A107, 93.
McQueen H.J., Sakaris P. and Bowles J. (1993), in *Proc. Int. Conf. on Advanced Composites*, ed. T. Chandra, TMS Warrendale, USA, 1193.
McQueen H.J., Myshlyaev M., Konopleva E. and and Sakaris P. (1998), *Canadian Metall. Quarterly*, 37, 98.
McQueen H.J., Konopleva E.V. and Sauerborn M. (1999), in *Synthesis of Lightweight Metals III*, TMS Warrendale, 121.
Miller W.S. and Humphreys F.J. (1989), in *Fundamental Relationships between Microstructure and Mechanical Properties of Metal Matrix Composites*, eds M.N. Gungor and P.K. Liaw, TMS Warrendale, 517.
Miller W.S. and Humphreys F.J. (1991), *Scripta Metall. Mater.*, 25, 33.
Mishra R.S., Bieler T.R. and Mukherjee A.K. (1997), *Acta Mater.*, 45, 561.
Mochida T., Taya M. and Obata M. (1991a), *JSME Int. J.*, 34, 187.

Mochida T., Taya M. and Lloyd D.J. (1991b), *Mater. Trans. JIM,* 32, 931.
Montheillet F., Jonas J.J. and Neale K.W. (1996), *Metall. Mater. Trans*, 27A, 232.
Mori T. Okabe M. and Mura T. (1980), *Acta Metall.*, 28, 319.
Mortensen A. and Jin. I. (1992), *Int. Mater. Rev.*, 37, 101.
Mukherjee A.K. (2002), *Mat. Sci. Eng.*, A322, 1.
Murthy S.V.S. and Nageswara Rao B. (1998), *Met. Sci. Eng*, A254, 76.
Nair S.V., Tien J.K. and Bates R.C. (1985), *Int. Met. Rev.*, 30, 257.
Nieh T.G. (1984), *Metall. Trans.*, 15A, 139.
Nieh T.G. and Wadsworth J. (1991), *Mat. Sci. Eng.*, A147, 129.
Nieh T.G. and Wadsworth J. (1995), *Scripta Metall. Mater.*, 32, 1133.
Nieh T.G., Imai T., Wadsworth J. and Kojima S. (1994), *Scripta Metall. Mater.*, 31, 1685.
Nishimura H., Sato T. and Yamamoto H. (1994), *J. Japan. Inst. Metals*, 44, 480.
Ozdemir I., Tekmen C. and Toparli M. (2003), *Z. Metallkde*, 94, 41.
Park D.-H., Ko B.-C and Yoo Y.-C. (2002), *J. Mat. Sci.*, 37, 1593.
Parker B.A. (1991), in *Proc. 2nd Aust. Forum on Metal Matrix Composites*, eds S. Bandyopadhyay and A.G. Crosky, UNSW Press, 83.
Petch N.J. (1953), *J. Iron Steel Inst. London*, 173, 25.
Pickens J.R., Langen I.J., England R.O. and Leibson M. (1987), *Metall. Trans.*, 18A, 303.
Pilling J. and Ridley N. (1989), *Superplasticity in Crystalline Solids*, Institute of Metals, London.
Poole W.J., Silvetti S.P., Embury J.D. and Kocks U.F. (1992), in *Proc. Recrystallization 92; Mat. Sci. Forum,* ed. M. Fuentes, 113–115, 551.
Poudens A. and Bacroix B. (1996), *Scripta Mater.*, 34, 847.
Poudens A., Bacroix B. and Bretheau T. (1991), in *Proc. 12th Int. Risø Symp.*, Risø National Laboratory, Roskilde, Denmark, 595.
Prasad Y.V.R.K. (1990). *Ind. J. Tech.*, 28, 435.
Qin Q. and McQueen H.J. (2000), *Canadian Inst. Mining Metall. Pet.*, 553.
Rack H.J. and Niskanen P.W. (1984), *Light Metal Age*, 42, 9.
Raj R. (1981), *Metall. Trans*, 12A, 1089.
Sarkar J. and Surappa M.K. (1993), in *Proc. Int. Conf. on Advanced Composites*, ed. T. Chandra, TMS, Warrendale, USA.
Sellars C.M. (1986), in *Proc. 7th Int. Risø Symp.*, Risø National Laboratory, Roskilde, Denmark, 167.
Sellars C.M. and Tegart W.G. McG. (1966), *Mém. Sci. Rev. Métall.*, 63, 731.
Shahani R.A. (1991), *Ph.D. Thesis*, University of Cambridge.
Shahani R.A. and Clyne T.W. (1991a), in *Proc. 12th Int. Risø Symp.*, Risø National Laboratory, Roskilde, Denmark, 655.
Shahani R.A. and Clyne T.W. (1991b), *Mat. Sci. Eng.*, A135, 281.
Shen Y.-L., Finot M., Needleman A. and Suresh S. (1995), *Acta Metall. Mater.*, 43, 1701.
Sheppard T. and Vierod R.P. (1987), *Mat. Sci. Tech*, 3, 285.
Shu J.Y. and Barlow C.Y. (2000), *Int. J. Plasticity*, 16, 563.
Sorensen N.J., Suresh S., Tvergaard V. and Needleman A. (1995), *Mat. Sci. Eng*, A197, 1.
Sparks C.N. and Sellar C.M. (1992), in *Proc. Recrystallization 92; Mat. Sci. Forum,* ed. M. Fuentes, 113–115, 557.
Spigarelli S., Evangelista E., Cerri E. and Langdon T.G. (2001), *Mat. Sci. Eng.*, A319, 721.
Spigarelli S., Cern E., Cavalleve P. and Evangelista E. (2002), *Mat. Sci. Eng.*, A327, 144.
Stanford-Beale C.A. and Clyne T.W. (1988), in *Proc. 9th Int. Risø Symp.,* Risø National Laboratory, Roskilde, Denmark, 479.

Stanford-Beale C.A. and Clyne T.W. (1989), *Comp. Sci. Tech.*, 35, 121.
Stewart A.T. and Martin J.W. (1975), *Acta Metall.*, 23, 1.
Stobbs W.M. (1973), *Phil. Mag.*, 30, 1073.
Sugumata M. (1996), in *Processing, Properties and Applications of Cast Metal Matrix Composites*, TMS, Warrendale, USA, 121.
Surappa M.K. (1993), *J. Mat. Sci. Lett.*, 12, 1272.
Taha A.S. (1996), *Interceram*, 45, 174.
Taya M. and Arsenault R.J. (1989), *Metal Matrix Composites: Thermomechanical Behaviour*, Pergamon Press, Oxford.
Taya M., Lulay K.E. and Lloyd D.J. (1991), *Acta Metall. Mater.*, 39, 73.
Tuler F.R., Beals J.T., Demetry C., Zhao D. and Lloyd D.J. (1988), in *Proc. Conf. on Cast Reinforced Metal Composites*, eds S.G. Fishman and A.K. Dhingra, Metals Park OH, ASM International, 321.
Van Aken D.C., Krajewski P.E., Vyletel G.M., Allison J.E. and Jones J.W. (1995), *Metall. Trans.*, 26A, 1395.
Vogelsang M., Arsenault R.J. and Fisher R.M. (1986), *Metall. Trans.*, 17A, 379.
Wang Y.Z., Cavalier P., Spigarelli S. and Evangelista E. (2001), *J. Mat. Sci. Lett.*, 20, 1195.
Watanabe H., Saitoh K. and Okaniwa S. (1990), *J. Japan Inst. Light. Metals*, 40, 1990.
Watanabe H., Mukai T., Nieh T.G. and Higashi K. (2000), *Scripta. Mater.*, 42, 249.
Watanabe H., Mukai T. and Higashi K. (2001), *J. Japan. Inst. Light. Metals*, 51, 503.
Wei Z., Zhang B. and Wang Y. (1994), *Scripta Metall. Mater.*, 30, 1367.
Wilson D.V. (1986), in *Formability and Metallurgical Structure*, eds A.K. Sachdev and J.D. Embury, TMS Warrendale, 8.
Withers P.J., Stobbs W.M. and Pedersen O.B. (1989), *Acta Metall. Mater.*, 37, 3061.
Xia X. and McQueen H.J. (1997), *App. Comp. Mater.*, 4, 333.
Xia X., Zhao J. and McQueen H.J. (1996), *Sci. Eng. Comp. Mater.*, 5, 215.
Yamada T., Tsuzuka T., Hirota M. and Kawach Y. (2000), *J. Japan. Inst. Metals*, 64, 838.
Yamada T., Tsuzuka T. and Sato H. (2001), *J. Japan. Inst. Metals*, 65, 946.
Yan F., Wu K., Wu G.L., Lee B.L. and Zhao M. (2003), *Mater. Lett.*, 57, 1992.
Yoo Y.-C. and Ko B-C. (2000), *J. Mat. Sci.*, 35, 4073.
Yu D., Munroe P.R., Bandyopadhyay S. and Mouritz A. (1994), *Scripta Metall. Mater.*, 30, 927.
Zaidi M.A. and Wert J.A. (1989), in *Treatise on Materials Science and Technology*, eds A.K. Vasudevan and R.D. Doherty, Academic Press Inc., 31, 137.
Zener C. (1948), Private communication to C.S. Smith, *Trans. AIME.*, 175, 47.
Zener C. and Hollomon J.H. (1944), *J. App. Phys.*, 15, 22.
Zhao D., Tuler F.R. and Lloyd D.J. (1992), *Scripta Metall. Mater.*, 27, 41.
Zhao J., Xia X. and McQueen H.J. (1996), *Canadian Inst. Mining Metall. Pet.*, 151.
Zhang W.L., Wang D.Z., Yao Z.K. and Gu, M.Y. (2001), *Mat. Sci. Tech.*, 17, 349.

7
Semi-solid processing of metallic alloys

A DAS and ZY FAN, Brunel University, UK

7.1 Introduction

Ever since the accidental origin of the concept of semisolid metal (SSM) processing in early 1971 by Spencer as a part of his doctorate research under the supervision of Professor Flemings at MIT, SSM processing has evolved over the last 32 years to be considered as the potential processing route for the future. SSM processing involves near net shape processing of metallic components in the semisolid state utilising a slurry consisting of non-dendritic solid evenly dispersed in the liquid matrix. The exciting rheological characteristics of such a slurry structure, namely thixotropic and pseudoplastic behaviour, allows it to be treated like a solid at rest but like a viscous fluid under pressure that forms the basis of SSM processing. In this chapter the concept and practice of SSM processing will be discussed. The basic rheological characteristics of semisolid metal are discussed in section 7.2 including rheology of low solid fraction slurries (pertaining to rheo-processing) and the deformation behaviour of high solid fraction slurries (pertaining to thixo-processing) is discussed in section 7.3. Section 7.4 discusses the evolution of microstructure in SSM processing including the characterisation of microstructure, solidification behaviour under forced convection and the mechanism of non-dendritic structural evolution in SSM processing. The alloy design criteria for SSM processing and the mechanical properties of processed material are discussed in the two next sections. An overview of the different SSM processing techniques is presented in section 7.7 followed by a brief discussion of the likely future trends to end the chapter.

7.2 Rheology of semi-solid metals

The first investigation of the rheology of stirred SSM slurries conducted at MIT by Spencer *et al.* (Spencer, 1972) on the Sn-Pb system showed that at a solid fraction higher than 0.2 it behaves like non-Newtonian fluid with apparent viscosity orders of magnitude less than that of unstirred dendritic slurry. The numerous extensive rheological studies on stirred SSM slurries that followed have identified the continuous cooling behaviour, which describes the viscosity evolution during continuous cooling at constant cooling rate and shear rate, along with detailed characterisation of pseudoplastic (shear thinning) and thixotropic behaviour.

The continuous cooling behaviour describes the effects of solid fraction, shear rate and cooling rate on the rheological behaviour of SSM slurries that is of particular relevance to SSM processing techniques such as rheocasting. Figure 7.1 shows an example of results obtained from the continuous cooling experiments on Sn-15Pb (all alloy compositions are expressed in weight % unless otherwise stated) alloy (Joly, 1976). Generally, for a given cooling rate and shear rate, the measured apparent viscosity increases with increasing solid fraction, slowly at low solid fraction and sharply at high solid fraction. At a given solid fraction, the apparent viscosity decreases with increasing shear rate and decreasing cooling rate. However, as has been pointed out by

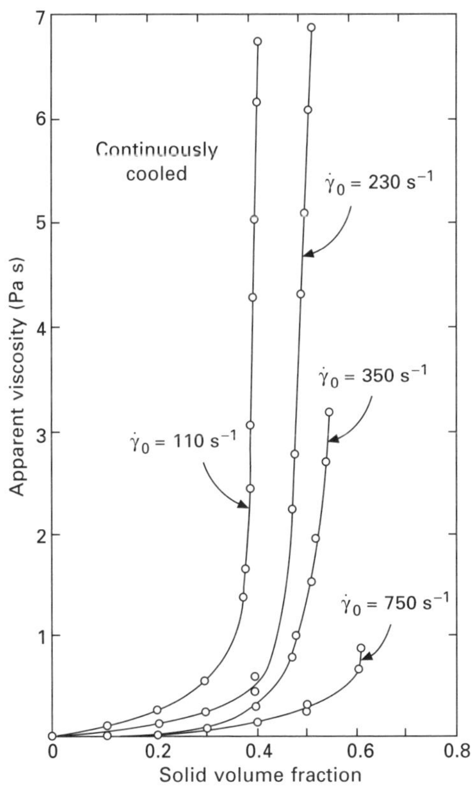

7.1 Apparent viscosity as a function of solid fraction for Sn-15 wt.% Pb alloy continuously cooled at 0.33 K min-1 and continuously sheared at different rates $\dot{\gamma}_0$ (after Joly, 1976).

Suéry *et al.* (1996), such experiments are more relevant to exploiting the solidification behaviour rather than studying the rheology of SSM slurries.

Experiments on the isothermal steady state, at which viscosity is a function of solid fraction and shear rate for a given alloy system, lead to a more precise rheological characterisation. For SSM slurry with a fixed solid fraction, the steady state viscosity has shown pseudoplastic behaviour decreasing with increasing shear rate and approaching an asymptotic value when the shear rate becomes infinite as shown in Fig. 7.2. It is now generally accepted that the steady state viscosity at a given shear rate depends on the degree of agglomeration between solid particles, which, in turn, is the result of a dynamic equilibrium between agglomeration and deagglomeration processes (Quaak, 1996). Another phenomenon associated with steady state behaviour is the presence of yield stress. Although not well understood at present (Barnes, 1985; Hartnett, 1989), it is generally agreed that many suspensions have a yield point at low shear rate resulting from the structure formation due to the dynamic interaction between solid particles. The yield phenomenon is generally inherent to the thixotropy, however, there are very little data of yield stress for SSM slurries.

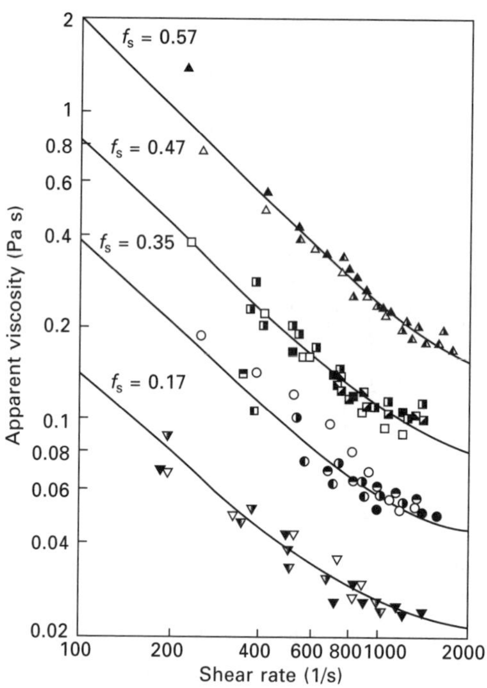

7.2 Steady state apparent viscosity as a function of shear rate for Sn-15 wt.% Pb alloy at different solid fractions f_s (after Turng, 1991).

The thixotropic behaviour of SSM slurry was initially demonstrated by measuring the hysteresis loops during a cyclic shear deformation (Joly, 1976; Martin, 1994). However, such a procedure is not sufficient to quantify the kinetics of agglomeration and deagglomeration processes. To overcome this shortcoming, special experimental procedures involving an abrupt shear rate jump or a shear rate drop were developed to characterise the kinetics of structural evolution (Turng, 1991; Martin, 1994; Modigell, 1998). An example of such thixotropic experiments is given in Fig. 7.3 for Sn-15Pb alloy obtained by Modigell (1998). It was found that the agglomeration process dominates after a shear rate drop whereas the deagglomeration process dominates after a shear rate jump (Martin, 1994; Mada, 1996a,b; Modigell, 1999].

Another group of interesting experiments associated with shear rate transient experiments are the so-called isostructure tests. If the shear rate transient is fast enough, the slurry structure will not change during the shear rate ramping process. A shear-thickening phenomenon was observed experimentally both in Sn-Pb alloys (Kumar, 1993; Modigell, 1999) and Al-based alloys by Kumar (1993). However, recent investigation in the transient behaviour of A357 alloy of a large range of solid fractions showed consistent shear thinning behaviour, the transient shear-thickening behaviour was not detected for transition times as low as 200 μs (De Figueredo, 2000).

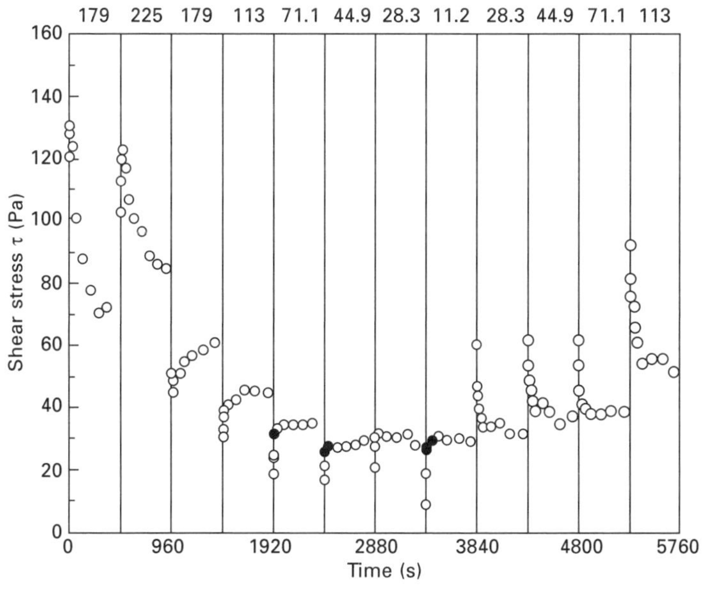

7.3 Results from shear rate transient experiment in Sn-15 wt.% Pb alloy with a solid fraction of 0.45. Top scale represents shear rate (s^{-1}) corresponding to the vertical section (after Modigell, 1998).

7.3 Deformation behaviour of SSM slurries with high solid fractions

It would be worthwhile to mention in brief the deformation characteristics of SSM slurries with high volume fraction solid due to its high relevance to thixoforming process. However, although thixoforming enjoyed being the most widely utilised form of SSM processing in the past, it is now being discarded in favour of the more promising rheo-based SSM processing technologies due to the potential advantages of the latter. SSM slurries with high solid fractions are characterised by the existence of a solid skeleton, which can be formed either by partial solidification at low temperature or by partial remelting at high solid fraction. The experimental results so far have identified the existence of large yield stress and the occurrence of solid phase cracking and liquid phase separation. Due to the formation of a solid skeleton, the solid phase responds to the macroscopically applied hydrostatic pressure by densification through either particle rearrangement or by slight deformation at the contacts between the solid particles (Martin, 1997). The reduction of the interstitial volume leads to the development of hydrostatic pressure in the liquid phase. Presence of the spatial gradient of hydrostatic pressure in the liquid phase will cause phase separation (Suéry, 2000). Conditions that promote segregation include low deformation rates (or low die filling rates), high-pressure gradients (e.g., in case of filling narrow and long channels), high fraction of solid and large and non-spherical particles.

7.4 Microstructural evolution during slurry preparation

The ideal microstructure for a semisolid slurry prior to the component shaping process would be an accurately specified volume fraction of fine and spherical solid particles uniformly dispersed in a liquid matrix. A primary goal of slurry preparation is to create a structure that will ensure the favourable rheological characteristics necessary to facilitate the subsequent component shaping process. Technically, this structure can be achieved by a number of different techniques. They can be divided categorically into two different approaches. The first approach (utilised in rheo-processing) involves partial solidification of a melt under forced convection induced by either electromagnetic or mechanical stirring, or partial solidification under the influence of an external field, such as ultrasonic vibration or pulsed electrical current. The second approach (used in thixo-processing) involves partial remelting of a solid feedstock material, which has been solidified earlier under specific conditions or worked thermomechanically. In this section the scientific understanding on microstructural evolution during the slurry preparation by those two approaches is reviewed.

7.4.1 Microstructural characterisation

For a fixed alloy composition, a complete description of the microstructure of semisolid slurry involves quantifying the volume fraction, the size, shape and distribution of the solid particles. Microstructural quantification has been performed using thermodynamic calculations or through various techniques such as thermal analysis, ultrasonic monitoring and measurement of electrical resistivity/magnetic permeability. However, quantitative metallography of quenched specimen appears to be the most popular technique although inadequate for measuring particle shape and distribution accurately due to the inherent problems of measuring 3D microstructure in 2D sections. Attempts have been made in the past to obtain 3D images by serial sectioning (Ito, 1992; Sannes, 1994; Wolfsdorf, 1997; Niroumand, 1998, 1999) although it is time consuming and suffers due to its low resolution (about 20 μm). Recently, back scattered diffraction (Arnberg, 1999) and X-ray microtomography (Verrier, 2000; Suéry, 2002) have been suggested for accurate identification of solid morphology and distribution. Qin (2001) used fractal dimension to describe the morphology of solid particles in semisolid slurry and successfully applied the concept to model the rheological behaviour of SSM slurries with non-spherical particles (Fan, 2002a).

7.4.2 Solidification behaviour under forced convection

The intensively used direct method of using transparent organic analogues to study conventional solidification events has little applicability in studying solidification under forced convection due to the blurred image caused by intensive stirring (Mullis, 1998). Therefore, the current understanding of the solidification behaviour under forced convention is entirely based upon examination of the final solidified microstructures. However, solidification is a complicated phenomenon and the final microstructure observed is a result of interplay between both nucleation and growth characteristics making it difficult, if not impossible, to understand the exact mechanism of morphological evolution under forced convection. Present understanding is further hindered due to the lack of theoretical investigations undertaken so far.

It has been conclusively established from experimental observation that solidification under melt-stirring produces non-dendritic structures (Flemings, 1991; Kirkwood, 1994; Fan, 2002b). Early rheological work (Spencer, 1972; Joly, 1976) on the Sn-Pb system identified the existence of a degenerated dendritic structure or rosette morphology in the semisolid state under shearing in a rotational viscometer. With prolonged stirring time, a more or less coarse spherical morphology containing entrapped liquid was obtained possibly by a ripening process. Increasing the shear rate accelerates this morphological transition and reduces the amount of entrapped liquid inside the solid particles (Joly, 1976). The rosette and degenerated dendritic morphology of solid particles in the semisolid slurry has been confirmed later in other alloys both under mechanical (Vogel, 1979; Apaydin, 1980; Van Dam, 1982) as well as magnetohydrodynamic (MHD) stirring (Flemings, 1991; Kirkwood, 1994; Fan, 2002b). For sheared Al-Cu alloys some of the observation indicated that the primary particles grow as rosettes until a certain limit beyond which solidification takes place with the appearance of more (new) particles (Vogel, 1979). Others observed no great influence of stirring on the density or size of the rosettes or cellular type particles grown under different cooling rates (Molenaar, 1985). The cell spacing was considerably greater than the secondary dendrite arm spacing from unstirred melt, indicating that stirring promotes crystal growth. On the other hand, solidification structure in a stirred Al-19Si alloy indicated a decrease in average particle diameter and increase in the particle density with increasing the shear rate (Smith, 1991).

It should be mentioned that in all investigations on melt stirring the flow conditions are generally characterised by largely laminar flow with relatively low shear rate achieved by using rod or simple impeller as stirrers or under magnetohydrodynamic flow conditions. Ji and Fan, on the other hand, (2000, 2001, 2002a, 2002b) studied the effect of turbulent flow on the solidification morphology of different alloy systems for the first time using a novel twin-screw rheomoulding (TSRM) machine. They found that under the intensive turbulent flow the solidification morphology consists of near monosize extremely fine solid particles, devoid of any entrapped liquid, evenly distributed throughout the microstructure. The particle morphology was shown to be spherical even at a very early stage of solidification (Ji, 2000). Further systematic study (Das, 2003) on the growth morphology under various fluid flow conditions using carefully selected stirring devices and controlled microstructural observation has revealed that laminar flow promotes a coarsened dendritic or rosette morphology depending on the intensity of shearing, while turbulent flow is crucial for the formation of extremely fine spherical particle morphology as shown in Fig. 7.4.

7.4.3 Structural evolution during partial remelting and isothermal holding

Thixo-processing, in general, utilises partial remelting of alloy billets (rather than solidification) to generate desirable nominal liquid fraction and globular (non-dendritic) solid morphology in the semisolid slurry. The reheating process needs to be optimised to achieve the most desirable slurry characteristics for thixoforming. Loue and Suéry (1995) studied the influence of thermomechanical history on the microstructural evolution of A357 alloy during partial remelting and isothermal holding and found that particle density

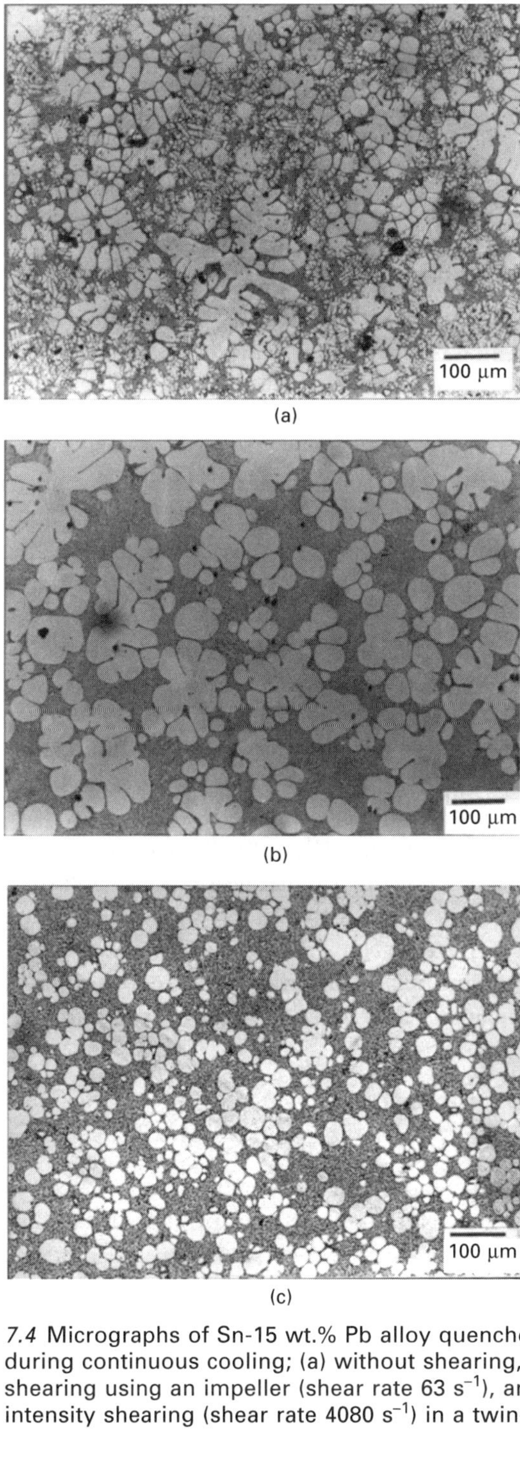

7.4 Micrographs of Sn-15 wt.% Pb alloy quenched from 204 °C during continuous cooling; (a) without shearing, (b) under melt shearing using an impeller (shear rate 63 s^{-1}), and (c) under high-intensity shearing (shear rate 4080 s^{-1}) in a twin-screw slurry maker.

for the initial globular structures decreases with increasing isothermal holding time, while that for initially dendritic structures remains fairly constant. They also found that longer solidification time and smaller initial grain size accelerate the coarsening kinetics. During isothermal holding in the semisolid state, regardless of the initial particle morphology, the solid phase always evolves towards a spherical morphology. However, the kinetics of such evolution is governed by the initial microstructure; the more complex the starting particle shape, the faster the kinetics for spherodisation and the longer the holding time required to obtain spherical particles (Blais, 1996).

Another phenomenon observed during microstructural coarsening is entrapped liquid inside the solid particles. Plastic deformation prior to reheating has been found to be an important factor to the coarsening kinetics. Cold working prior to partial remelting of the alloy billet allows the most rapid globularisation of the solid phase (Loue, 1995; Yunhua, 1998) but too much plastic deformation and too high a temperature will lead to grain coarsening. Tzimas and Zavaliangos (2000) compared the reheating behaviour of three different initial microstructures obtained by magnetohydrodynamic (MHD) stirring, stress induced and melt activated (SIMA) process and spray casting. They concluded that at medium liquid fraction, the microstructure of spray-cast and SIMA consists of discrete equiaxed grains uniformly dispersed in the liquid matrix, while the corresponding microstructure of MHD alloys exhibit extensive agglomerates consisting of incompletely spheroidized grains. Their results also demonstrated that the reheated MHD microstructures are less equiaxed compared with SIMA and spray-cast alloys even after five minutes soaking in the semisolid state.

Recent work by the present authors (Das, 2002b) has suggested that billets cast at liquidus temperature following optimal shearing in a twin-screw machine produces very fine and uniform microstructure with globular microstructure as shown in Fig. 7.5. Reheating such a microstructure was shown to require minimal remelting to produce an ideal slurry structure consisting of evenly distributed fine spherical particles without any entrapped liquid (Fig. 7.5).

7.4.4 Mechanisms for the formation non-dendritic structure

Any proposed mechanism of morphological evolution under forced convection has to address the fundamental characteristics of experimentally observed microstructure, namely, origin of the non-dendritic solid morphology and particle refinement. It should be noted that prolonged isothermal holding will spherodise any solidification morphology (including dendrites) in order to reduce the total interfacial energy, even under complete diffusion control. Such ripening process is operative once the microstructure reaches the equilibrium solid fraction, and is different from the non-dendritic evolution

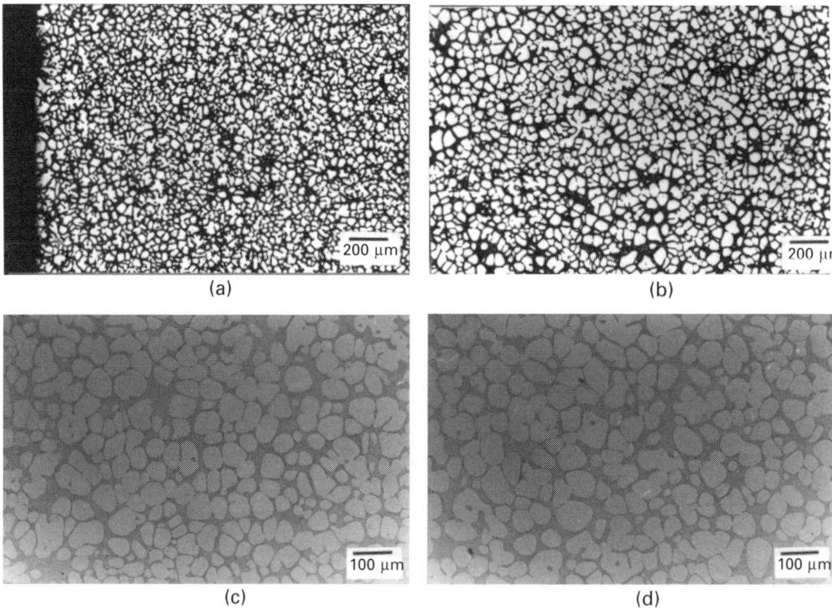

7.5 Micrographs of Sn-15 wt.% Pb alloy cast at the liquidus temperature following shearing (at 2080 s^{-1} for 60 s) in the twin-screw machine from (a) the edge and (b) the centre of the billet. Corresponding micrographs following remelting for 10 minutes at 190 °C in a resistance-heating furnace is shown in (c) from the edge and (d) from the centre of the specimen.

of the solid under forced convection before the solid fraction reaches its equilibrium value. Several mechanisms have been proposed as varied as dendrite arm fragmentation, dendrite arm root remelting and growth-controlled mechanisms.

From the initial boundary layer simulation work on growth under melt convection it was concluded that stirring destabilises the solid liquid interface for both low and high mobility interfaces and increases the dendrite tip velocity and decreases the tip radius (Vogel, 1977a; Cantor, 1977). In view of the apparent contradiction with observed experimental results it was suggested that the non-dendritic microstructure observed during SSM processing might be a result of reduced concentration gradients arising out of overlapping diffusion fields of growing particles. In the absence of any observable difference in the nucleation characteristics with or without melt stirring, Doherty and co-workers proposed the dendrite fragmentation theory, as schematically illustrated in Fig. 7.6, to account for the observed microstructural features (Vogel, 1977b; Lee, 1983; Doherty, 1984). It states that melt stirring induces plastic bending of the dendrites creating high and low angle grain boundaries. The former is wetted by the melt producing

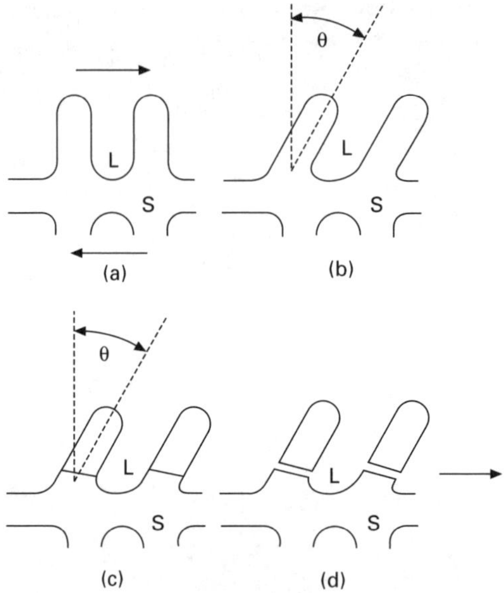

7.6 Schematic illustration of the dendrite fragmentation theory (after Doherty, 1984): (a) original dendritic growth, (b) bending of dendrite arms due to shear, (c) formation of grain boundaries due to migration of dislocations, (d) fragmentation of arms due to wetting of high angle grain boundaries by the melt.

dendrite fragmentation and explains increased particle density in stir-cast structures while low angle or coincidence-site boundaries remain in the microstructure.

It was argued that dendrites cannot fracture near the solidification temperature but secondary arms could, however, remelt at the root due to solute enrichment and thermo-solutal convection (Hellawell, 1996). Crystal multiplication was suggested to be operative through a continuous nucleation mechanism in the absence of a distinct recalescence where each volume element of the liquid passes through different temperature zones periodically. According to the author, vigorous stirring prevents the establishment of stable diffusion fields for continued dendrite evolution, and therefore, smooth rounded non-dendritic shapes are expected in the semi-solid slurries. Molenaar (1985) on the other hand, discarded dendrite fragmentation theory and proposed that the growth type is cellular from experimental observation in intermediate and fast cooled and sheared Al-Cu alloys. Based on a further mathematical analysis the authors proposed that in metal alloys with low Prandtl number the thermal boundary layer is hardly affected by stirring whereas the hydrodynamic boundary layer is significantly reduced and mass transport is dominated by convection (Molenaar, 1986). Consequently, uniform liquidus temperature distribution ahead of the interface (due to reduced solute peak) coupled with the very small negative temperature gradient pertains to an ideal condition for cellular growth of isolated particles in the bulk melt. Dendrite fragmentation theory has been increasingly questioned recently as serial sectioned stir-cast structures seem to suggest that apparently fragmented particles in 2d microstructure may belong to more complex 3d single crystals in reality (Niroumand, 1999).

From experimental investigation using different stirrer geometry to induce different fluid flow patterns in the melt (Das, 2003) and supported by further Monte Carlo simulation (Fig. 7.7) (Das, 2002a), the present authors have proposed an entirely growth-based evolution of microstructure in sheared semisolid slurries. According to the authors the nature of fluid flow in the melt governs the diffusion layer geometry around the growing solid, and therefore, determines the growth morphologies as shown in Fig. 7.8. A low-intensity laminar flow is unlikely to modify the diffusion geometry to a large

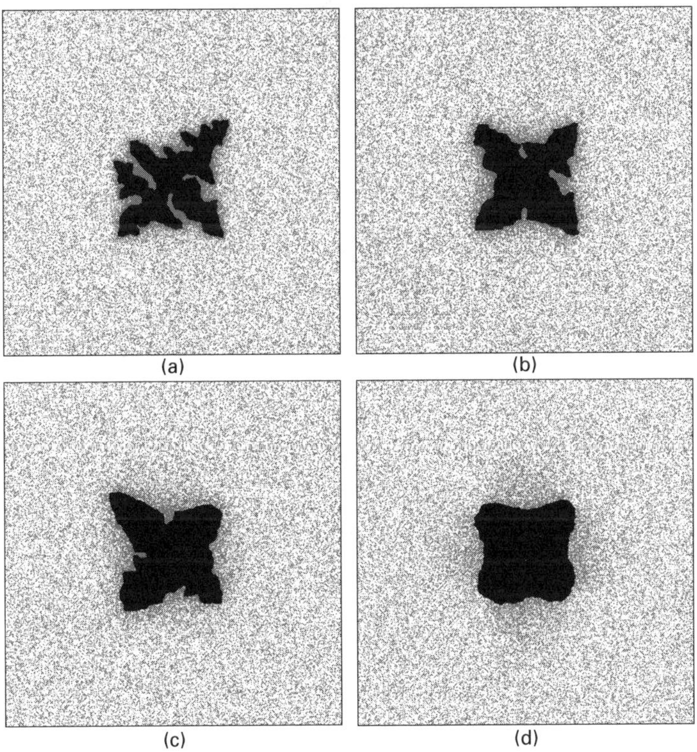

7.7 Particle morphology generated through Monte Carlo simulation for solidification under (a) no melt convection, (b) low intensity laminar flow, (c) high intensity laminar flow, and (d) turbulent flow in the melt. Greyscale intensity represents the solute concentration in the melt.

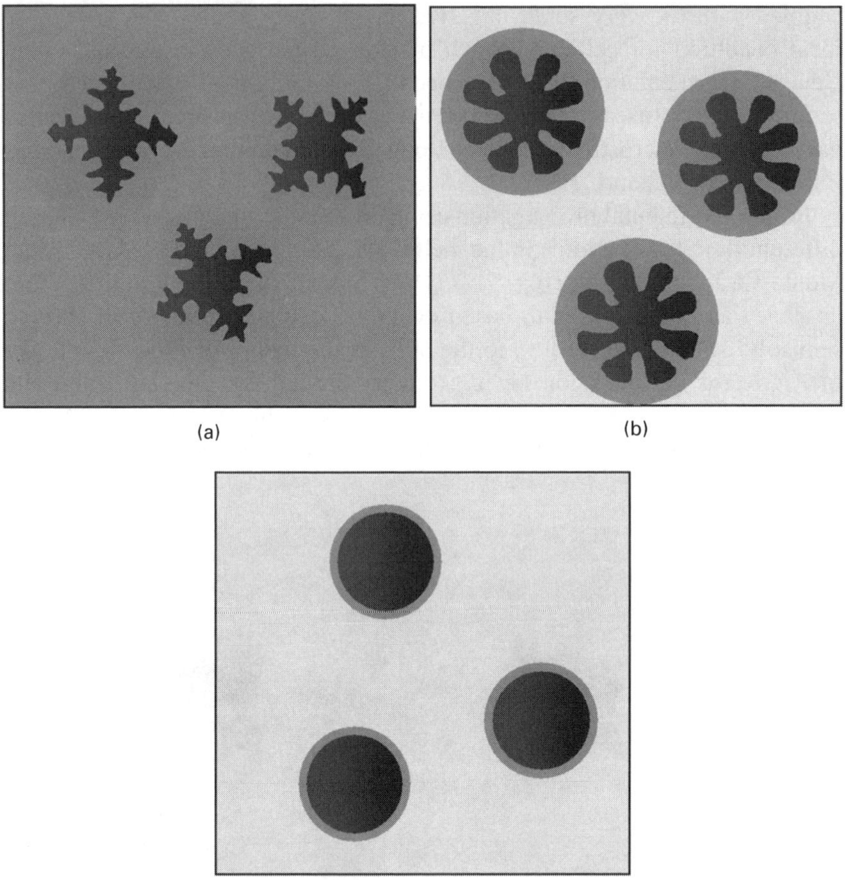

(a) (b) (c)

7.8 Schematic illustration of the effect of nature of melt flow on the solutal diffusion boundary layer (dark grey), and consequently, the morphology of the solid under (a) no melt convection, (b) laminar flow and (c) turbulent flow in the melt. Light grey region represents completely homogenised liquid due to melt convection.

extent where the secondary arms of the dendrite would not experience any relative liquid motion. However, increased intensity of laminar flow will reduce the thickness of diffusion boundary layer around a growing particle, and therefore, modify the growth morphology of the solid. They have shown that particles growing in the bulk liquid will in fact be rotating under a laminar flow and under such circumstances growth morphology will resemble rosettes observed under experimental conditions. Under high intensity turbulence in the melt a compact and spherical growth morphology of the solid from the initiation of solidification is suggested due to the almost complete elimination of diffusion layer around the solid and thus preventing any constitutional undercooling phenomena. The experimentally observed grain refinement under high shear rate and high intensity of turbulence was explained by the homogenisation of both temperature and composition fields in the TSRM machine. Although the actual nucleation rate in the rheomoulding process may not be increased, all the nuclei formed will survive due to the uniform temperature field resulting in an increased effective nucleation rate. In addition, the intensive mixing action may disperse the clusters of potential nucleation agents increasing the number of effective nucleation sites.

On the other hand, the driving force for such morphological evolution in the semisolid state due to partial remelting is the reduction of the interfacial energy between the solid and liquid phases. The holding time has to be carefully determined to ensure complete morphological transition from dendritic (or rosette) to spherical particles without encountering excessive grain growth. The kinetics of semisolid thermal transformation (SSTT) of initially dendritic structure has been examined by a number of investigators, for example, Seconde (1984); Yang (1994); Chen (1995); Hong (2000); Tzimas (2000). The general findings are summarised here.

With an increase in temperature, partial remelting has been found to initiate at the grain boundaries followed by an apparent decrease in the proportion of the eutectic phase. During partial remelting and the subsequent isothermal holding in the semisolid state, coarsening first proceeds predominately through coalescence of dendritic arms. As the dendritic arms of the same cell have a perfectly matching crystallographic orientation, the dendrite arms coalescence is supposed to be extremely rapid. This mechanism of rapid coalescence causes substantial entrapment of the liquid phase at the inter-dendritic region, while in the case of the electromagnetic stirred alloys the coalescence of short dendritic arms leads to a more spherical morphology. After this rapid coalescence stage, much slower coarsening by diffusion of solid atoms from areas of high curvature to areas of low curvature can take place. As long as the particles are not spherical, coarsening will lead to an increase in the mean free path in the solid phase, but the number of particles remains constant until, eventually, the solid phase becomes spherical. Further coarsening will then take place through Ostwald ripening through the dissolution of the small globules and the particle density will then decrease.

7.5 Alloy development for SSM processing

The alloys commonly used for SSM processing are limited to a few casting alloys and only a limited number of trials have been carried out on wrought alloys (Salvo, 1996). In addition, SSM processing, as a new forming technology, is currently utilised without taking advantage of its unique features and is judged by sub-optimal trials using existing alloys, which are optimised either for casting or for forging (Apelian, 2000). For those reasons, there has been a growing need for the development of new alloys specifically optimised for SSM processing to extract maximum benefit in terms of an enlarged processing window, improved mechanical properties, and cost benefits offered by SSM processing.

The first commercial attempt at alloy development especially suited for thixoforming was carried out by Pechiney (Garat, 1998). The properties of the modified Al-6Si-1Cu-Mg alloy after thixoforming match or exceed those of permanent mould cast A356 alloy under T6 condition. The other approach to alloy design for SSM processing is based on existing wrought alloys for improving processability and maintaining the good combination of mechanical properties. Efforts in this direction have been mainly concentrated on Al-Mg-Si systems with increased Mg and Si contents and the minor addition of other alloying elements (Eagler, 2000). It was found that such alloys have sufficient thixoformability and good combinations of mechanical properties. In addition, hot cracking was eliminated. Other preliminary work on alloy development focused primarily on self-hardening alloys aimed at reducing heat-treatment cost (Eagler, 2000).

Unlike the conventional trial and error based experimental alloy development approach that is both time and cost intensive, alloy development can now tremendously benefit from thermodynamic calculations. The availability of phase diagram calculation methods such as the CALPHAD approach, combined with the immensely improved thermodynamic databases now available, can be a very powerful tool to guide the development of new alloys. Generally, thermodynamic calculations can be used to select the major alloying elements followed by experimental work to select the minor alloying elements towards the final alloy composition. To design alloys specifically for SSM processing a number of scientific and technological factors need to be considered. Based on prior experimental work and computational thermodynamic calculations (Witulski, 1998; Maciel Camacho, 2003; Liu, 2003a, 2003b), such basic considerations include:

- optimisation of the solidification range, ΔT_{S-L}, to have better castability and greater resistance to hot tearing of the liquid alloy
- minimal temperature sensitivity of solid fraction, df_s/dT, such that for a given alloy composition solid volume fraction (v_s) remains stable under processing temperature fluctuation giving rise to a larger processing window and consistent rheological characteristics
- greater potential for age hardening by careful selection of alloy compositions ensured by a large solid solubility difference between SSM temperature and ageing temperature to utilise the more cost effective T5 heat treatment
- good castability or fluidity of the remnant liquid in the semisolid slurry that is subjected to the laws of conventional casting through careful alloying addition during alloy design

- rheological properties in the semisolid state as demanded by SSM processing, by selecting proper morphology, size and distribution of the solid phase in the liquid matrix. A reduced tendency for particle agglomeration that strongly affects the rheology will facilitate SSM processing and can be achieved through addition of alloying elements.

Figure 7.9 presents a prediction of potential alloy compositions for SSM processing from the Al-rich corner of the commercially important Al-Si-Mg

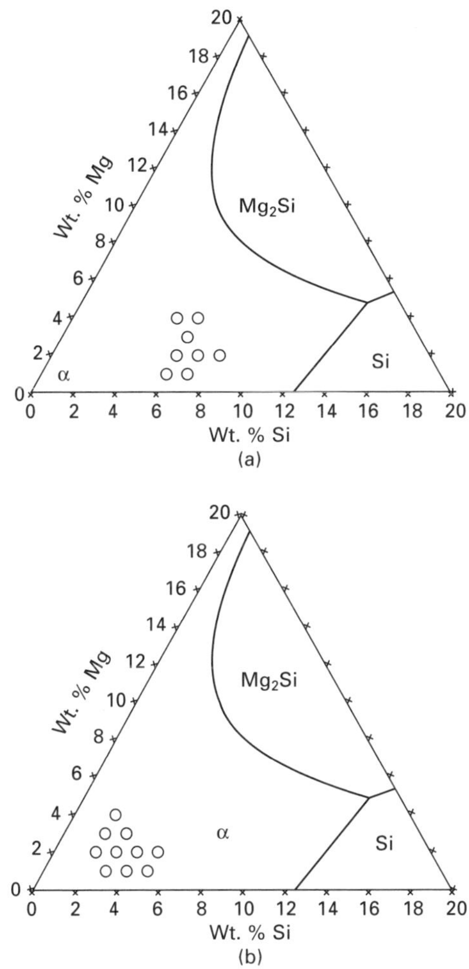

7.9 Computational thermodynamics prediction of promising alloy compositions (represented by open circles) in the Al-rich corner of the Al-Mg-Si ternary system, suitable for SSM processing in the (a) rheo-route (at f_S = 0.3) and (b) thixo-route (at f_S = 0.6). Liquidus projections are superimposed as solid lines.

Table 7.1 Thermodynamic characteristics of Al-Si-Mg and Mg-Al-Zn alloys designed for SSM processing in comparison with those of commercial alloys based on the same system

Alloys	$f_S = 0.3$		$f_S = 0.6$	
	$\Delta T_{S\text{-}SS}$	$\lvert df_S/dT \rvert$	$\Delta T_{S\text{-}SS}$	$\lvert df_S/dT \rvert$
Aluminium alloys designed for rheo-route $f_S = 0.3$				
Al-6Si-1Mg	46.9	0.014	–	–
Al-6Si-2Mg	42.2	0.013	–	–
Aluminium alloys designed for thixo-route $F_S = 0.6$				
Al-4Si-2Mg	–	–	34.6	0.008
Al-5Si-2Mg	–	–	20.5	0.006
Commercial aluminium alloys				
356.0	30.5	0.012	6.8	~ 0.21
357.0	36.4	0.012	11.9	~ 0.13
6061	57.7	0.075	51.8	0.034
6066	83.5	0.039	71.6	0.016
Magnesium alloys designed for rheo-processing $f_S = 0.3$				
Mg-10Al-1Zn	142.1	0.014	–	–
Mg-11Al-1Zn	147.1	0.012	–	–
Commercial magnesium alloys				
AZ91	121.6	0.016	95.4	0.008
AZ63	174.0	0.018	148.3	0.008

system based on the CALPHAD technique (Liu, 2003b). The thermodynamic characteristics of some newly designed Al-based and Mg-based alloys are compared in Table 7.1 with those of the conventional cast and wrought alloys based on the same system. Generally, the new alloys have a more suitable solidification range, reduced temperature sensitivity of solid fraction and enhanced potential for age hardening, and therefore improved processability and increased potential for property enhancement compared with the conventional alloys. It should be noted that the present directions for alloy development for SSM processing also involve either modification of the existing cast or wrought alloys to improve strength and ductility while maintaining castability in the former and improved fluidity and temperature sensitivity of solid fraction and reduced tendency for hot tearing in the latter.

7.6 Mechanical properties

Although SSM processing, especially Rheo-processing, is expected to play a very significant role in the manufacturing of future lightweight high-strength components (e.g., automobile components), the evaluation of mechanical properties of SSM processed alloys is still very limited. The majority of the results available concern thixo-processing and are primarily confined to cast

Al-alloys, with less attention being given to wrought alloys. The data for Mg-based alloys are particularly scarce. The available tensile property data in the literature show that the strengths of SSM processed Mg-alloys compare well, but are not superior to those of conventional die-casting, although the ductility values are improved. Mechanical property data for SSM processed aluminium cast alloys can be found in Brown (1992); Kiuchi (1994); Kirkwood (1996); Bhasin (1998); Chiarmetta (2000); Tsutsui (2002).

It has been observed that the mechanical properties of thixoformed A356 alloy in the fully heat treated conditions (T6) are superior to those obtained by sand or permanent mould casting but comparable to those achieved by squeeze casting; the improved ductility of thixoformed specimens are particularly noteworthy. Such improvement of both strength and ductility can be attributed to significantly reduced porosity and refined microstructure coupled with increased chemical homogeneity in the SSM processed component. Similar results have been observed for cast alloys A357. However, thixoformed wrought alloys do not quite achieve the highest strength and ductility that they are potentially capable of in the wrought form as suggested in the literature (Chiarmetta, 1996; Fink, 1998; Davis, 1993; Tausig, 1998, 2000; Kaufmann, 2000a).

This appears to originate from the presence of defects such as hot cracks, residual porosity, oxide inclusions, and liquid segregation in the thixoformed wrought alloys. This emphasises again the need for new alloys developed specifically for SSM processing. An interesting comparative study between the mechanical properties of thixo- and rheo-cast A357 in the T5 condition has recently been presented by Stampel S.p.A, Italy (Giordano, 2002). It was observed that although average yield strength is roughly the same, UTS and elongation is higher for the rheocasting in addition to lower strength variation in the components. Such encouraging results were attributed to reduced micro defects, higher stability and greater consistency in rheocasting over thixocasting and substantiate the fact that rheo-based technologies are the future for SSM processing.

In spite of the stringent requirements in the automobile industry and the potential for SSM processing in that sector, fatigue and creep data on SSM processed material is even sparser as compared to the tensile data. Gabathuler (1993) studied the fatigue strength of SSM formed alloy. The fatigue strength of SSM formed A356 at high cycles ($N > 10^6$) have shown substantial improvement over samples produced by high-pressure die casting (HPDC), PMC and sand casting. Similar enhancement in fatigue strength for SSM formed samples over samples produced by other processing routes has also been reported for A357 and modified A319 alloys (Bergsma, 2000).

7.7 Technologies for component shaping

7.7.1 Rheocasting

The process of production of non-dendritic semisolid slurry by shearing during solidification has been identified as rheocasting. Direct transfer of the semisolid slurry into a die or mould to produce a finished product has also been identified as rheocasting or stir casting. The technology that originated at MIT was based on mechanical stirring (Mehrabian, 1972; Flemings, 1976). Melt agitation is commonly generated by means of augers, impellers, or multipaddle agitators mounted on a central rotating shaft (as shown in Fig. 7.10). Although rheocasting was identified as the production technology at the very beginning of semisolid processing research, it has not been

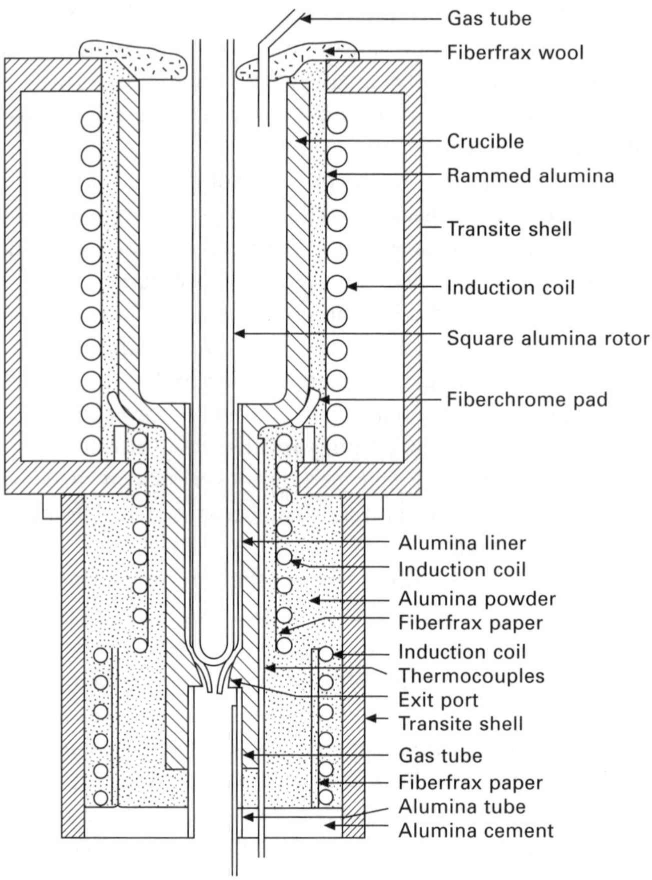

7.10 Schematic illustration of a high-temperature continuous rheocaster utilising mechanical stirring (after Flemings, 1976).

commercialised to any great extent so far. The slurries produced under such conditions do not have adequate thixotropic characteristics for a successful direct shaping by either the casting or forging route due to the usually coarse rosettes observed in the microstructure (Sumartha, 1998) apart from the problems associated with contamination of the melt through oxidation and chemical reaction with the stirring system. To overcome the difficulties associated with direct mechanical stirring, a MHD stirring process was developed by ITT in the USA and described in a series of patents originally held by Alumax Inc (Kenney, 1988). In this technique local shear is generated by rotating electromagnetic fields within the continuous casting mould, and the stirring is deep in the liquid, previously filtered and degassed, so that contamination is virtually eliminated.

Electromagnetic stirring can be achieved through three different modes: vertical flow (Blais, 1996; Pluchon, 1995), horizontal flow (Gabathuler, 1992) and helical flow (Garat, 1998) modes, with the helical flow being ultimately a combination of vertical and horizontal stirring modes. In the horizontal flow mode the motion of the solid particles takes place in a quasi-isothermal plane so that mechanical shearing is probably the dominant mechanism for spheroidisation. In the case of vertical flow mode, the dendrite located near the solidification front are recirculated to the hotter zone of the stirring chamber and partially remelted, and therefore, the thermal process is dominant over mechanical shearing (Hellawell, 1996; Garat, 1998). However, MHD stirring usually gives rise to degenerated equiaxed dendrites, which are inherently unsuitable for direct component shaping unless they are held at semisolid temperatures for a sufficiently long time. The other possible reason for the failure of commercialisation might be insufficient process control and low productivity associated with the original version of rheocasting. Therefore, MHD stirring has established itself as a widespread feedstock production practice for thixocasting rather than rheocasting.

7.7.2 New rheocasting

One of the 'slurry-on-demand' processes developed to overcome the difficulties associated with melt stirring is the new rheocasting (NRC) process, which eliminates the necessity for melt stirring, developed and patented by UBE Industries Ltd and described first by Adachi (1998). In the NRC process, as sketched in Fig. 7.11, liquid metal slightly above its liquidus temperature is poured into specially designed steel crucibles. In the crucible semisolid slurry is formed by controlled cooling through a mechanism similar to liquidus casting and distribution of the solid nuclei throughout the melt. By controlled air blowing the slurry temperature (therefore, solid fraction) is brought to the desired semisolid region within a few minutes after pouring. The temperature gradient in the crucible and the slug temperature is homogenised by high-

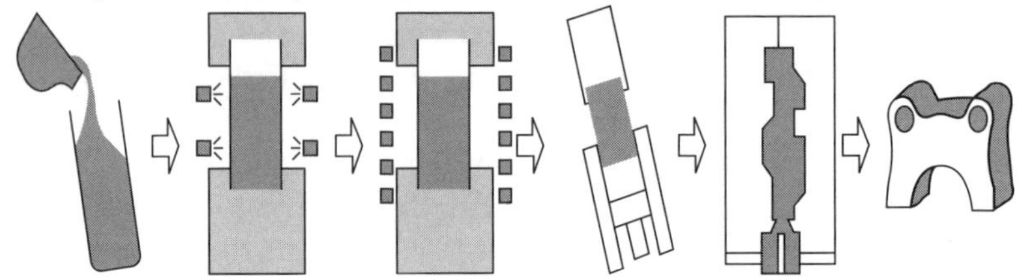

7.11 Schematic illustration of the new rheocasting process.

frequency induction heating before transferring the slurry into the sleeve of the vertical SQC machine, where it is cast into its final shape.

So far the NRC process has been evaluated by processing cast Al-alloys (Kaufmann, 2000b; Wabusseg, 2000), wrought Al-alloys (Kaufmann, 2000a) and Mg-alloys (Potzinger, 2000). Since conventional alloys can be used, NRC offers a significant cost advantage over thixoforming and SQC processes (Hall, 2000; Giordano, 2002). It also allows the use of a wider range of alloys; of particular interest are the Mg-alloys, where a thixotropic feedstock material is not readily available on the market at the present time. An interesting recent comparison of the new rheocasting process against thixocasting (on an industrial scale) by Stampel S.p.A has shown that new rheocasting is significantly advantageous due to lower cost, better mechanical property of components, lower metal losses, higher production rate and improved scrap recycling (Giordano, 2002).

7.7.3 Rheodiecasting

The major drawback of direct mechanical stirring or MHD stirring to produce slurry for SSM processing is the inhomogeneous microstructure and the coarse and complex morphology of the solid particles. Moreover, liquid entrapment in the solid particles implies a lowering of the effective liquid fraction in the slurry as compared to the equilibrium amount predicted at the processing temperature. This lowering of liquid fraction due to entrapment inside the solid particles of the semisolid slurry may adversely affect castability during the processing of components.

By realising that turbulent flow is more effective than laminar flow on microstructural refinement and growth of solid as compact and spherical particles (Ji, 2000, 2002a; Das, 2002a, 2003), Fan and co-workers have developed the twin-screw rheodiecasting process (Ji, 2001; Fan, 2002b), schematically illustrated in Fig. 7.12. The heart of the system is the twin-screw slurry maker, which consists of two closely intermeshing, co-rotating and self-wiping twin-screw assemblies capable of rotating at extremely high speed. The liquid alloy feed is supplied from the holding furnace to the

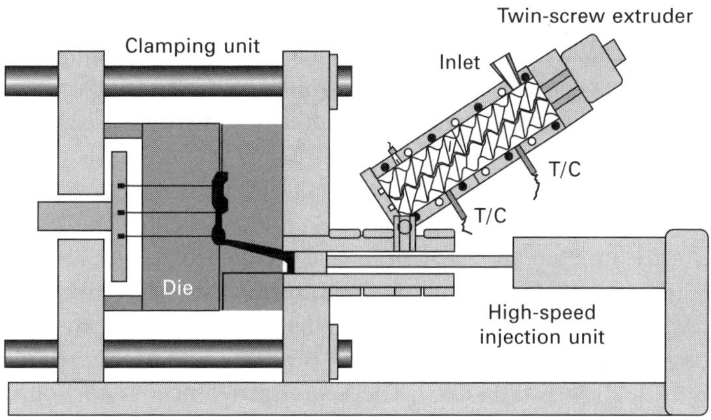

7.12 Schematic illustration of the rheodiecasting process employing the twin-screw slurrymaker.

twin-screw slurrymaker and intensively sheared under high shear rate, high intensity of turbulence, and cyclic variation of shear rate for a predetermined interval under an environment of protective gas (Ji, 2001).

As a consequence of such fluid flow characteristics and accurate temperature control, the temperature and composition fields inside the barrel are extremely uniform and produce a slurry structure with an extremely fine near monosize homogeneous distribution of solid particles. The slurry is fed into a conventional cold-chamber die-casting machine to produce near-net shape components. The major advantages of the process include ideal slurry structure, accurate process control and ability to automation among others. Moreover, the slurry-maker assembly can be directly integrated into an existing high-pressure diecasting facility, and therefore, require minimal modification in the existing casting infrastructure to adapt for SSM processing. So far this technology has been tested successfully using a model Sn-Pb system as well as with commercial and specially designed Mg and Al-based alloys.

Recently, the twin-screw process has also been extended to a variety of rheo-based processing technologies such as the rheomixing process for processing the usually immiscible liquid alloys (Fang, 2002), a one step rheo-extrusion method to produce extruded products directly starting with liquid alloy (Roberts, 2002), and a twin-screw low-superheat casting process (Das, 2002b).

7.7.4 New MIT process

It has now been realised that in order to obtain an ideal SSM slurry, the ability to achieve copious nucleation is far more effective than relying upon the mechanical action of stirring. Therefore, the recent advances on rheo-processing (such as the new rheocasting processes) have been primarily based on rapid heat extraction from the melt at or near the liquidus temperature. Based on this concept of rapid and controlled heat extraction and convection, Flemings and co-workers have developed a new method of Rheocasting and licensed to Idra press SpA (Yurko, 2002). The process is schematically illustrated in Fig. 7.13. The liquid alloy is held in the container isothermally just above the liquidus temperature. A rotating coated and water-cooled copper rod is inserted into the melt for a short duration to drop the temperature of the liquid below the liquidus creating nucleation events at the stirrer surface and distributing them throughout the melt. The stirring rod is withdrawn and the liquid is cooled to the desired processing temperature followed by casting of the semisolid slurry. The fast effective heat extraction combined with the convection created by the brief stirring was shown to be more effective compared to the long mechanical stirring and isothermal coarsening originally developed at MIT in creating an ideal semisolid slurry with non-dendritic and globular particles. Moreover, the new process has been demonstrated to produce particles without any entrapped liquid, thus ensuring larger quantities of effective liquid fraction available for better fluidity of the

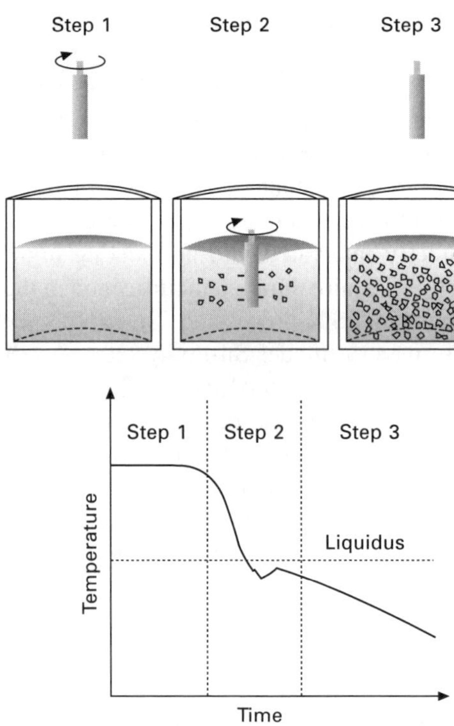

7.13 Schematic illustration of the new MIT semisolid rheocasting process (after Yurko, 2002).

slurry. Increased stirring speed and stirring time once the melt temperature has dropped below the liquidus temperature had minimal impact on the final microstructure. Initial processing run with A356 and A357 alloys seemed to suggest that the SSM slurry produced by the new MIT process showed better microstructure and fluidity compared to MHD processed slurry.

7.7.5 Melt mixing approach

Another implementation of the rapid heat extraction near the liquidus temperature of the melt is the recent melt mixing approach developed at the WPI, MA, USA. In this method two liquids of similar or slightly different composition is held above the liquidus temperature in separate crucibles and then allowed to flow into and mix inside a flow dispenser with patterned channels to induce turbulence in the mixing melts. The liquid is rapidly cooled below the liquidus inside the dispenser and results in copious nucleation following which the semisolid slurry is poured into a mould and adjusted for the required volume fraction and slurry structure by proper temperature control. The method has been shown to produce non-dendritic slurry structure mixing A356 alloy melt at similar or slightly different temperature (Findon, 2002) or by mixing hyper- and hypo-eutectic compositions of Al-Si alloy to achieve refined particle size through spontaneous heat extraction on mixing (Saha, 2002).

7.7.6 Thixoforming

Thixoforming is a general term used to describe the near-net shape forming processes from a partially remelted non-dendritic alloy slug within a metal die. If component shaping is performed in a closed die, it is referred to as thixocasting, while if shaping is achieved in an open die, it is called thixoforging, as schematically illustrated in Fig. 7.14. In the two-step process a specially processed feedstock is partially remelted and homogenised in the semisolid stage to produce a thixotropic slurry structure with appropriate solid fraction followed by transferring the semisolid slug in a forging die or shot chamber for component processing. Thixoforming necessitates the use of suitable feedstock requiring minimal holding at the semisolid temperature to produce a non-dendritic slurry structure, although feedstock preparation is rarely an integral part of thixoforming itself. The most commonly used feedstock preparation involves: MHD stirring (as discussed earlier in rheocasting section), SIMA process involving cold working (Young, 1983) or warm working (Elias Boyed, 1988; Kirkwood, 1989) of the feedstock billet to introduce recrystallisation and grain boundary melting during the reheating stage to produce fine globular slurry structure, spray casting (Annavarapu, 1995; Blais, 1996; Underhill, 1995) and casting near the liquidus temperature (Tausig, 1998; Xia, 1998; Das, 2002b).

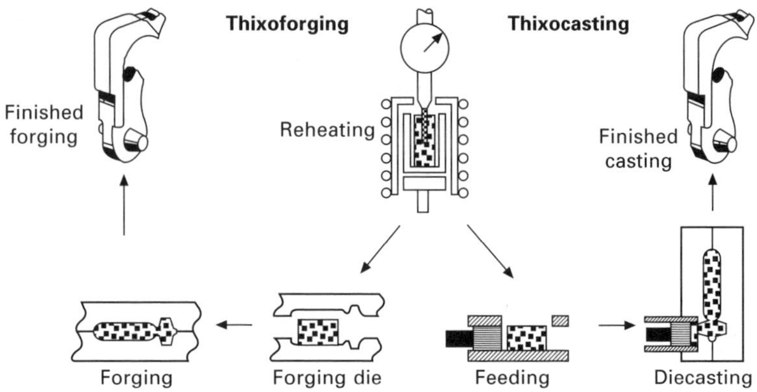

7.14 Schematic illustration of the thixoforming process.

Reheating to the semisolid state is a particularly important phase in the thixoforming process (Garat, 1998) to produce a semisolid slug with controlled solid fraction of fine and spherical particles uniformly dispersed in a liquid matrix of low melting point. The accuracy and uniformity of heating temperature is essential to prevent slug instability and slug handling difficulties due to high temperature or impartial melting resulting in die filling problems arising from too low a temperature. In addition, as discussed previously, the alloys currently used not being optimised for SSM processing, small variations in temperature can lead to a large difference in solid fraction resulting in vastly different rheological characteristics of the slug. Finally, the heating duration has to be optimised; too long a heating time will cause microstructural coarsening, while too short a heating time will lead to incomplete spheroidisation of the solid particles leading to mould filling problems. Induction heating is normally used due to precise and fast heating, although a convection furnace is also used in some cases.

The forming process takes place either with casting (thixocasting) or with forging (thixoforging). At the present time, thixocasting through horizontal cold chamber die casting is the dominant process. A smooth laminar mould filling is the crucial step for the forming process that can be achieved by an optimised shot profile tailored for specific alloys and their physical conditions and by proper design of gating system and die cavity and the correct choice of die temperature. The main advantage of thixoforming is the freedom from handling liquid metal and the ability to automate the process using approaches similar to those employed in forging and stamping. However, thixocasting is finding progressively lower acceptance due to the high cost of feedstock, inhomogeneity of structure and composition, high amount of metal loss and low recycling rate of scrap (Fan 2002b).

7.7.7 Thixomoulding

Thixomoulding is a relatively new process for the production of near-net shaped components from Mg-alloys in a single integrated machine (Pasternak, 1992), as shown schematically in Fig. 7.15. The raw materials for thixomoulding are Mg-alloy chips of 2–5 mm in size obtained during metal working of conventional solid Mg-alloys. A volumetric metering device feeds the Mg chips into an electrically heated plasticising and conveying unit where they are partially melted and transformed under continuous shear force into semisolid slurry. The core of this unit is the screw, which performs both a rotary and translational movement. Upon entering the unit at the feed throat, the chips are forced to pass from the heating zone to the front of the screw while the screw retracts. Once the plasticising volume corresponds to the weight of the part to be moulded, the screw advances at high speed and injects the materials into a mould. A non-return valve keeps the materials from flowing back from the front of the screw to the infeed zone. In order to prevent the Mg-alloy from oxidising and igniting as it is heated, an argon atmosphere is usually maintained at the infeed displacing the air between the magnesium chips.

The advantage of thixomoulding is that it effectively combines slurry making and component forming operations into an one-step process leading to high efficiency and energy management, and the elimination of liquid metal handling that is particularly advantageous for processing Mg-alloys. A number of thixomoulding machines are now in operation for production of Mg castings, particularly electronic housings. At present, it appears that this process is limited to relatively low solid fraction and to Mg-alloys for thin-wall components.

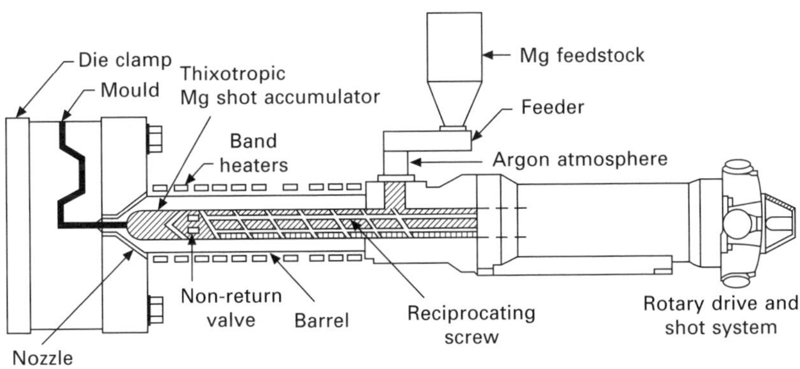

7.15 Schematic illustration of the Dow Thixomoulder (after Pasternak, 1992).

7.8 Future trends

From the inception of the concept of SSM processing, there has been encouraging progress in developing new techniques, a quest for fundamental understanding of the process and in identifying the areas for the exploitation of such processing techniques. Undoubtedly there is a tremendous potential for the development and application of SSM processing especially in the areas of high-performance lightweight materials such as in the automobile sector. However, it appears that significant efforts are still necessary for achieving the full potential and successful exploitation of SSM processing both in terms of fundamental understanding and process and material development.

Till now, thixo-processing has been the dominant route for SSM processing. Not only is this a two-step process, but also it is heavily dependent on the availability of quality feedstock material. Expensive feedstock, inhomogeneous microstructure, lower production yield, larger material loss and reduced scrap recycling rate makes thixoprocessing progressively uncompetitive compared to one-step rheoprocessing routes. This opinion was reflected strongly in the recent SSM conferences held in Turin, Italy in September 2000 (Chiarmetta, 2000) and Tsukuba, Japan in September 2002 (Tsutsui, 2002). It appears that significant efforts have been directed towards developing new rheo-based direct single-step SSM processing techniques that can integrate slurry making and component-shaping operation into one single process. The essential requirement in rheo- (also called 'slurry-on-demand') based SSM processing technology is to produce an ideal slurry microstructure comprising fine and spherical particles without any entrapped liquid distributed uniformly in a liquid matrix such that the necessary rheological properties and solid fraction can be obtained with reliability, and efficient integration of slurry making into a component shaping technology with accurate process control.

Understanding the mechanisms for the formation of globular structure is crucial to the successful slurry making technologies. Moreover, there seem to be two different approaches utilised in the slurry making process. In the first process melt agitation by mechanical means (stirrers) or external fields (electromagnetic fields) have been utilised. Contrary to earlier belief, present investigations seem to suggest that dendrite fragmentation due to shearing is less likely to contribute towards the evolution of non-dendritic slurry structure. It has been suggested by the present authors that morphological evolution can be an entirely growth phenomenon depending on the flow pattern in the liquid. Controlled solidification experiments and rigorous simulation is necessary to understand the exact nature of flow created in the melt by melt stirring, and investigate the effect of such flow patterns on the heat and mass diffusion geometry in the liquid and semisolid melt in order to address the influence of stirring on the nucleation and growth behaviour in the melt.

Recent work (Ji; 2000, 2001, 2002a,b; Das, 2003) using the twin-screw machine has clearly identified the importance of turbulent flow on the formation of fine and spherical solid particles. However, further work is required to quantify turbulent flow and to understand the physical interaction between turbulent flow and the solidification process. There appears to be an increased interest in generating the ideal slurry structure through non-agitation methods. The recent resurgence of interest in methods primarily based on low superheat casting appears to suggest that future development in slurry generation will concentrate on developing simpler methods based on increased effective nucleation rate rather than relying upon the modification of growth of solid particles.

A better understanding of the physical nature of the liquid metals and alloys above their liquidus may help to identify physical methods to interfere more effectively and positively with the nucleation process and eventually to develop more efficient solidification technologies. It is also necessary to understand the solidification behaviour of the remaining liquid in the SSM slurry. The secondary solidification may have little significance to slurry generation itself, but it has great implications on the mechanical properties of the final products. So far, there has been little attention paid to secondary solidification.

Currently, the alloys used for SSM processing are mainly cast alloys with a limited number of wrought alloys. However, wrought alloys have narrow processing windows for available SSM technologies and thus limit further realization of the full potential of SSM processing. Therefore, there is a great demand for new alloys designed especially to utilise the maximum benefit of SSM processing in terms of a large processing window, greater control and superior properties. Considering that SSM processing has been stated to have tremendous potential for lightweight high-strength alloys, such as for the automobile and aerospace industries, there is an increasing need for designing new alloys for improved strength and ductility to match or exceed those of wrought alloys and alloys with superior creep and fatigue properties at elevated temperatures.

Understanding the rheological behaviour has a direct influence on the successful development and effective control of SSM processes. However, due to the complexity of the rheology of SSM processing and the experimental difficulties associated with high temperature and high chemical reactivity, the rheology of SSM slurry is not well understood and there is still a scarcity of accurate and well-defined viscosity data. Future efforts should be directed to the development of more reliable techniques for rheological characterisation of SSM slurries, understanding the effects of slurry structure on the rheological behaviour and the development of rigorous physical models for an improved scientific understanding and phenomenological models for process simulation. Apart from this, the flow characteristics of the semisolid slurry inside the die also require better understanding.

One of the stated benefits of SSM based casting processes is the reduced turbulence created inside the die cavity due to increased viscosity of the slurry compared to molten alloys. This is of special significance in high-pressure diecasting usage where reduced turbulence may significantly eliminate porosity and thereby render the components heat treatable. However, at the same time, it is necessary to ensure proper filling of different cross-sectional components and eliminate the possibility of segregation of liquid and solid components of the slurry. The difference in rheological behaviour and flow characteristics of semisolid slurry with molten alloy requires better understanding in order to assist suitable designing of die and gating and runner system (that could be significantly different from the conventional design criterion for casting of liquid alloys) for successful production of defect-free finished or semifinished components.

In spite of the interest shown in SSM processing, there is a scarcity of mechanical property data. Considering the potential for automobile, aerospace and structural usage, there is a growing need for exhaustive evaluation of SSM processed materials and their comparison against components produced under conventional processing techniques. Of special consideration is the fatigue and creep behaviour of SSM processed materials, which is lacking at the moment. It is desirable to understand the effects on mechanical properties of microstructural parameters, such as volume fraction, size, morphology and distribution of the primary phase, and of the phase arrangement in the matrix. So far, there has been very limited attention devoted to this aspect of SSM processing research. More investigations should be undertaken both theoretically and experimentally to establish the relationships between SSM formed microstructures and the mechanical properties of the final parts. This is a complex task, which may involve process development, alloy design and schedules for heat treatment.

7.9 Sources of further information

There have been significant efforts directed towards developing and understanding semisolid processing technology since its origin in 1971 that cover a broad area of interest from fundamental understanding of rheological characteristics, microstructure formation, nucleation, to the development of technologies, industrial adaptation and efforts to develop semisolid processing for light metals and alloys. It is beyond the scope of this chapter to provide an overall picture of the different aspects of SSM processing. Fortunately, three comprehensive reviews have been published on the subject of semisolid processing of metallic alloys and composites by Flemings (1991), Kirkwood (1994) and finally by Fan (2002b). The readers are directed to these excellent review articles for a greater understanding of the subject. Furthermore, the potential of semisolid processing has resulted in seven international conferences so far solely dedicated to this field, organised every alternate year. The proceedings of these conferences (Brown, 1992; Kiuchi, 1994; Kirkwood, 1996; Bhasin, 1998; Chiarmetta, 2000; Tsutsui 2002) are worth reading, considering the enormous knowledge they contain covering every aspect of semisolid processing.

7.10 References

Adachi M. (1998), in 'Proc. Japanese Die Casting Assoc.', JD98-19, 123–128.

Annavarapu S. and Doherty R.D. (1995), *Acta Metall. Mater.*, 43, 3207–3230.

Apaydin N., Prabhakar K.V. and Doherty R.D. (1980), *Mater. Sci. Eng.*, 46, 145–150.

Apelian D. (2000), in Chiarmetta (2000), 47–54.

Arnberg L., Bardal A. and Saud H. (1999), *Mater. Sci. Eng.*, A262, 300–303.

Barnes H.A. and Walters K. (1985), *Rheologica Acta*, 24, 323–326.

Bergsma S.C., Kassner M.E., Evangelista E. and Cerri E. (2000), in Chiarmetta (2000), 319–324.

Bhasin A.K., Moore J.J., Young K.P. and Midson S. (eds), (1998), 'Proceedings of Fifth International Conference on Semisolid Processing of Alloys and Composites', Colorado School of Mines, Golden, Colorado.

Blais S., Loue W. and Pluchon C. (1996), in Kirkwood (1996), 187–192.

Brown S.B. and Flemings M.C. (eds), (1992), 'Proceedings of the Second International Conference on the Semisolid Processing of Alloys and Composites', M.I.T., Cambridge, MA.

Cantor B. and Vogel, (1977), *J. Cryst. Growth*, 41, 109–123.

Chen C.P. and Tsao C.Y.A. (1995), *J. Mater. Sci.*, 30, 4019–4026.

Chiarmetta G.L. (1996), in Kirkwood (1996), 204–207.

Chiarmetta G.L. and Rosso M. (eds), (2000), 'Proceedings of sixth International Conference on Semi-Solid Processing of Alloys and Composites', Turin, Italy.

Das A., Ji S. and Fan Z. (2002a), *Acta Mater.*, 50, 4571–4585.

Das A., Ji S. and Fan Z., (2002b), in Tsutsui (2002), 689–694.

Das A. and Fan Z. (2003), *Mater. Sci. Tech.*, 573–580.

Davis J.R. (1993), *Aluminium and Aluminium Alloys*, ASM. Metals Park, OH.

De Figueredo A.M., Kato A. and Flemings M.C. (2000), in Chiarmetta (2000), 477–482.

Doherty R.D., Lee H.-I. and Feest E.A. (1984), *Mater. Sci. Eng.*, 65, 181–189.

Eagler S., Hartman D. and Niedick I. (2000), in Chiarmetta (2000), 483–488.

Elias Boyed L., Kirkwood D.H. and Sellars C.M. (1988), in 'Proc. 2nd World Basque Cong. on New structural materials', Spain, Servico Central de Publicaciones del Gobierno Vasco, 285–295.

Fang X., Fan Z., Ji S. and Hu Y. (2002), in Tsutsui (2002), 695–700.

Fan Z. and Chen J.Y. (2002a), *Mater. Sci. Tech.*, **18**, 258–267.

Fan Z. (2002b), *Inter. Mater. Rev.*, 47, 49–85.

Findon M. de Figueredo Apelian D. and Makhlouf M.M. (2002), in Tsutsui (2002), 557–562.

Fink R. and Witulski T. (1998), in Bhasin (1998), 557-564.

Flemings M.C., Riek R.G. and Young K.P. (1976), *Mater. Sci. Eng.*, 25, 103–117.

Flemings M.C. (1991), *Metall. Trans.*, 22A, 957–981.

Gabathuler J.P., Barras D., Krähenbühl Y. and Weber J.C. (1992), in Brown (1992), 33–46.

Gabathuler J.P., Hubbr H.J. and Erling J. (1993), in 'Proc. Int. Conf. on Aluminium alloys: New processing technologies', Ravenna, Italy, Associazione Italiana di Metallurgia, 169–180.
Garat M., Blais S., Pluchon C. and Loue W.R. (1998), in Bhasin (1998), 199–213.
Giordano P. and Chiarmetta G.L (2002), in Tsutsui (2002), 665–670.
Hall K., Kaufmann H. and Mundl A. (2000), in Chiarmetta (2000), 23–28.
Hartnett J.P. and Hu R.Y.Z. (1989), *J. Rheology*, 33, 671–679.
Hellawell A. (1996), in Kirkwood (1996), 60–65.
Hong T.W., Kim S.K., Ha H.S., Kim M.G., Lee D.B. and Kim Y.J. (2000), *Mater. Sci. Tech.*, 16, 887–892.
Ito Y., Flemings M.C. and Corne J.A. (1992), in *Nature and properties of semi-solid materials*, (ed. J.A. Sekhar and J.A. Dantzig), 3-17, Warrendale, PA, TMS Publ.
Ji S. and Fan Z. (2000), in Chiarmetta (2000), 723–728.
Ji S., Fan Z. and Bevis M.J. (2001), *Mater. Sci. Eng.*, A299, 210–217.
Ji S. and Fan Z. (2002a), *Met. Mater. Trans.*, A33, 3511–3520.
Ji S., Fan Z., Fang X. and Song S.H. (2002b), in Tsutsui (2002), 683–688.
Joly PA and Mehrabian R, (1976), *J. Mater. Sci.*, 11, 1393–1418.
Kaufmann H., Wabusseg H. and Uggowitzer P.J. (2000a), in Chiarmetta (2000), 457–462.
Kaufmann H., Wabusseg H. and Uggowitzer P.J. (2000b) *Alluminium*, 76, 70–75.
Kenney M.P., Courtois J.A., Evans R.D., Farrior G.M., Kyonka C.P., Koch A.A. and Young K.P. (1988): *Metals Handbook*, 15, Metals Park, OH, ASM. International.
Kirkwood D.H. and Kapranos P. (1989), *Metall. Mater.*, 5, 16–19.
Kirkwood D.H. (1994), *Inter. Mater. Rev.*, 39, 173–189.
Kirkwood D.H. and Kapranos P. (eds), (1996), 'Proceedings of Fourth International Conference on Semisolid Processing of Alloys and Composites', The University of Sheffield, England.
Kiuchi M. (ed.), (1994), 'Proceedings of the Third International Conference on the Semisolid Processing of Alloys and Composites', Institute of Industrial Science, University of Tokyo, Tokyo, Japan.
Kumar P., Martin C.L. and Brown S.B. (1993), *Metall. Trans.*, 24A, 1107–1116.
Lee H.-I., Doherty R.D., Feest E.A. and Titchmarsh J.M. (1983), in *Solidification Technology in Foundry and Casthouse*, London, Metals Society.
Liu Y.Q., Das A. and Fan Z. (2003a), *Mater. Sci. Tech.*, in press.
Liu Y.Q., Das A. and Fan Z. (2003b), *Metall. Mater. Trans.*, accepted for publication.
Loue W.R. and Suéry M. (1995), *Mater. Sci. Eng.*, A203, 1–13.
Maciel Camacho A., Atkinson H.V., Kapranos P. and Argent B.B. (2003), *Acta Mater.*, 51, 2319–2330
Mada M. and Ajersch F. (1996a), *Mater. Sci. Eng.*, A212, 157–170.
Mada M. and Ajersch F. (1996b), *Mater. Sci. Eng.*, A212, 171–177.
Martin C.L., Kumar P. and Brown S.B. (1994), *Acta Metall. Mater.*, 42, 3603–3614.
Martin A.L., Favier D. and Suéry Y.M. (1997), *Int. J. Plasticity*, 13, 215–235, 237–259.
Mehrabian R. and Flemings M.C. (1972), *Tran. A.F.S.,* 80, 173–182.
Modigell M., Koke J. and Petera J. (1998), in Bhasin (1998), 317–326.
Modigell M and Koke J, (1999), *Mech. Time-Depend. Mater.*, 3, 15–30.
Molenaar J.M.M., Salemans F.W.H.C and Katgerman L. (1985), *J. Mater. Sci.*, 20, 4335–4344.
Molenaar J.M.M., Katgerman L., Kool W.H. and Smeulders R.J. (1986), *J. Mater. Sci.*, 21, 389–394.
Mullis A.M., Battersby S.E. and Fletcher H.L. (1998), in Bhasin (1998), 233–240.

Niroumand B. and Xia K. (1998), in Bhasin (1998), 637–644.
Niroumand B. and Xia K. (1999), *Mater. Sci. Eng.*, A283, 70–75.
Pasternak L., Carnahan R., Decker R. and Kilbert R. (1992), in Brown, (1992), 159–169.
Pluchon C., Loue W.R., Menet P.Y. and Garat M. (1995), *Light Metals 1995*, TMS, 1233–1242.
Potzinger R., Kaufmann H. and Uggowitzer P.J. (2000), in Chiarmetta (2000), 85–90.
Quaak C.J., Katgerman L. and Kool W.H. (1996), in Kirkwood (1996), 35–39.
Qin R.S. and Fan Z. *Mater* (2001), *Sci. Tech.*, 17, 1149–1152.
Roberts K., Fang X., Ji S. and Fan Z. (2002) in Tsutsui (2002), 677–682.
Saha D., Apelian D. and Dasgupta R. (2002), in Tsutsui (2002), 323–328.
Salvo L., Suéry M., de Charentenay Y. and Loue W. (1996), in Kirkwood (1996), 10–15.
Sannes S., Gjestland H., Amberg L. and Solberg J.K. (1994), in Kiuchi (1994), 75–84.
Seconde J.F. and Suéry M. (1984), *J. Mater. Sci.*, 19, 3995–4006.
Smith D.M., Eadyl J.A., Hogan M. and Irwin D.W. (1991), *Metall. Trans.*, 22A, 575–584.
Spencer D.B., Mehrabian R. and Flemings M.C. (1972), *Metall. Trans.*, 3, 1925–1932.
Suéry M., Martin C.L. and Salvo L. (1996), in Kirkwood (1996), 21–29.
Suéry M. and Zavaliangos A. (2000), in Chiarmetta (2000), 129–135.
Suéry M. (2002), in Tsutsui (2002), 31–34.
Sumartha Y., de Figueredo A.M. and Flemings M.C. (1998), in Bhasin (1998), 57–67.
Tausig G. and Xia K. (1998), in Bhasin (1998), 473–480.
Tausig G. (2000), in Chiarmetta (2000), 489–494.
Tsutsui Y., Kiuchi M. and Ichikawa K. (eds), (2002), 'Proceedings of seventh International Conference on Semi-Solid Processing of Alloys and Composites', Tsukuba, Japan.
Turng L.S. and Wang K.K. (1991), *J. Mater. Sci.*, 26, 2173–2183.
Tzimas E. and Zavaliangos A. (2000), *Mater. Sci. Eng.*, A289, 217–227.
Underhill P.R., Grant P.S., Bryant D.J. and Cantor B. (1995), *J. Mater. Syth. Proc.*, 3, 171.
Van Dam J.C. and Mischgosfsky F.H. (1982), *J. Mater. Sci.*, 17, 989–993.
Verrier S., Josserond C., Salvo L., Suéry M., Cloetens P. and Ludwig W., (2000), in Chiarmetta (2000), 423–428.
Vogel A. and Cantor B. (1977a), *J. Cryst. Growth*, 37, 309–316.
Vogel A. (1977b), *D Phil thesis*, Univ. Sussex.
Vogel A., Doherty R.D. and Cantor B. (1979), in *Solidification and casting of metals*, 518–525, London, The Metals Society.
Wabusseg H., Kaufmann H. and Uggowitzer P.J. (2000), in Chiarmetta (2000), 777–782.
Witulski T., Morjan H., Niedick I. and Hirt G. (1998) in Bhasin (1998), 353–360.
Wolfsdorf T.L., Bendeer W.H. and Woohees P.W. (1997), *Acta Mater.*, 45, 2279–2295.
Xia K. and Tausig G. (1998), *Mater. Sci. Eng.*, A246, 1–10.
Yang Y.S. and Tsao C.Y.A. (1994), *Scripta Metall. Mater.*, 1994, 30, 1541–1546.
Young K.P., Kyonka P. and Courtois J.A. (1983), U.S. Patent 4, 415, 374.
Yunhua W., Mingfang Z., Zhiqiang G. and Huaqin S. (1998), in Bhasin (1998), 693–698.
Yurko J.A., Martinez R.A. and Flemings M.C. (2003), in Tsutsui (2002), 659–664.

8
Superplasticity in structural materials

N RIDLEY, University of Manchester, UK

8.1 Introduction

There are two main types of superplastic (SP) behaviour: fine grain or micrograin superplasticity, and internal stress superplasticity. The present chapter is concerned only with fine grain superplasticity. Superplastic materials are polycrystalline solids that have the ability to undergo abnormally large tensile strains prior to failure. In metallic systems for deformation in uniaxial tension, elongations of ≥ 200% would be indicative of superplasticity. However, a number of metallic materials may exhibit elongations of 500–1000% or more, while Higashi[1] has reported an elongation of ~8000% for an aluminium bronze. This behaviour is related to the observation that at elevated temperatures the flow stress, σ, of a SP material is very sensitive to the rate of deformation. The characteristic equation which describes SP behaviour is usually written:

$$\sigma = k\dot{\varepsilon}^{m} \qquad 8.1$$

where k is a constant which incorporates the dependency of temperature and grain size, $\dot{\varepsilon}$ is strain rate and m is the strain rate sensitivity of flow stress.

For significant SP behaviour m would be greater or equal to 0.3 and for the majority of SP materials m is 0.4–0.8. The presence of a neck in a material subject to tensile straining leads to a locally high strain rate and, for a high value of m, to a sharp increase in the flow stress required to sustain plastic flow (i.e. strain rate hardening) within the necked region. Thus, high m confers a high resistance to neck development and leads to the high tensile strains observed for SP materials.

Superplasticity is a characteristic of materials which can be processed to develop a fine grain size, usually 10–15 μm or less, when they are deformed under specific conditions within the strain rate range 10^{-5}–$10^{-1}s^{-1}$ at temperatures >0.4 T_m. The low flow stresses, often <10 MNm^{-2}, combined with the relatively high uniformity of plastic flow has led to increasing commercial interest in the superplastic forming (SPF) of components, often complex in shape, from sheet matcrials using techniques similar to those developed for the gas pressure bulge forming of thermoplastics.

For many years studies of superplasticity were confined to metallic alloys but more recently SP behaviour characterised by high tensile ductility has been reported for intermetallics, ceramics, and both metal and ceramic based composites. However, while many materials may exhibit superplasticity, commercial SPF is currently concerned with a relatively narrow range of alloys based mainly on Al or Ti.

8.2 Mechanical aspects of superplasticity

It can be seen from eqn 8.1 that if the relationship between stress and strain rate is plotted logarithmically then the slope of the plot is *m* where

$$m = \frac{\partial(\log \sigma)}{\partial(\log \dot{\varepsilon})} \qquad 8.2$$

Many SP materials exhibit a sigmoidal variation of flow stress with strain rate, as shown in Fig. 8.1a for AA7475, while in Fig 8.1b it is seen that at 516 °C the *m* value passes through a maximum (~ 0.8) at a strain rate of ~3 $\times 10^{-4}s^{-1}$.[2] A value of $m > 0.3$ delineates the superplastic regime (Region II). Both the high (Region III) and low (Region I) strain rate ranges have *m* values of 0.1–0.3.

In Region III, deformation is generally considered to involve conventional recovery controlled dislocation creep (power law creep). The activation energy for flow is similar to that for lattice diffusion and the strain rate is relatively insensitive to grain size. Deformation in this region leads to the observation of slip lines and to the development of high dislocation densities. Crystallographic texture increases and significant grain elongation occurs during deformation.

The origin of Region I is uncertain. It has been suggested that the decrease in strain rate sensitivity with decreasing strain rate results from a threshold stress, σ_0, or alternatively is due to microstructural instability (grain growth hardening). It has been noted for some materials that at low strain rates *m* can increase and take values close to unity (Region 0), implying diffusion creep.

In Region II, *m* is a maximum leading to high tensile strains. It is widely believed that in Region II strain is accumulated by the motion of individual grains or groups of grains relative to each other by sliding and rotation. Equiaxed grains tend to remain equiaxed or, if not equiaxed, become more so during deformation. Grains have been observed to change their neighbours and to emerge at the free surface from the interior. Texture tends to become less intense, while the reverse is true in Region III.

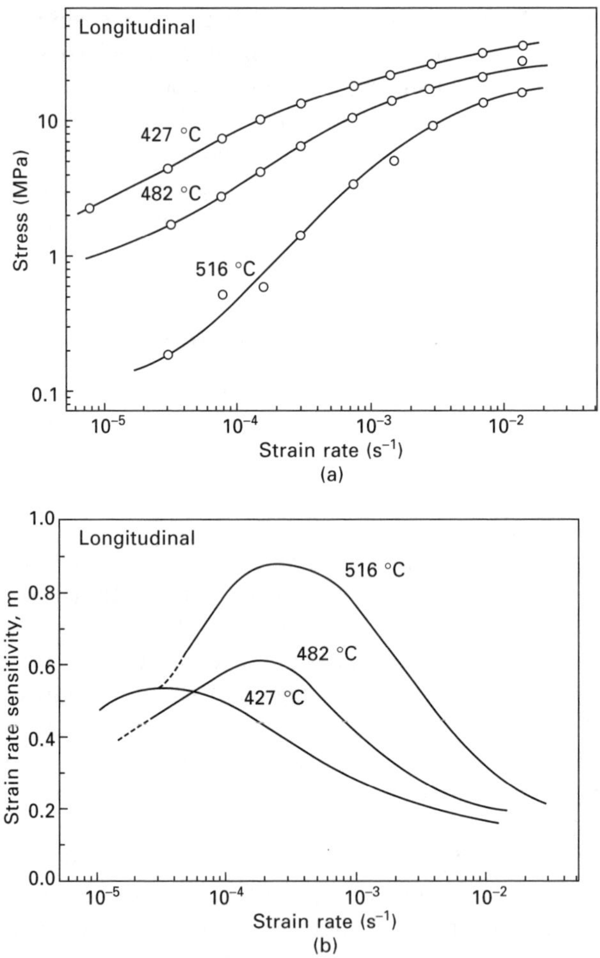

8.1 (a) Stress versus strain rate, (b) m versus strain rate, for superplastic AA7475 (from ref. 2).

Sliding between two grains cannot occur without the necessity for accommodation at the intersection with a third grain. The most probable accommodation mechanisms involve diffusion and/or the glide and climb of dislocations. A number of attempts has been made to develop theoretical models capable of predicting both the mechanical and topological features of SP deformation, although none has been entirely successful.[3–6] These models have been subject to review.[7–11]

Mechanical behaviour is sensitive both to temperature and grain size. In general, increasing temperature decreases the flow stress particularly for Regions I and II (Fig. 8.1a). Maximum *m* has been found to increase with increasing temperature and decreasing grain size, and the strain rate for maximum *m* moves to higher values (Fig. 8.1b). The strain rates at which superplasticity is normally observed lie in the range between $10^{-4}\ s^{-1}$ and $5 \times 10^{-3} s^{-1}$. These rates are considerably less than those usually involved in conventional hot and cold shaping processes ($>10^{-2} s^{-1}$) and, as a consequence, there has been considerable interest in high strain rate superplasticity (HSRS), which will be outlined later (section 8.4.1).

8.3 Characterisation of superplastic materials

8.3.1 Determination of strain rate sensitivity

Optimum conditions of strain rate and temperature are those which lead to maximum *m*, or to maximum tensile strain. In an ideal material where the microstructure remains constant, the true value of flow stress, σ, can be measured by carrying out tensile tests at a range of constant strain rates, and measuring the steady state loads. From eqn 8.1, the slope of a logarithmic plot of true stress versus true strain rate gives *m*. However, it is difficult to obtain a meaningful flow stress for most engineering materials from a constant strain rate test because the flow stress will increase with strain due to the effects of grain growth. It is important when comparisons are made to determine the flow stress at constant structure. For this purpose step strain rate or strain rate jump tests were developed by Backofen *et al.*[12]

To measure the range of strain rates over which a material might exhibit SP behaviour a series of step strain rate tests is carried out at various constant temperatures. At a given temperature a specimen is deformed at a constant strain rate (or constant crosshead velocity) until a steady state load is reached. After a strain rate jump, which may be up or down, a new steady state load is allowed to develop. Several methods have been proposed by Hedworth and Stowell[13] for determining the relevant true stresses at each strain rate, from which the *m* value over the range of strain rates experienced by the test piece can be obtained. Procedures for measuring *m* have been given by Pilling and Ridley.[9]

8.3.2 Tensile testing

This involves the measurement of tensile elongations for a range of constant strain rates. At a given temperature these tests would give information on elongations to failure, the extent of hardening during SP flow, flow stress levels and hence the pressures required for SPF. Metallographic examination of the test pieces would indicate microstructural changes associated with deformation such as grain growth and cavitation (see section 8.7).

8.3.3 Biaxial cone testing

Although information obtained from uniaxial tensile testing can reflect behaviour during biaxial SP forming, attempts to produce a standard test for SP formability have led to the development of the biaxial cone test.[14,15] Using gas pressure, sheet is bulged into a conical die of constant angle (~57°) designed to give a constant average strain rate by maintaining constant pressure conditions after the expanding sheet has made contact with the sides of the die.

The parameters measured include radius of curvature and sheet thickness at the pole, and the height of the cone. Material flow stress, strain and strain rate, obtained from the test provides useful information for the modelling of SPF. The cone test is also used as an acceptance test for SP sheet material in that the cone is required to rise to a predetermined height when subjected to a given pressure-time cycle. Figure 8.2 shows a cone test for AA7475 in which thinning at the cone apex has led to failure.

8.3.4 Constitutive relationships

If superplasticity is regarded as a type of creep behaviour then eqn 8.1 can be rewritten as:

$$\dot{\varepsilon} = k_1 \sigma^n \qquad 8.3$$

where, n, the stress exponent is equal to $1/m$. Equation 8.3 can be restated in an expanded temperature dependent form as:

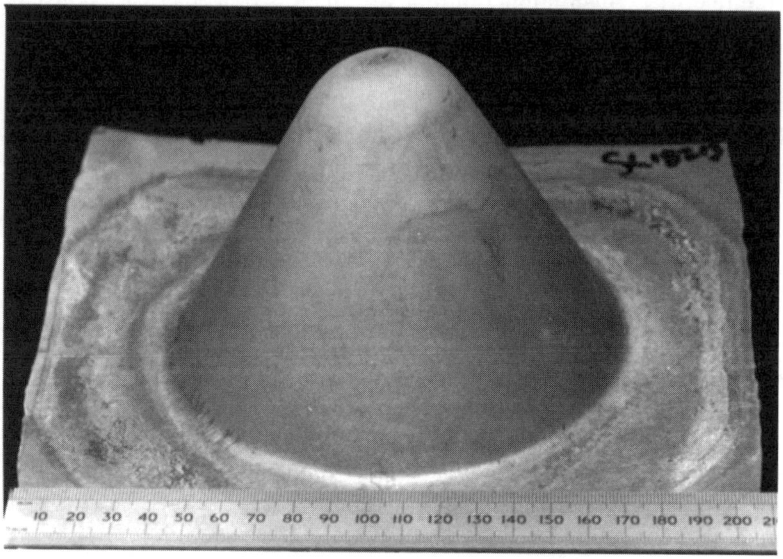

8.2 Cone test on AA7475 sheet showing failure at cone apex.

$$\dot{\varepsilon} = A\frac{DGb}{\mathbf{k}T}\left(\frac{b}{d}\right)^{\mathrm{p}}\left(\frac{\sigma}{G}\right)^{\mathrm{n}} \qquad 8.4$$

This is the constitutive relationship commonly used to describe elevated temperature deformation, in which D is the appropriate temperature dependent diffusion coefficient (= D_0 exp-Q_s/RT) where D_0 is the frequency factor, Q_s is the activation energy and R the gas constant; G is shear modulus, b is Burgers vector, **k** is Boltzmann's constant, T is absolute temperature, d is grain size, p is the exponent of inverse grain size (usually 2 or 3 for Region II). Deformation is driven by the deviatoric (shear) component of the applied stress field characterised by σ, or by $\sigma^* = \sigma - \sigma_0$, where σ^* is the effective stress and, σ_0, a threshold stress. A is a mechanism-dependent constant.

To establish a constitutive relationship for a material, which is an important component of any process model of SPF, it is necessary to determine values for the materials parameters m, p, Q_s and σ_0. Procedures for determining these parameters are available.[9,16]

For constant stress and temperature it follows from eqn 8.4 that:

$$\dot{\varepsilon} \propto \frac{1}{d^{\mathrm{p}}} \qquad 8.5$$

Since the value of p lies between 2 and 3, the marked effect that a reduction in grain size could have on strain rate is apparent, i.e., an order of magnitude reduction in grain size could lead to an increase of 10^2–10^3 in the SP strain rate. Alternatively, for a given strain rate, a reduction in grain size could enable SPF to be carried out at lower temperatures, although this is likely to lead to an increase in flow stress.

8.4 Types of superplastic materials

To exhibit superplasticity a material must be capable of being processed to give a fine grain equiaxed microstructure which will remain stable during deformation. The processing should also lead to a predominance of high angle boundaries (lattice misorientation >15°), in order that grain boundary sliding and grain rotation, characteristic of superplasticity, can occur. There are two main types of SP microstructure: pseudo-single phase and microduplex.

8.4.1 Pseudo-single phase materials

The former group includes Al alloys and these would normally contain <10% volume of second phase. They are processed to develop a distribution of fine precipitates (dispersoids) so that on recrystallisation the alloy will have a fine grain size which will be maintained during SPF. This is due to the pinning effect of dispersoids on high angle grain boundaries (Zener pinning).[17]

Closely spaced small particles may also inhibit recrystallisation and restrict subgrain growth.[18]

While numerous Al alloys have been shown to exhibit superplasticity, some of current and recent commercial interest are listed in Table 8.1, along with ambient temperature tensile properties. These alloys can be subdivided into two groups: those which are recrystallised prior to SPF, and those which develop a SP microstructure during the early stages of hot forming. The former group includes the 7000 series alloys, e.g., AA7475, and also AA5083, while the latter include AA2004 (SUPRAL). The Al-Li alloys such as AA8090 and AA2090 may be processed by either route.[19] Several pseudo-single phase Mg alloys have been thermomechanically processed to develop SP microstructures, in which the grains are stabilised by various fine precipitates,[10,20] although most of the current activities appear to be still on a laboratory scale.[21]

Much effort has been put into the development of new and faster forming SP materials. This has led to the observation of high strain rate superplasticity (HSRS), which is defined in JISH 7007 by the Japanese Standards Association[22] as superplasticity at $\dot{\varepsilon} \geq 10^{-2}s^{-1}$. HSRS is associated with ultrafine grain sizes and is consistent with the predictions of eqn 8.5. Most HSRS materials are Al-based, either alloys or Al alloy composites, many with ceramic whisker or particulate reinforcement. Although interesting, these materials are produced in only small amounts by expensive processing and currently are not used commercially for SPF.

8.4.2 Microduplex materials

Grain growth can be limited by having a microstructure containing approximately equal volume fractions of two or more chemically and structurally different phases. Examples of duplex alloys which can be

Table 8.1 Compositions of SP aluminium alloys, typical room temperature tensile properties and optimum SPF temperatures

Alloy	Composition Wt%	Temper	0.2% PS (MPa)	UTS (MPa)	El%	SPF temperature (°C)
2004	Al-6 Cu-0.4 Zr	0	150	250	10	460
2004	Al-6 Cu-0.4 Zr	T6	315	420	9	460
5083	Al-4.5 Mg-0.7 Mn-0.1 Zr	0	140	290	23	500–520
7475	Al-5.7 Zn-2.3 Mg-1.5 Cu-0.2Cr	T76	500	550	10	515
8090	Al-2.4 Li-1.2 Cu-0.7 Mg-0.12Zr	T6	350	460	5	520–530
2090	Al-2.5 Cu-2.3 Li-0.12Zr	T6	340	450	5	520

thermomechanically processed to develop fine grain SP microstructures include α/β Ti alloys, α/β Cu alloys, ultra-high carbon steels, α/γ stainless steels, nickel-based alloys, and eutectics such as Al-Cu, Al-Ca, Al-Ca-Zn, Pb-Sn and Zn-Al eutectoid.[9]

Of these materials α/β Ti alloys are of appreciable commercial significance for SPF, with most attention being concentrated on Ti-6Al-4V(Ti-6/4). Other Ti alloys are attempting to make inroads into the dominant position occupied by SP Ti-6/4, but this is difficult given the vast experience that design engineers have of this material. There is also some commercial activity in the SPF of α/γ stainless steels, nickel-based alloys (e.g. IN 718) and Zn-Al eutectoid.

8.4.3 Superplastic ceramics

Fine grain ceramic materials such as the pseudo-single phase, dispersion strengthened and toughened, yttria stabilised zirconia (Y-TZP) have been processed to develop superplasticity at elevated temperatures. Elongations of up to several hundred per cent can be obtained, compared with the very low plastic strains normally associated with these materials at room temperature.[23,24]

8.5 Processing and microstructure

Procedures used to develop SP microstructures in Al alloys by static recrystallisation or by strain enhanced dynamic recrystallisation will be outlined, using AA7475 and AA2004, respectively, to illustrate the principles involved. Specific reference will also be made to the increasingly dominant AA5083. For the development of microduplex SP Ti alloys the processing of α/β Ti-6Al-4V (wt%) will be considered. Some reference will also be made to other SP materials which, although not of current commercial significance, have interesting properties.

8.5.1 Aluminium alloys

Static recrystallisation

AA7475. Al-Zn-Mg-Cu-Cr (Table 8.1)

This alloy is a high-strength heat treatable material which finds structural applications mainly in aerospace. The processing route involves the production of fine grain microstructures by rapid heating of heavily warm or cold worked material containing a bimodal distribution of precipitates. These consist of overaged M-phase and T-phase particles (1–2 μm) and Cr-rich dispersoids, which are typically 0.1–0.2 μm in diameter. Rapid heating to 480 °C results in a large number of recrystallisation nuclei in the locally deformed regions adjacent to the coarse precipitates, i.e., particle stimulated nucleation, and thus to the development of a small grain size.[2,25]

At 480 °C the large precipitates dissolve, but the small Cr-rich precipitates inhibit grain growth following recrystallisation, and during subsequent SPF. The well documented Rockwell processing route is shown schematically in Fig. 8.3. The material would normally have a grain size of 10–15 μm and be capable of sustaining large tensile strains, e.g., 500–1000% elongation under optimum deformation conditions (~515 °C; $2 \times 10^{-4} s^{-1}$) at a flow stress <10 MPa (see Fig. 8.1). The slow strain rate can lead to relatively long forming times (~2 hours).

AA5083 Al-Mg-Mn (Table 8.1)

This alloy, which is essentially non-heat treatable, has medium strength, excellent cold forming and welding behaviour, is relatively inexpensive and can be processed to give a moderate degree of superplasticity. This attractive combination of characteristics has led to AA5083 being the dominant Al alloy currently used in the production of parts by SPF, where high strength is not essential.

After casting and processing to sheet the alloy contains two types of particles: constitutive particles in the size range 1–5 μm which typically contain Al, Mn, Si and Fe, and Al_6Mn particles in the size range 0.2–0.8 μm. After heavy cold work (70–80%), the larger particles act as nuclei for recrystallisation, while the finer Al_6Mn particles (or other dispersoids) stabilise the microstructure. The cold worked material recrystallises rapidly at temperatures in the range 350–550 °C to give grain sizes of 10–15 μm

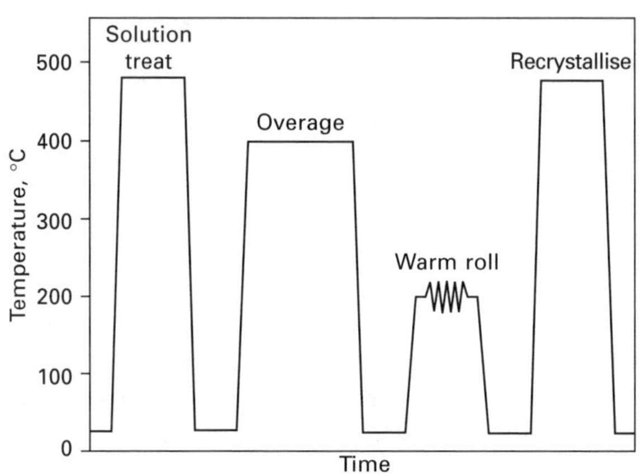

8.3 Schematic of processing route to a fine grain microstructure in AA7475.

(Fig. 8.4a). Tensile elongations of 300–350% can be obtained at strain rates of 10^{-3} s^{-1} at about 500 °C, with m~0.5.[26]

It has been noted that AA5083 shows significant strain hardening during SP flow which is not consistent with grain growth hardening. There is strong evidence that for this material strain is accumulated by solute-drag dislocation creep which occurs independently of grain boundary sliding.[27] This leads to significant grain elongation in the straining direction.

Dynamic recrystallisation

SUPRAL: Al-6Cu-0.4Zr wt%

The processing route has been described by Watts *et al.*[28,29] The alloy is rapidly solidified from a high superheat (~780 °C) to avoid the formation of coarse $ZrAl_3$ precipitates and to retain Zr in solution. Fine $ZrAl_3$ particles (~50 nm) are precipitated on ageing at 360 °C, have a volume fraction of ~0.003, and are homogeneously distributed. The alloy is solution treated at ~500 °C and hot rolled to break down the as-cast structure. The material is subsequently warm/cold worked to ≥ 80% reduction, when recovery and recrystallisation of the heavily dislocated structure is prevented by the pinning action of the $ZrAl_3$ precipitates.

The material can be formed at 460 °C at a strain rate ≥ 10^{-3} s^{-1} and is capable of substantial strains to failure of ~1000%. During forming, continuous dynamic recrystallisation leads to the evolution of a fine SP microstructure of grain size ~5 μm. (Fig. 8.4b). The mechanism of microstructural evolution has been the subject of investigation and discussion.[30,31]

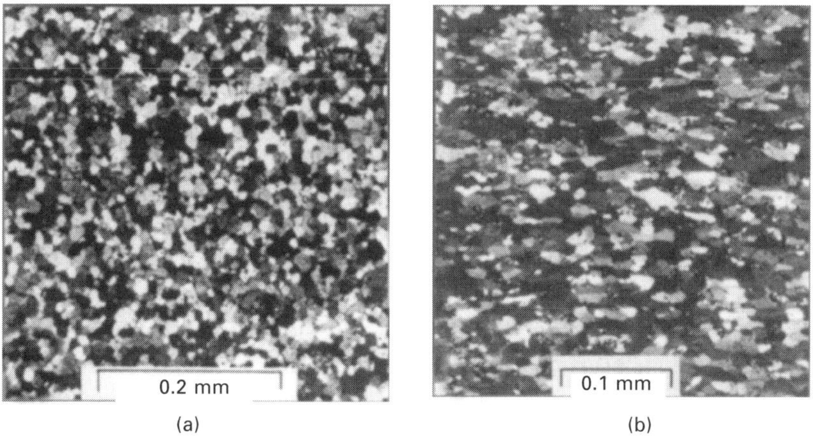

8.4 Fine grain microstuctures developed in (a) AA5083 by static recrystallisation at 520 °C and (b) AA2004 by dynamic recrystallisation; 450 °C, ε = 1.39 (300%).

Al-Li-based alloys

Because of the beneficial effects of Li additions on density and stiffness, considerable effort has been devoted to the design of a range of Al-Li alloys, several of which were processed to develop excellent SP formability. For AA8090 sheet material processed by the SUPRAL route, tensile elongations >1000% were reported for temperatures of 520–530 °C and a strain rate of ~$5 \times 10^{-4}s^{-1}$.[9] Despite optimistic forecasts, the low take-up of Al-Li alloys by aerospace companies has resulted in these materials currently finding few commercial applications, although the Russian alloy coded 1420, based on the wt% composition Al-5Mg-2Li-0.12Zr, is still the subject of much attention.[32] Recently Airbus announced that in order to save weight the company will use Al-Li alloys in the freighter version of the Airbus 380, which is planned to have its maiden flight in the first half of 2007, and subsequently in the Airbus 380 passenger aircraft.[33]

Al-based HSRS materials

HSRS was initially reported by Nieh *et al.*[34] in the mid-1980s and has been intensively studied since the early 1990s. The subject has been reviewed by Mabuchi and Higashi.[35] A combination of thermomechanical processing and powder metallurgy has been used to produce alloys with grain sizes of < 3 μm. In addition the techniques of physical vapour deposition, mechanical alloying, or consolidation of amorphous and nano-crystalline powders result in increasingly fine sub-micron grain sizes, capable of giving SP strains to failure at increasingly high strain rates up to 10^3s^{-1}. Data from Higashi[36] for Al alloys, are shown in Fig. 8.5.

Nano-phase microstructures produced in IN9021: Al-4.0Cu-1.5Mg-1.1C-0.8O_2 wt%, by mechanical alloying are stable at elevated temperatures because of the presence of ~ 5% volume of very fine oxides and carbides. For this alloy and many other HSRS materials, the optimum temperature for maximum SP elongation is above a temperature at which partial melting occurs. It is believed that liquid phase helps with the accommodation of grain boundary sliding, but a liquid phase is not always necessary for a material to exhibit HSRS.[35]

Very small grain sizes (≤1 μm) have been produced by equal-channel angular extrusion (ECAE).[32] A die is used which contains two channels of equal cross-section that intersect near the centre of a die. A bar of material which fits the channels is pushed through the die and undergoes straining by shear. By control of bar rotation, temperature and the number of passes, considerable redundant strains can be developed leading to fine equi-axed recrystallised microstructures. The fine microstructures lack thermal stability at normal SPF temperatures but can be deformed at appreciably lower

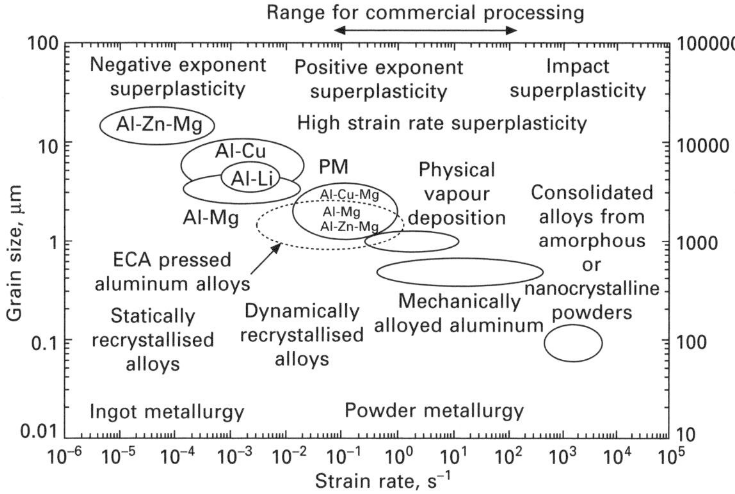

8.5 Variation of optimum strain rate with grain size for a range of Al alloys processed by different routes (from ref 36, courtesy of K. Higashi).

temperatures (i.e. low temperature superplasticity) and at high strain rates to give high tensile elongations. Figure 8.6 shows stress-strain curves for a Russian Al-Mg-Li-Zr alloy at 350 °C. The test at 10^{-2} s^{-1} was discontinued at 1180% elongation.[32]

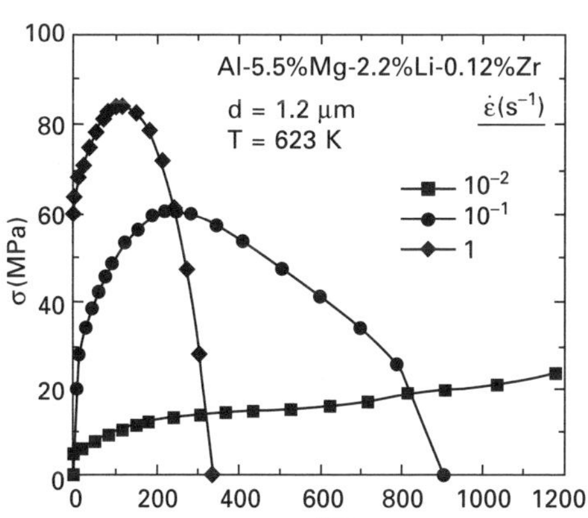

8.6 True stress versus true strain for tensile tests at 523 K (350 °C) on Al-Mg-Li-Zr alloy processed by ECAE (from ref. 32, courtesy of R.Z. Valiev).

α/β titanium alloys

The compositions of important duplex Ti alloys that have been processed to give SP behaviour are listed in Table 8.2. The table also includes the maximum temperatures at which these alloys are likely to be used. The as-cast structure is broken down by forging at a temperature of 100–150 °C above the β-transus (~1000 °C). Secondary processing involving forging/extrusion/plate or sheet rolling, is carried out below the β-transus to obtain a microstructure of fine equiaxed α-phase (grain size ~5/10 μm) in a matrix of transformed β-phase. Since the mill product contains residual cold work, it recrystallises on heating to the SPF temperature. The kinetics and mechanisms that control the development of the fine microstructure have been discussed recently by Semiatin *et al.*[37]

Superplastic behaviour in Ti-6/4 is dependent on grain size, its distribution and aspect ratio, the relative proportions of the α and β phases, and on the texture and inhomogeneity of the material.[38] The α and β phases have very different deformation characteristics. The β-phase has a high self-diffusivity and the rate of grain growth in alloys with a large volume fraction of β-phase is high. Hence, a substantial volume fraction of α-phase is required to minimise grain growth.

For Ti-6/4, SP flow can occur over a range of temperatures in the α + β field from 860–930 °C, but optimum behaviour is observed for a microstructure containing 40–50% β-phasc[38,39] (Fig. 8.7). Thc optimum strain ratc is ~ $4 \times 10^{-4} s^{-1}$, with m in the range 0.5–0.8 and the flow stress < 10 MPa, leading to tensile elongations ~1000%. The duplex microstructure of partially diffusion bonded sheets of Ti-6/4 showing interfacial voids is seen in Fig. 8.8. SPF is carried out in an inert gas environment, as oxygen and nitrogen both react

Table 8.2 SPF temperatures and maximum service temperatures for some α/β titanium alloys

Composition (wt%)	SPF temperature (°C)	Service temperature (°C)
Ti-4.5Al-3V-2Fe-2Mo (SP700)	750–800	~250
Ti-6Al-4V	880–920	300
Ti-4Al-4Mo-2Sn-0.5Si (IMI 550)	875–900	400
Ti-6Al-2Sn-4Zr-2Mo (Ti-6242)	850–940	500
Ti-5.8Al-4Sn-3.5Zr-0.7Nb-0.5Mo-0.3Si-0.05C (IMI 834)	950–990	600
Ti-14Al-19Nb-3V-2Mo (Super α_2)	960–990	650

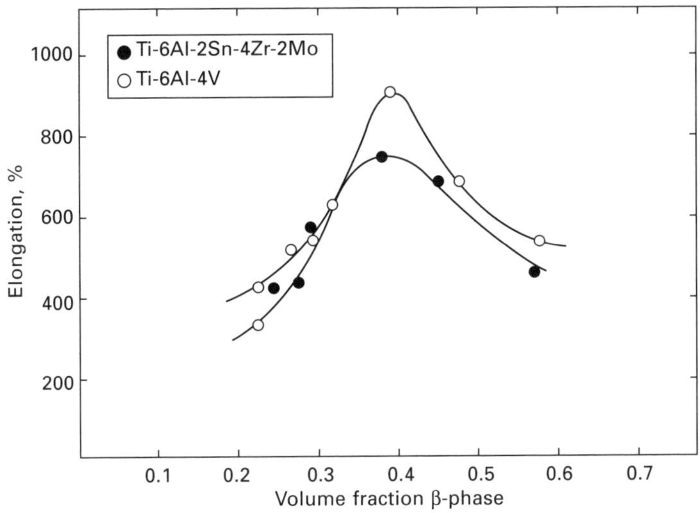

8.7 Effect of volume fraction of β-phase on the elongation to failure of two SP α/β Ti alloys for optimum deformation conditions.[36]

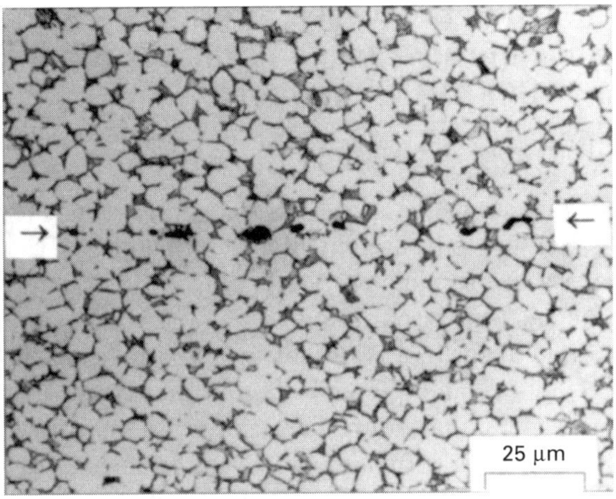

8.8 Duplex microstructure of partially diffusion bonded sheets of Ti-6/4 showing interfacial voids. 910 °C, 2.1 MPa (300 psi), 15 minutes.

with titanium leading to a thin brittle surface layer of α-phase (termed α-case) which has to be chemically removed.

Efforts have been made to reduce the SPF temperatures for α/β Ti alloys as this would save energy costs, lower grain growth which adversely affects mechanical properties, and reduce the wear and damage which SPF dies experience. It is possible to lower the forming temperature by reducing the β-phase transus temperature with small additions of elements such as Fe, Co, Ni and hydrogen.[40]

The Ti alloy SP700 contains ~4.5%Al with small additions of the β-stabilising elements V, Mo and Fe (Table 8.2). The alloy can be processed to give a fine duplex microstructure with typically a 3 μm α-phase grain size. It shows considerable SP formability at temperatures below 800 °C with tensile elongations up to ~2500% at 750 °C at a strain rate of $10^{-3}s^{-1}$, with an excellent range of ambient temperature properties.[41] However, the commercial use of a new alloy or an alloy of modified composition may be a slow process as it must go through a lengthy qualification procedure before acceptance, particularly for aerospace applications.

Swale[42] has pointed out that producers of Ti-6/4 sheet are able, by careful processing, to reduce the mean grain size to ~1 μm, such that the alloy could be deformed at ~760 °C. The availability of this material would reduce energy costs, problems of die wear/damage, grain growth, and eliminate α-case formation.

8.6 Superplastic forming (SPF) and quick plastic forming (QPF) processes, products and mechanical properties, and diffusion bonding (DB) and SPF/DB technology

8.6.1 Sheet-forming techniques

Superplastic forming is essentially a sheet-forming process and the low flow stresses enable gas pressure to be used to drive deformation. Advantages are that SPF is a near net-shape forming process, multiple parts can be produced in one operation, there is little or no spring back, large plastic strains can be obtained, and only one major tool is usually required rather than a pair of accurately matched tools, or multiple tools. For Al alloys the die sets can be relatively inexpensive, e.g., cast iron, because of the moderate forming temperatures required. SPF is currently applied to the shaping of Al alloys such as AA5083, AA7475 and AA2004 (SUPRAL) at temperatures in the range 430–520 °C.

For α/β Ti alloys, particularly Ti-6/4, SPF temperatures range from 860 to 920 °C, with dies often made from cast heat-resisting austenitic steels. However, in an attempt to reduce the high cost and long manufacturing time required to produce these dies, there has been a move to the use of ceramic dies set into a creep-resistant backing material which supports most of the forming stresses, as described by Sanders.[43]

Simple female forming

To form the sheet, the periphery of a heated blank is rigidly clamped to provide a gas-tight seal. The dies are located between heated platens in a SPF press which supplies a force to keep the dies closed during forming. On application of gas pressure to cause the material to stretch into the die cavity (Fig. 8.9a), the constraint provided by the clamped edge results in a stress system that varies across the sheet. The differential stress system leads to a variation in thickness strain, with maximum thinning occurring at the pole. Although a thickness variation would occur if $m = 1$, the differential thinning is smallest for alloys having the highest m value, and grows with increasing dome height.[44]

Once the pole of the bulged sheet contacts the die surface it is locked against the tool by friction and forming pressure, and this inhibits further thinning in this region. Continuing deformation leads to progressively more of the unsupported regions making contact with the die. Since the corners of the die are the last to fill, the greatest strain occurs in these regions. During forming the rate of application of pressure is such that the strain rates induced in the sheet are maintained in the SP range.

Drape forming

A heated and clamped sheet is stretched into a female die in which one or more male tools are located (Fig. 8.9b).[45] The polar region of the bulging sheet will make early contact with the male tool and the continued application of pressure will drape the sheet over the male tool as it bulges into the female annular regions. The choice of either female forming or drape forming could be influenced by whether the internal or external dimensions of the part are

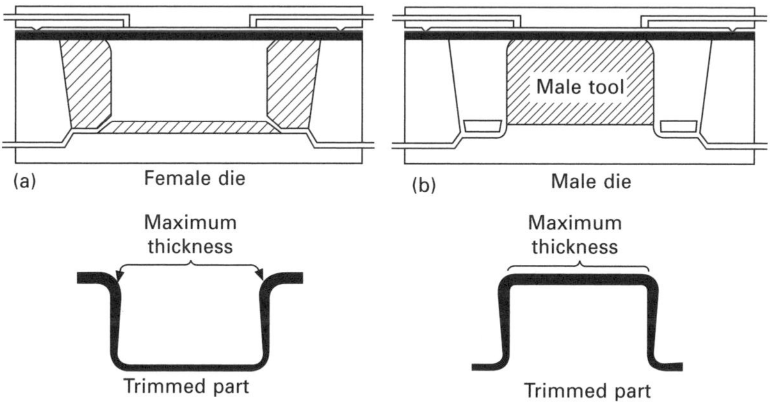

8.9 Diagrammatic illustration of (a) female forming (b) drape forming (from ref. 45, courtesy of JM Story).

the most critical. As seen in Fig. 8.9, if the outside shape is specified as the critical dimension, then female forming will be used. When the inside shape is critical, then drape forming will be used. If a number of male tools are placed within the female forming tool several similar parts, or different parts of a given component, can be produced at the same time.

Back-pressure forming

In back-pressure forming both sides of the sheet are pressurised. This produces a hydrostatic pressure capable of suppressing cavitation. Gas control creates a positive pressure differential enabling forming to be carried out. Back pressure can be applied to both female and drape forming when structural applications demand that cavitation be held below critical levels to avoid degradation of service properties (section 8.7).

Male forming (bubble forming)

The combined use of gas pressure and tool movement enables deeper parts of more uniform thickness to be made. Male forming is applied mainly to 2004 Al alloys. The sheet is first blown into a bubble away from the tool. The tool is then moved into the bubble and the pressure is reversed forcing the bubble to collapse onto the tool.

8.6.2 Forming equipment

The two main types of sheet-forming equipment include the presses developed by Superform AluminiumUK/ Superform USA, which are currently used to form alloys 5083 and 2004, and the platen type presses used both for titanium alloys and Al alloys. Aspects of the configuration and operation of SPF presses have been discussed by Laycock,[46] while the design and manufacture of hydraulic presses for SPF has been described by Whittingham.[47]

8.6.3 Procedures for reducing forming cycle time

The need to reduce vehicle weight in order to improve fuel economy has caused the automobile industry to look at Al alloys as a replacement material for vehicle body construction. Traditional SPF processes have been used for many years to produce automotive panels for specialist models, but these are cost effective only for the limited quantities involved, i.e., up to ~ 5000 parts per annum.[48] A new manufacturing strategy has to be developed if SPF is to be used for high-volume production. The developments necessary for SPF to remain cost effective have been discussed by Friedman and Luckey,[49] and relate to material costs and reduction in forming cycle time.

Dunwoody[50] has described a novel press construction for the SPF of automotive body parts from AA5083. The press consists of two levels with electrically heated moving platens measuring 3 m × 2.2 m on each side at each level. During operation, as the moving platens with formed parts move out of the press so the platens with pre-heated tools and sheets move in. Up to three forming tools may be placed on each platen, allowing six parts to be formed in the same cycle. This double-decked press is used by Form Tech AG, Switzerland, to produce inner body panels for the Maybach cars made by Daimler ChryslerAG.

The principles of quick plastic forming (QPF), which is essentially a pseudo-SPF process developed by General Motors Corporation, have been outlined by Schroth.[51] QPF is based on eight years of research during which every aspect of conventional SPF was examined, and modified where appropriate to maximise the rate of production of dimensionally accurate panels to meet the requirements of an automobile mass production line. It uses gas pressure to hot form AA5083 parts, the large rear door (decklid) inner panel for the Chevrolet Malibu Maxx (~100,000 vehicles per annum) being the first product. Many aspects of QPF are dealt with in a series of papers included in the conference volume given in reference 51.

QPF is carried out in an automated purpose-built 'forming cell'. To reduce forming time and strain in the cell, the blank sheet is preheated and pre-formed outside the cell, transferred by robot to the forming press which contains a heated tool, formed, and extracted by spring-actuated pads to a position above the tool face. It is then transferred by robot to a cooling fixture. The temperature used, ~450 °C, is less than that in conventional SPF and the forming pressure is higher (~3 MPa). QPF is capable of producing panels continuously with a forming cycle of about six minutes per panel.

8.6.4 Simulation and control of SPF

SPF of Al sheet has usually been based on a trial and error approach, combined with simple calculations and considerable practical experience. However, if maximum benefit is to be gained from SPF then some form of numerical simulation of the process is desirable. Numerical analyses have been developed but the finite element (FE) method has emerged as the most powerful technique. Wood and Bonet have reviewed the numerical analysis of SPF.[52]

The use of FE analysis to simulate SPF is a relatively complex topic but to be successful requires accurate constitutive equations. The equations can be introduced into a FE simulation package in order to determine the variation of the shape of the part with time, the stress and the strain rate, the evolution of grain size, the final thickness distribution and the pressure-time forming cycle.

8.6.5 Some applications of parts produced by SPF

Aluminium alloys

The SP aluminium alloys 7475, 2004 clad and unclad, and 5083 are used to produce a wide range of parts for civil and military aircraft (Fig. 8.10), with parts also being used on rockets, the Cassini space probe and the space station. For primary structures, AA7475 is used. Parts on civil and/or military aircraft include engine and baggage bay doors, avionics enclosures, air intake skins, ejector seat panels, engine cowlings, hydraulic and mechanical housings, wing and tail leading-edge panels. Interior parts include ceiling lighting panels, cabin surfaces and floor panels.

The rail market is important in Europe, where large numbers of parts produced from AA5083 find applications on commuter trains (Fig. 8.11). External parts include end panels fitted to London Underground trains, while internal fittings include window surrounds, ventilation, roof and door panels. More than 2,500 seats comprising one-piece shells produced by SPF and covered with fire resistant fabrics have been fitted in the Heathrow Express to Central London.

In architecture and building products made from AA5083 are used for external and internal cladding. For external cladding, ribbed panels can enhance

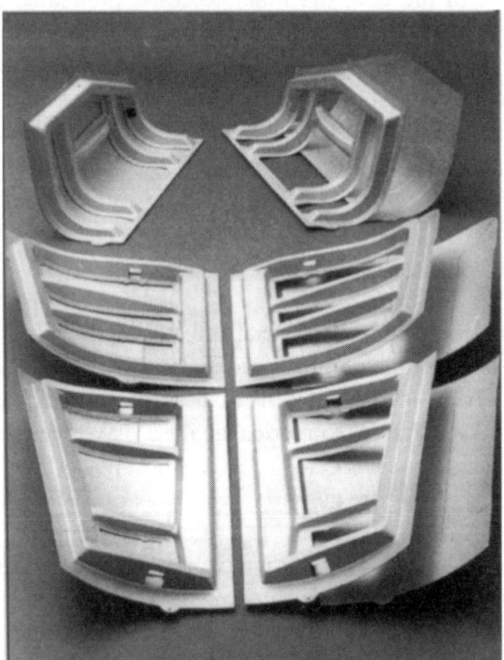

8.10 Access door components for British Aerospace Hawk aircraft; AA7475 (courtesy of Superform Aluminium).

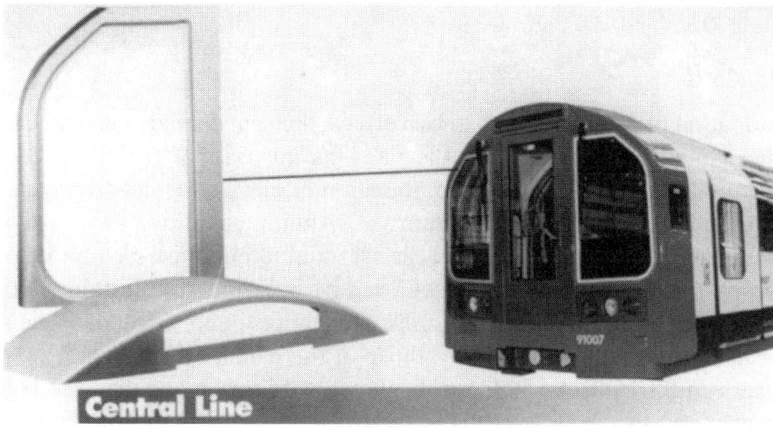

8.11 End panels for London Underground trains; AA5083 (courtesy of Superform Aluminium).

both structural and aesthetic performance, and also act as a rain screen. Clad buildings include the Financial Times print works in London, Gatwick Airport N. terminal, and the roof of the Charlety Stadium, Paris, which makes use of 14,000 panels.

SPF is used to form numerous parts from AA5083 for specialist and prototype automobiles, with QPF now being used to produce a component for a mass production vehicle. About 90% of the exterior surface of the new 2005 Ford GT40, which is a powerful sports car, is made from SPF, as are the entire bodies of the Roadster and Esperante models produced for the American sports car maker Panoz. In the UK much of the bodywork of the Morgan sports car is produced by SPF, while a number of parts are made for Aston Martin, Rolls-Royce and Bentley cars.

Titanium alloys

A large number of components are produced mainly from Ti-6/4 for civil and military aerospace applications. These include fan blades for civil turbofan engines (see section 8.8), large nacelle structures, engine intake lip skins, leading edges of helicopter blades, wing access panels (2-sheet DB/SPF), and escape hatches (4-sheet DB/SPF). In some cases complex components previously fabricated from a large number of parts are produced by SPF, or DB/SPF, leading to considerable savings in manufacturing and materials costs, as in Fig. 8.10.

8.6.6 Post-SPF properties

Aluminium alloys

SP deformation of Al alloys leads to two effects that can degrade mechanical properties. Firstly, and potentially the most serious, is that cavitation can occur during SPF, and secondly, a combination of elevated temperature and SP strain can lead to an increase in grain size. Aluminium alloy 7475 is used for primary aerospace structures and it is essential to minimise cavitation to avoid property degradation. This is achieved by back pressure forming and provided a SP strain equivalent to 150% elongation is not exceeded, then post-formed properties are similar to those of the parent metal sheet.

Neither AA2004 nor AA5083 are back pressure formed as they are not used for primary structures. AA2004 is highly superplastic but forming strains are usually limited to an equivalent SP elongation of ~250% to avoid significant loss of properties. For AA5083 the forming strain is limited to ~100% equivalent SP elongation because the material does not have very high reserves of superplasticity. Cavitation, and hence loss of properties, is not appreciable at small strains, but as SP strain approaches 200% falls in properties become more significant (Fig. 8.12).

Titanium alloys

SPF of Ti alloys leads to some loss of tensile properties. The reasons for this are:

- the as-received sheet materials are usually in a 'mill annealed' form in

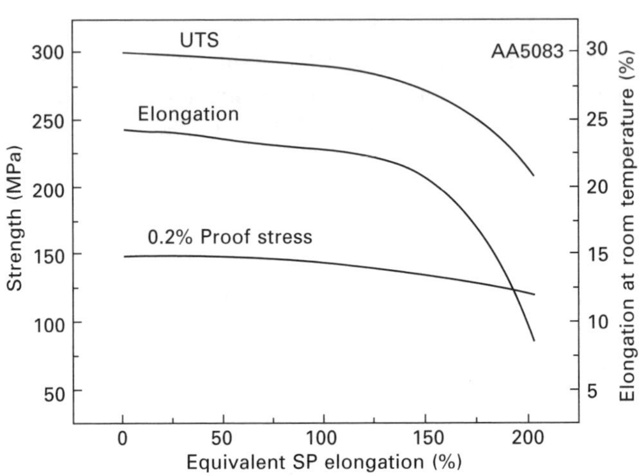

8.12 Post SPF tensile properties of AA5083 at room temperature (courtesy Superform Aluminium).

which they contain residual cold work which is removed on heating to the SPF temperature
- the low optimum strain rate (~4 × $10^{-4}s^{-1}$) leads to long forming times, and grain growth
- rolling texture may be removed or modified by SPF.

For Ti-6/4, thermal cycling reduces the proof stress (PS) by 8–12%, while a SP strain of ~0.9 (~150% elongation) reduces it a further 2%. In some alloys it is possible to restore tensile properties by a post-SPF heat treatment. The effect of SPF on mechanical properties has been discussed by Partridge *et al.*[53] and Inglebrecht.[54]

8.6.7 Diffusion bonding

Diffusion bonding is a joining process which involves minimal macroscopic deformation of the parts being bonded. Joining may occur entirely within the solid state or may involve isothermal melting and re-solidification of a thin interfacial region of transient liquid phase (see section 8.8). Bonding is carried out at temperatures which are usually close to those for optimum SP flow. The object of the process is to develop a microstructure in the bond region similar to that of regions remote from the join with parent metal properties. Figure 8.8 shows interfacial voids in the bond line of partially bonded sheets of Ti-6/4.

The bonding process brings together two cleaned surfaces. Mechanisms involved in joining of the two surfaces involve: (i) instantaneous plastic collapse of contacting asperities on the application of pressure to give a planar array of interfacial voids; removal of these by (ii) time-dependent plastic collapse involving creep and/or superplasticity, and (iii) diffusion processes. Models have been developed which predict the time to give 100% contact between bonding surfaces. Some models assume that bonding occurs under plane strain conditions,[55–57] although the model of Pilling[58] assumes isostatic conditions. The models for the most part show reasonable agreement with the limited experimental data available.

8.6.8 DB/SPF technology

DB/SPF technology is readily applicable to Ti alloys such as Ti-6/4 since the Ti lattice is capable of taking into solution the surface oxide and other contaminants which would prevent the formation of a sound bond. A widely used method for bonding involves the static compression of contacting surfaces by using gas pressure as in a platen press. Figure 8.13 shows a section through a four-sheet structure. The core configuration is determined by the stop-off pattern (regions where bonding is inhibited) on the surfaces of the initial sheet materials. Bond quality is assessed using metallographic examination, lap shear and other mechanical tests.

Diffusion bonding of Al alloys is difficult because of the presence of stable and tenacious oxide layers. However, procedures which remove or disrupt the oxide layers have been developed. These have been summarised by Huang *et al.*[59] Most methods are capable of producing high-quality bonds on a laboratory scale, although several would be difficult to scale up to produce production parts. However, even if good bonds are produced, their strengths have not yet proved to be high enough to withstand the peeling stresses induced during SPF to give an expanded cellular structure as seen in Fig. 8.13 for Ti6/4.

8.7 Problems associated with SP and SPF

8.7.1 Miscellaneous problems

While SP alloys have numerous attractions for use in tensile sheet forming, there are also some problems which could detract from their more extensive use. For Al alloys these problems relate to the high costs of superplastic sheet, the relatively slow forming rates, and cavitation which occurs during SP flow. For Ti alloys, problems are associated with the high cost of SP sheet and the high temperatures at which SPF is carried out. Some of these problems have been addressed in section 8.6. The present section will consider cavitation.

8.7.2 Failure and cavitation of superplastic alloys

Failure

When a superplastic material fails during tensile flow it is either the result of unstable plastic flow or a consequence of the growth and coalescence of cavities. Unstable plastic flow normally results in the material pulling down to a fine point at failure, while cavitation leads to a pseudo-brittle failure,

8.13 Section through 4-sheet cellular structure of Ti-6/4 produced by DB/SPF.

due to the tearing of ligaments between cavities (Fig. 8.14). The presence of cavities in a commercially formed part may also have an adverse effect on its service behaviour.

Cavitation

Since many materials cavitate during SP flow, the literature on the subject is substantial and has been reviewed.[60,61] Metallographic observations on Al alloys show that cavities are most likely to develop at grain boundary particles.[62] A cavity located on a grain boundary, whether nucleated or pre-existing, may grow by diffusion and/or plastic deformation of the surrounding matrix. Many studies show that cavity growth during SP deformation is dominated by matrix plastic flow. For plastic strain control, the rate of cavity growth increases linearly with cavity size and is independent of strain rate within the SP region.[63] This leads to the relationship:

$$C_v = C_0 \exp(\eta\varepsilon) \qquad 8.6$$

where C_v is the volume fraction of cavities at strain ε, and C_0 is a constant; η is the cavity growth rate parameter and is given by the expressions

$$\eta = \frac{3}{2}\left(\frac{m+1}{m}\right)\sinh\left[2\left(\frac{2-m}{2+m}\right)\left(\frac{K}{3} - \frac{P}{\sigma_e}\right)\right] \qquad 8.7$$

and

$$\left(\frac{K}{3} - \frac{P}{\sigma_e}\right) = \frac{\sigma_m}{\sigma_e} \qquad 8.8$$

K is a constant whose value depends on the deformation geometry and the extent of grain boundary sliding, P is the imposed pressure, σ_m the mean stress and σ_e is equal to the uniaxial flow stress. If 50% of SP strain is

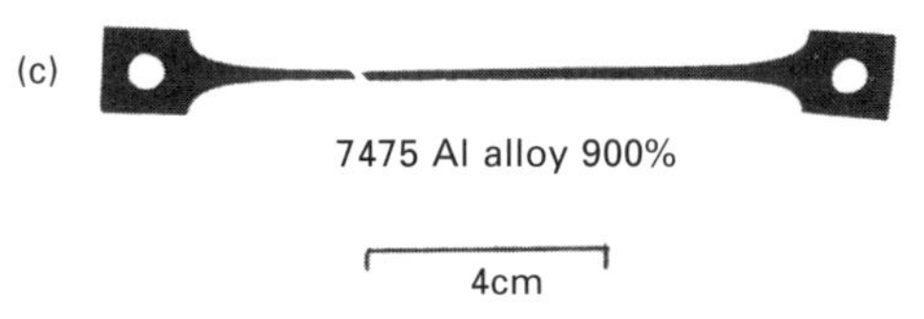

8.14 Shadowgraphs of failed SP alloys showing unstable plastic flow in (a) Ti-6/4 and (b) Ti-14Al-19Nb-3V-2Mo wt% (super alpha-2) and (c) pseudo-brittle fracture in AA2004.

attributed to grain boundary sliding then $K = 1.5$ and 2.25 for uniaxial and biaxial straining, respectively, and 2.7 for plane strain deformation.[64]

It can be seen from eqns 8.6 and 8.7 that the application of an imposed pressure, P, will reduce η, the cavity growth rate parameter, and hence the level of cavitation for a given strain. For the same strain, cavitation will be higher for biaxial and plane strain deformation than for uniaxial flow. Figure 8.15 shows the effect of imposed pressure (or back or hydrostatic pressure) on cavitation of AA7475 deformed in equi-biaxial tension to different strains,[65] and Fig. 8.16 illustrates the effect of pressure on microstructure.

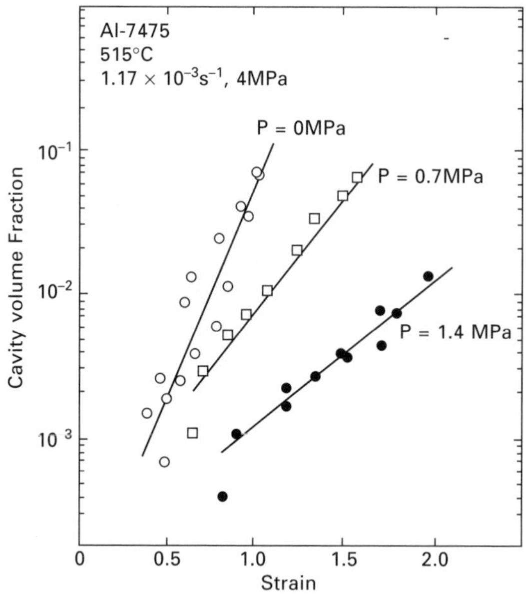

8.15 Effect of imposed pressure on cavitation in Al alloy 7475. Biaxial deformation.

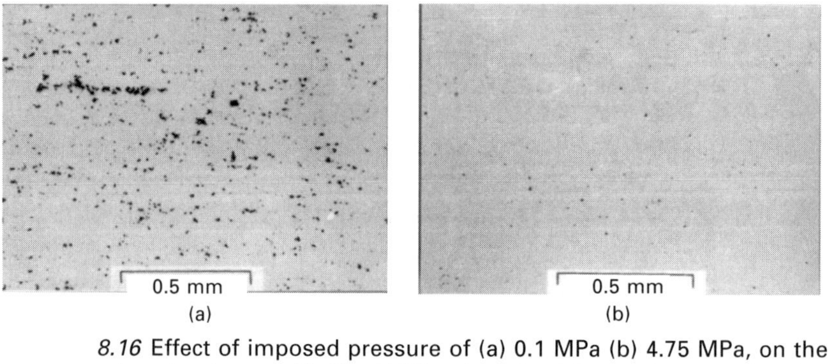

8.16 Effect of imposed pressure of (a) 0.1 MPa (b) 4.75 MPa, on the microstructure of Al alloy 7475 uniaxially deformed to a strain of 1.7 (450%).

Control of cavitation

To minimise cavitation during SP forming, there must be rigorous control of processing to ensure the development of a material with a fine stable grain size comprised predominantly of high angle boundaries. Coarse precipitates and impurity particles should be minimised and dispersoids should be uniformly dispersed.[66] Cavitation can be controlled by various procedures including annealing and/or the application of hydrostatic pressure, either prior to, during, or after SPF.[67] However, the most effective and practical way of controlling cavitation is to superimpose a hydrostatic pressure during SPF. Work on Al alloys deformed in uniaxial, equibiaxial or plane strain conditions, has shown that increasing imposed pressure reduces the rate at which the overall volume of cavities increase with strain, and decreases the level of cavitation for a given strain (Fig. 8.15).

It can be seen from equations 8.6 to 8.8 that to prevent cavity growth during SPF it is necessary for $P \geq K\sigma_e/3$, i.e., for $P \geq 0.5\sigma_e$ for uniaxial tensile deformation (although $P > \sigma_e/3$ is also quoted in the literature), $P \geq 0.75\sigma_e$ for equibiaxial bulging, and $P \geq 0.9\sigma_e$ for plane strain. These predictions are broadly in agreement with observation. It can be seen from equation 8.7 that the ratio P/σ_e is important in determining the level of cavitation. In commercial forming practice, P is unlikely to exceed ~ 4 MPa for technical reasons so the flow stress, σ_e, (dependent on grain size and strain) should not be too high if cavitation is to be prevented. However, even if the criterion $P > K\sigma_e/3$ is not met, any level of pressure will be beneficial in its effect on cavitation (see Fig. 8.15).

Titanium alloys such as Ti-6/4 are remarkably resistant to cavitation. The Ti alloys take impurities such as oxides, carbides and nitrides, into solid solution so they are essentially free from hard particles. At their usual deformation temperatures, ~900 °C, the alloys usually contain about 40% by volume of β-phase which is able to fully accommodate grain boundary sliding.

8.8 Case study: the role of DB/SPF in the manufacture of wide chord fan blades

On modern 'high bypass ratio' turbofan engines, the fan makes a major contribution to thrust and efficiency.[68] At take off it generates about 75% of the total engine thrust. The fan is located at the front of the engine and must be capable of developing sufficient power for aircraft safety after impact by birds and debris, such as ice and pebbles, during take off. On the very rare occasions when detachment of a fan blade occurs, it is vital that it is contained within the engine structure. The fan blades must resist fatigue stresses since they are subjected to low cycle fatigue during each flight and high cycle fatigue stresses from air intake disturbances. Hence, aerodynamics and mechanical integrity, along with weight, cost and noise, are the primary considerations that have led to the design and development by Rolls-Royce of light weight wide chord fan blades for their turbofan engines. The fan blades are hollow, made from a titanium alloy, and are used in fans for civil engine applications in the thrust range 10,000 Kg (10 tonnes) to 50,000 Kg (50 tonnes).

In the 'high bypass ratio' turbofan engine, about a third of the air drawn through the fan passes into the engine where it is progressively compressed and heated. It reacts with aviation fuel in the ignition chamber, and the resultant hot gases are exhausted through a turbine at the rear of the engine and provide additional thrust. The turbine is fitted with nickel alloy blades and operates at temperatures in excess of 1,350 °C. It is mechanically linked to the fan and the rest of the compressor, and controls the speed of fan rotation. The greater fraction of bypass air flows between the engine and the double-skinned outer casing (nacelle), and provides propulsion and sound insulation.

8.8.1 Properties required of a fan blade

It is instructive to consider what material properties would be required for the manufacture of a fan blade for an aeroengine. These would include: high strength and toughness; good ductility; good fatigue behaviour; low density; corrosion resistance; ease of fabrication and economic cost of manufacture. Since the fan is at the cool end of the engine it admits air at temperatures ranging from –50 °C to 80 °C, so that creep is not likely to be a serious problem.

Low alloy steels, ferritic stainless steels, aluminium alloys or carbon-fibre composites would satisfy some of these property requirements, but not all. However, only titanium alloys can be processed to develop the high strength/weight ratios combined with the other required properties. Table 8.3 lists mechanical properties of Ti-6/4.

Table 8.3 Typical room temperature properties for annealed Ti 6/4 alloy, mechanically processed in the α/β phase field

UTS (MPa)	980
YS (MPa)	940
Elongation %	16
Reduction in area %	45
Fracture toughness (MPa $m^{1/2}$)	60
10^7 fatigue limit (MPa)*	+/–494
Density	4.46

*Axial loading: smooth specimens, K_t = 1.0

8.8.2 Evolution of fan blade design

Conventional fan blades were of narrow chord and were manufactured from Ti-6/4 alloy forgings with mid-fan snubbers to counteract vibrational instability. These blades were used on the RB211 engines from the late 1960s to ~1988. However, the snubbers impede air flow and reduce fan and engine efficiency. To eliminate the need for mid-span supports, Rolls-Royce pioneered the development of the wide chord fan blade. The chord width was increased by 40% to give blade stability and the number of blades per fan was reduced by about one-third (Fig. 8.17). However, a solid blade consistent with the new design would give an unacceptable increase in the rotation (blade and disc) and static (bearings and containment) component weights. The solution to the problem was to design a hollow blade with an internal core structure which separates and supports the titanium alloy 'skins'. This new design led to improved fan efficiency and reduced fuel consumption. The first-generation blades were developed during the late 70s and early 80s, entered service in 1984, and their fabrication required the development of novel joining and forming technologies. Figure 8.18 is a schematic of a wide chord fan blade and the location of the blade in the fan disc slot.

8.8.3 First-generation wide chord fan blades

The manufacture of this fan blade, which has a non-load-bearing Ti honeycomb core and external Ti alloy skins, has been discussed by Fitzpatrick and Broughton.[69] The external concave and convex panels are produced from flat, hot rolled Ti-6/4 plate by a combination of forging, hot isothermal

(a) (b)

8.17 Rolls Royce turbofans with (a) conventional solid Ti-6/4 fan blades showing mid-span supports (b) wide chord Ti-6/4 hollow fan blades (courtesy of Rolls-Royce plc).

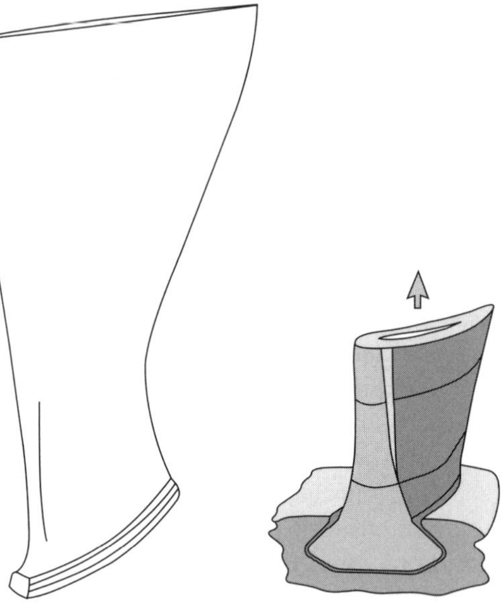

8.18 Schematic of (a) wide chord fan blade (b) location of blade in fan disc slot.

forming and chemical machining processes, which give the panels their aerodynamic shape and form an internal cavity. The internal core is produced by machining a honeycomb slab fabricated from crimp-rolled and resistance welded Ti foil. The panels and honeycomb are then joined by transient liquid phase diffusion bonding. Many different joining systems were evaluated before the Cu-Ni system was identified as the most promising.

The interlayers for joining are attached to the panel surfaces as thin foils. Heating to elevated temperatures causes solid state diffusion between the copper, nickel and the titanium alloy surfaces, leading to a critical Ti-Cu-Ni composition which melts at the bonding temperature. Further interdiffusion leads to isothermal solidification to form the bond and, after a longer time, to a high-strength joint with properties comparable to the fan blade panel material. Figure 8.19 shows the microstructure developed at the panel-panel interface and it can be seen that the transient liquid region has solidified to give a relatively fine acicular basket weave structure.[69]

To establish fan blade quality and to monitor and control the manufacturing process, conventional non-destructive techniques such as X-ray radiography were supplemented by ultrasonic inspection. These tests were combined with an extensive assessment of the component using destructive techniques including mechanical testing, microstructural examination, with further tests being carried out both on engines and rigs to examine fatigue behaviour, bird

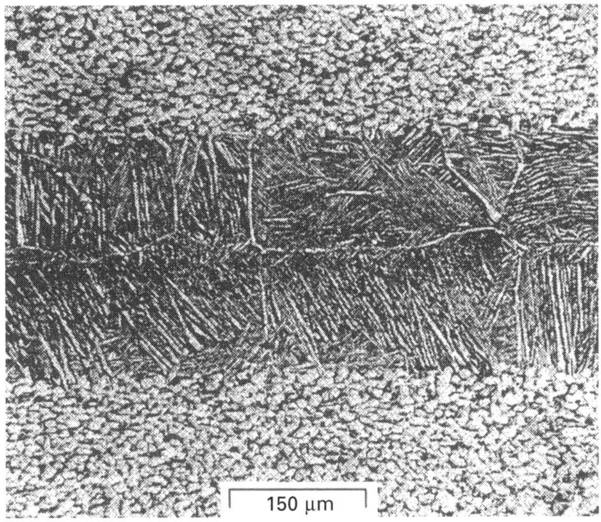

8.19 Transient liquid bond in first-generation wide chord fan blade showing basket weave microstructure (courtesy of Rolls-Royce plc).

ingestion and containment. The work showed that the manufacturing procedures developed led to a high-quality, consistently reliable fan blade.

The first commercial application of these blades was in the RB211-535E4 used to power the Boeing 757 in 1984. The technology was later applied to the V2500 engine used in the Airbus 320 family (1988) and the McDonnell Douglas MD90, and to the RB211-524G/H engines for the Boeing 747 and 767 aircraft. Between 1984 and 2003 more than 100,000 first-generation fan blades have seen service.

8.8.4 Second-generation wide chord fan blades

Research continued during the late 1980s and early 1990s to develop an alternative fan blade for the larger engines being planned, with weight and cost being the prime considerations.[70,71] The second-generation design incorporates a superplastically formed corrugated core which largely supports its own centrifugal load. This results in a fan blade which is 15% lighter than its first-generation equivalent. Its manufacture makes use of DB/SPF, referred to in section 8.6.

The fan blades are manufactured as an assembly of external Ti alloy panels and a central Ti alloy membrane sheet. The cleaned external panels have their mating faces coated with a barrier layer (or stop-off) in a predetermined pattern before being assembled with the cleaned internal membrane. All these operations take place in an environmentally controlled facility for operational discipline and the prevention of particle contamination within the joints of the components.

The DB pack is superplastically formed between precision dies to develop both the internal corrugated core and the external aerodynamic shape. Further machining and surface processing complete the production sequence. As with the first-generation wide chord fan, a rigorous programme of testing confirmed that the DB/SPF manufacturing route led to a consistently high-quality product. Figure 8.20 is a schematic of sections through the first- and second-generation wide chord fan blades. Engines with this type of fan blade include the Trent 700 on the Airbus 330 in 1995, the Trent 800 on the Boeing 777 in 1996, and the Trent 500 on the four-engined Airbus 340 in 2002. More than 25,000 of these blades have been produced to date.

8.8.5 Other developments[72]

Swept fan concept

This is the third-generation wide chord fan blade and is being developed for the fan system of the Trent 900 engine, which at 116″ (~3 m) diameter is the largest fan produced by Rolls-Royce. The blades are produced by the DB/SPF route and during prototype testing have shown increased air flow and efficiency compared with second-generation fan blades. At the same time, they have demonstrated a reduced sensitivity to bird strikes (due to its high

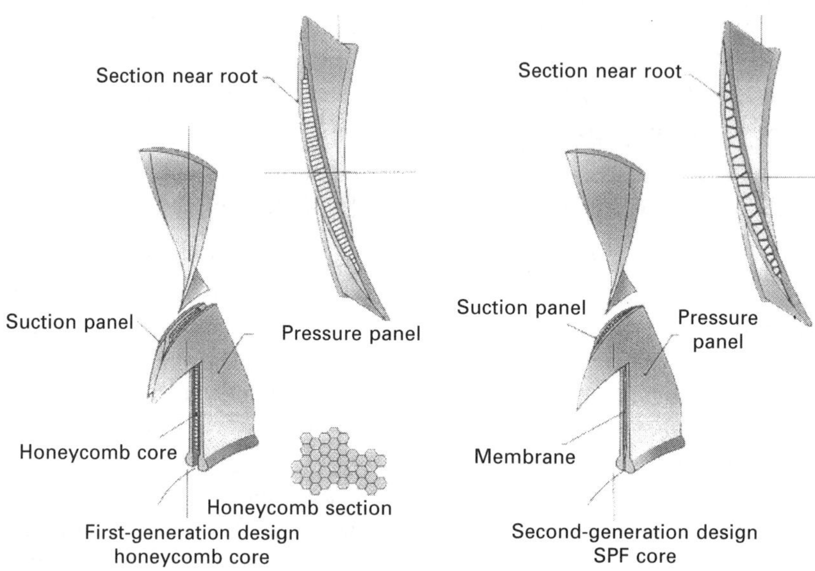

8.20 Schematic of sections through first- and second-generation wide chord fan blades (courtesy of Rolls-Royce plc).

tip stagger). This engine will power the four-engined double-decked Airbus 380 due to enter service in 2006.

Military aircraft applications

In addition to developing very large fans, Rolls-Royce is developing smaller fans for military aircraft engines to be used in the joint strike fighter (JSF) programme for the F136 aircraft. The hollow blade technology is being combined with linear friction welding to create a lightweight bladed disc or blisk, by eliminating the need for conventional blade root fixing, thus achieving further significant weight savings.

8.8.6 Summary

A commercial and performance need existed for the development of lightweight fans for the high bypass ratio jet engines manufactured by Rolls-Royce. Several approaches were identified including the use of carbon fibre material and the manufacture of hollow fan blades from Ti alloys. Innovative solutions were developed for both types of material but the hollow titanium fan blade was demonstrated to be superior following a rigorous engine and component testing programme. Innovation has continued in recent years with more complex component development, e.g., swept fans and military engines, as well as continuing the drive for more cost effective manufacturing processes. This case study illustrates the need to develop new technologies and materials systems for identified practical applications, with primary drivers being a reduction in production costs and improved engine efficiency.

8.9 Conclusion and future trends

The number of components produced by SPF, particularly of Ti and Al alloys, continues to grow, but the overall range of applications tends to remain in the fields of aerospace, rail transport, automobiles and architecture (see section 8.6.4). While SPF is not yet a mainstream manufacturing process and caters for what have been termed niche markets, the recent development by GM of the pseudo-SPF process, QPF, is likely to have a substantial knock-on effect within the automotive industry. Sanders[73] has recently forecast that within a few years most automobiles built in the USA will have at least one part produced by SPF, and within a decade a pseudo-SPF process will be the dominant means of producing car panels. It has also been noted by Sanders[74] and by Grimes[75] that for many years there has been a massive imbalance between the amount of research effort, as indicated by the number of research publications, directed towards new SP materials, compared with that directed towards manufacturing technology. This imbalance needs reversing.

The α/β Ti alloys with their unique combination of mechanical properties, along with the ability to combine DB and SPF, have secure markets in aerospace. However, Ti alloy sheet materials are expensive to produce and require expensive equipment for SPF. It has been suggested that if the price of sheet material could be reduced by 33%, then the number of aerospace applications for sheet assemblies could be increased by a factor of three.[74]

Since DB/SPF of Ti-6/4 is normally carried out at 900 °C or more, the dies and platens used have to resist these temperatures. Service problems encountered include surface oxidation, oxide adherence, and low cycle thermal fatigue which arises from stresses induced during cooling. There is a need for the development of new alloys which could withstand these problems and have high temperature strength up to 1000 °C, be resistant to creep, easy to cast and machine, and be able to be repaired by welding. Alternatively the use of ceramic dies might provide a solution to these problems.

Barnes[48] has discussed trends and prospects for the SPF of Al alloys. He has pointed out the need to increase the number of qualified Al alloys that can be shaped by SPF. The 6000 series alloys (Al-Mg-Si), which are heat treatable, would be a useful addition to the present SP alloys, provided they could be processed to develop suitable microstructures.

A simplified view of the manufacture of Al alloy components by conventional SPF is that it takes a relatively expensive sheet of material, and forms it relatively slowly into a complex shape using relatively inexpensive tooling. The cost-effective niche for SPF would be the production of between 50–5,000 parts of a given product. For SPF to become a higher volume process would require a reduction in materials costs and/or forming time. This points to the benefits of a lower cost, faster forming alloy. Grimes and his collaborators[76,77] have taken up the challenge and are involved in a project to develop a HSRS Al alloy. Work on an Al-Mg-Zr alloy has given encouraging results.[76]

Higashi[78] has reported that in Japan a 5083 alloy sheet has been processed to give high elongations at 10^{-1}–10^{-2} s^{-1}. He has listed a number of commercial Mg alloys that can be processed to develop superplasticity, and this reflects a likely area of future commercial activity.

8.10 Sources of further information

The literature on superplasticity is substantial and papers describing original work frequently appear in *Materials Science* and other related journals. A search on the World Wide Web using the words ‘Superplasticity’ or ‘Superplastic Forming’ will yield many hundreds of reference sources. There are three books of relatively recent origin.[9,10,20] International conferences have been held triennially since 1982 and are now designated International Conference on Superplasticity in Advanced Materials, or ICSAM. The most recent meeting in the series, ICSAM-03, was held in Oxford, UK, in 2003. The ICSAM proceedings are published by Trans Tech Publications, Switzerland, in Materials Science Forum. Additional conferences are also held from time to time in the UK, Europe, USA, Japan and China (see references). At Ufa in Russia there is an Institute dedicated to the study of superplasticity.

Much of the SPF of Ti alloy components for civil and military aircraft is done in-house by aerospace companies. SPF of Ti alloys is also carried out by such contractors as Aeromat International plc, Welwyn Garden City, UK, (www.aeromet.co.uk), and by Form Tech GmbH, Weyhe, Germany, (email:werner.beck@formtech.de). The leading producers of components made from Al alloys are Superform Aluminium, Worcester, UK, (www.superform-aluminium.com) and Superform USA, Riverside, California, (www.superform-aluminium.com/susa), although General Motors of Detroit, USA, are now using their QPF process to produce body panels. FormTec AG, CH-3965 Chippis, Switzerland, also produce automotive body panels (email: barry dunwoody@formtecag.com). Custom Metalforms Ltd, Poole, Dorset, UK, (Tel:+44(0)1202 381300) uses SPF of Zn-Al eutectoid alloy to make a range of products.

8.11 Acknowledgements

The author is grateful to Dr G.A. Fitzpatrick, formerly with Rolls-Royce plc, Barnoldswick, Lancs, for invaluable discussions on the wide chord fan blade, and to A.J. Barnes, Superform USA, and R.J. Butler and R.J. Stracey of Superform Aluminium, UK, for helpful discussions on the SPF of aluminium alloys.

8.12 References

1. Higashi K., private communication, University of Osaka Prefecture, Japan, 1988.
2. Hamilton C.H., Bampton C.C. and Paton N.E., 'Superplasticity in high strength aluminium alloys', *Superplastic Forming of Structural Alloys*, Warrendale, PA, TMS, 1982.
3. Ball A. and Hutchison M.M., 'Superplasticity in the aluminium-zinc eutectoid', *Metal. Sci. J.*, 1969 **3** 1–7.
4. Mukherjee A.K., 'The rate controlling mechanism in superplasticity', *Mater. Sci. Engin.*, 1971 **8** 83–89.
5. Gifkins R.C., 'Mechanisms of superplasticity', *Superplastic Forming of Structural Alloys*, San Diego, TMS, 1982.
6. Ashby M.F. and Verrall R.A., 'Diffusion-accommodated flow and superplasticity', *Acta Metall.*, 1973 **21** 149–163.
7. Kashyap B.P. and Mukherjee A.K., 'On the models for superplastic deformation', *Superplasticity*, Grenoble, CNRS Paris, 1985.
8. Langdon T.G., 'Experimental observations in superplasticity', *Superplastic Forming of Structural Alloys*, San Diego, TMS, 1982.

9. Pilling J. and Ridley N., *Superplasticity in Crystalline Solids*, London, Institute of Metals, 1989.
10. Nieh T.G., Wadsworth J. and Sherby O.D., *Superplasticity in Metals and Ceramics*, Cambridge University Press, 1997.
11. Mukherjee A.K., 'Superplasticity in metals, ceramics and intermetallics', *Plastic Deformation and Fracture of Materials, Volume 6,* New York, VCH Publishers Inc., 1993.
12. Backofen W.A., Turner I.R. and Avery D.H., 'Superplasticity in an Al-Zn alloy', *Trans. ASM,* 1964 **57** 980–990.
13. Hedworth J. and Stowell M.J., 'The measurement of strain rate sensitivity in superplastic alloys', *J. Mater. Science*, 1971 **6** 1061–1069.
14. McKay T.L., Sastry S.M.L. and Yalton C.F., 'Metallurgical characterisation of superplastic forming', *AFWAL-TR-80-4038*, 1980.
15. Goforth R.E., Chandra N.A. and George D., 'Analysis of the cone test to evaluate superplastic forming characteristics of sheet metal', *Superplasticity in Aerospace*, Phoenix, TMS, 1988.
16. Ridley N., 'Superplastic Forming', *Handbook of Aluminum, Vol 1, Physical Metallurgy and Processes,* New York, Marcel Dekker, 2003.
17. Zener C., quoted by C.S. Smith in 'Grains, phases, interfaces: an interpretation of microstructure', *Trans. AIME*, 1948 **175** 15–51.
18. Humphreys F.J. and Hatherley M., *'Recrystallisation and related annealing phenomena',* 2nd edn, Oxford, Elsevier, 2004.
19. Grimes R., 'Superplasticity in aluminium-lithium-based alloys', *Superplasticity*, Grenoble, CNRS Paris, 1985.
20. Kaibyshev O.A., *Superplasticity of alloys, intermetallides and ceramics*, New York, Springer Verlag, 1992.
21. Mohri T., Mabuchi M. and Nakamura M., *et al.*, 'Microstructural evolution and superplasticity in rolled Mg-9Al-1Zn', *Mater. Sci. Engin.*, 2000 **A290** 139–144.
22. *Glossary of terms used in metallic superplastic materials*, JISH 7007, Japanese Standards Association, 1995.
23. Wakai F., Sakaguchi S. and Matsuno Y., 'Superplasticity of yttrra-stabilised tetragonal zirconia ZrO_2 polycrystals', *Adv. Ceram. Mater.*, 1986 **1** 259–263.
24. Nieh T.G. and Wadsworth J., 'Superplasticity and superplastic forming of ceramics' *Mater. Sci. Forum*, 1994 **170–172** 358–368.
25. Humphreys F.J., 'The nucleation of recrystallisation at second phase particles in aluminium', *Acta Metall.*, 1977 **25** 1323–1344.
26. Imamura H. and Ridley N., 'Superplastic and recrystallisation behaviour of a commercial Al-Mg alloy 5083', *Proc. ICSAM-91*, Osaka, JSRS, 1991.
27. Li F., Roberts W.T. and Bate P.S., 'Superplasticity and the development of dislocation structures in an Al-4.5% Mg alloy', *Acta Mater.*, 1996 **44** 217–233.
28. Grimes R., Baker C., Stowell M.J. and Watts B.M., 'Development of superplastic aluminium alloys', *Aluminium*, 1975 **51** 720–723.
29. Watts B.M., Stowell M.J., Baikie B.L. and Owen D.G.E, 'Superplasticity in Al-Cu-Zr alloys, Part 1: Material preparation and properties', *Metal Sci.*, 1976 **10** 189–197.
30. McNelley T.R. and McMahon M.E., 'Recrystallisation and superplasticity in Al alloys', *Superplasticity and Superplastic Forming*, Warrendale, TMS-AIME, 1988.
31. Ridley N., Cullen E. and Humphreys F.J., 'Effect of thermomechanical processing on evolution of superplastic microstructures in Al-Cu-Zr alloys', *Mater Sci. Technol.*, 2000 **16** 117–124.

32. Valiev R.Z. and Islamgaliev, 'Microstructural aspects of superplasticity in ultrafine-grained materials', *Superplasticity and superplastic forming 1998'* Warrendale, TMS, 1998.
33. Kingsley-Jones, M., 'A380 weightsaving effort to focus on aluminium lithium', *Flight International*, 2004 26 Oct.–1 Nov. **9**.
34. Nieh T.G., Gilman P.S. and Wadsworth J., 'Extended ductility at high strain rates in a mechanically alloyed aluminium alloy', *Scripta Metall.*, 1985 **19** 1375–1378.
35. Mabuchi M. and Higashi K., 'The processing, properties, and applications of high strain-rate superplastic materials', *JOM*, 1998 June 34–39.
36. Higashi K., 'Positive exponent superplasticity in metallic alloys and composites', *Superplasticity; 60 years after Pearson*, London, Institute of Materials, 1995.
37. Semiatin S.L., Seetharaman V. and Weiss I., 'Hot working of titanium alloys – an overview', *Advances in Science and Technology of Titanium Alloy Processing*, Warrendale, TMS, 1997.
38. Paton N.E. and Hamilton C.H., 'Superplasticity in titanium alloys', *5th Intl. Conf. on Titanium,* Munich, Oberursel, DGM, 1984.
39. Cope M.T., Evetts D.R. and Ridley N., 'Superplastic deformation characteristics of two microduplex titanium alloys', *J. Mater. Sci.*, 1986 **21** 4003–4008.
40. Leader J.R., Neal D.F. and Hammond C., 'The effect of alloying additions on the superplastic properties of Ti-6Al-4V', *Metall. Trans.*, 1986 **17A** 93–106.
41. Wisbey A., Williams B.C., Ubhi H.S *et al.*, 'Superplasticity in the titanium alloy SP 700 with low SPF temperature', *Superplasticity: 60 years after Pearson*, London, Institute of Materials, 1995.
42. Swale W., 'Superplastic forming of titanium – a user friendly process', *Proc. Euro-SPF 2004, Third European Conf. on SPF,* Toulouse, Cepadues-Editions, 2004.
43. Sanders D.G., 'A production system using ceramic die technology for superplastic forming', *Mater. Sci. Forum*, 2004 **447–448** 153–158.
44. Cornfield G.C. and Johnson R.H., 'The forming of superplastic sheet metal', *Int. J. Mech. Sci.*, 1970 **12** 491–497.
45. Story J.M., 'Part selection criteria and design considerations for the use of superplastically formed aluminium for aerospace structures', *Superplasticity in Aerospace II,* Warrendale PA, TMS-AIME, 1990.
46. Laycock D., 'Superplastic forming of sheet metal', *Superplastic forming of structural alloys,* Warrendale, PA, TMS, 1982.
47. Whittingham R., 'The design and manufacture of superplastic forming/diffusion bonding presses', *Mater. Sci. Forum*, 2000 **357–359** 29–34.
48. Barnes A.J., 'Industrial applications of superplastic forming: trends and prospects', *Mater. Sci. Forum*, 2001 **357–359** 3–16.
49. Friedman P.A. and Luckey S.G., 'On the expanded usage of superplastic forming of aluminium sheet for automotive applications', *Mater. Sci. Forum*, 2004 **447–448** 199–204.
50. Dunwoody B.J., 'Series production of automotive body panels in 5083-SPF using a new press concept', *Mater. Sci. Forum*, 2004 **447–448** 205–210.
51. Schroth J.G., 'General Motors' Quick Plastic Forming Process', *Advances in superplasticity and superplastic forming,* Warrendale, PA, TMS, 2004.
52. Wood R. and Bonnet, 'A review of the numerical analysis of superplastic forming', *J. Mater. Process Technol.*, 1996 **60** 45–53.
53. Partridge P.G., McDarmaid D.S., Bottomley I. and Common D., 'The mechanical properties of superplastically formed Ti and Al alloys', *Superplasticity* (AGARD-LS-168), Neuilly-sur-Seine, France, AGARD, 1989.

54. Ingelebrecht C.D., 'Tensile properties of titanium alloys after superplastic strain', *Designing with titanium*, London, Institute of Metals, 1986.
55. Derby B. and Wallach E.R., 'Theoretical model for diffusion bonding', *Metal. Sci.*, 1982 **16** 49–56.
56. Guo Z.X. and Ridley N., 'Modelling of the diffusion bonding of metals', *Mater. Sci. Technol.*, 1987 **3** 945–954.
57. Pilling J., Livesey D.W., Hawkyard J. and Ridley N., 'Solid state bonding of superplastic Ti-6Al-4V', *Metal. Sci.*, 1984 **18** 117–122.
58. Pilling J., 'The kinetics of isostatic diffusion bonding in superplastic materials', *Mater. Sci. Eng.* 1988 **100** 137–144.
59. Huang Y., Ridley N., Humphreys F.J. and Cui J.-Z., 'Diffusion bonding of superplastic 7075 aluminium alloy', *Mater. Sci. Engin.*, 1999 **A266** 295–302.
60. Ridley N. and Wang Z.C., 'The effect of microstructure and deformation conditions on cavitation in superplastic materials', *Mater. Sci. Forum*, 1997 **233–234** 63.
61. Chokshi A.H., 'The influence of grain size on cavitation in superplasticity', *Mater. Sci. Forum*, 1997 **233-234** 89.
62. Bae D.H. and Ghosh A.K., 'Cavity formation and early growth in a superplastic Al-Mg alloy', *Acta Mater.*, 2002 **50** 511–523.
63. Hancock J.W., 'Creep cavitation without a vacancy flux', *Metal. Sci.*, 1976 **10** 319–325.
64. Pilling J. and Ridley N., 'Cavitation in superplastic alloys and the effect of hydrostatic pressure', *Res. Mechanica*, 1988 **23** 31–63.
65. Pilling J. and Ridley N., 'Effect of hydrostatic pressure on cavitation in aluminium alloys', *Acta Metall.*, 1986 **34** 669–679.
66. Stowell M.J., 'Cavitation in superplasticity', *Superplastic forming of structural alloys,* Warrendale, PA, TMS, 1982.
67. Varloteaux A., Blandin J.J. and Suery M., 'Control of cavitation during superplastic forming of high strength aluminium alloys', *Mater. Sci. Technol.*, 1989 **5** 1109–1117.
68. *The Jet Engine*, 5th edn revised, Derby, Rolls-Royce Technical Publications Dept., 1996.
69. Fitzpatrick G.A. and Broughton T., 'The Rolls-Royce wide chord fan blade', *Titanium Products and Applications*, San Francisco, CA, Ti Development Assoc., 1986.
70. Fitzpatrick G.A. and Cundy J.M., 'Rolls-Royce's wide chord fan blade – the next generation', *Titanium '92 Science and Technology*, Warrendale, PA, TMS, 1993.
71. Fitzpatrick G.A., 'Diffusion bonding and superplastically formed fan blades for Rolls-Royce's Trent Engines', *New Manufacturing Trends for Aeronautical Industries*, San Sebastian, Spain, HEGAN, 2000.
72. Fitzpatrick G.A., 'Further developments in the Rolls-Royce diffusion bonded and superplastically formed fan blade', *New Manufacturing Trends for Aeronautical Industries*, Bilbao, Spain, HEGAN, 2002.
73. Sanders D.G., 'The history and current state-of-the-art in airframe manufacturing using superplastic forming technologies', *Advances in superplasticity and superplastic forming,* Warrendale, PA, TMS, 2004.
74. Sanders D.G., 'Superplastic forming manufacturing technology moves into the twenty-first century', *Mater. Sci. Forum*, 1999 **304–306** 805–812.
75. Grimes R., 'Superplastic forming: evolution from metallurgical curiosity to major manufacturing tool', *Mater. Sci. Technol.*, 2003 **19** 3–10.

76. Dashwood R.J., Grimes R., Harrison A.W. and Flower H.M., 'The development of high strain rate superplastic Al-Mg-Zr alloy', *Mater. Sci. Forum*, 2001 **357–359** 339–344.
77. Grimes R., Dashwood R.J. and Flower H.M., *et al*, 'Progress towards high superplastic strain rate aluminium alloys' *Mater. Sci. Forum*, 2004 **447–448** 213–220.
78. Higashi K., 'Recent advances and future directions in superplasticity', *Mater. Sci. Forum*, 2001 **357–359** 345–356.

Index